Methods in Molecular Biology™

Series Editor
John M. Walker
School of Life Sciences
University of Hertfordshire
Hatfield, Hertfordshire, AL10 9AB, UK

For further volumes:
http://www.springer.com/series/7651

Microtubule Dynamics

Methods and Protocols

Edited by

Anne Straube

Centre for Mechanochemical Cell Biology, Warwick Medical School, University of Warwick, Coventry, UK

Humana Press

Editor
Anne Straube
Centre for Mechanochemical Cell Biology
Warwick Medical School
University of Warwick
Coventry, UK
anne@mechanochemistry.org

Please note that additional material for this book can be downloaded from
http://extras.springer.com

ISBN 978-1-4939-6185-6 ISBN 978-1-61779-252-6 (eBook)
DOI 10.1007/978-1-61779-252-6
Springer New York Dordrecht Heidelberg London

Printed on acid-free paper

Humana Press is part of Springer Science+Business Media (www.springer.com)

Preface

Microtubules are at the heart of cellular self-organization and the target of the most widely used anticancer drugs. The dynamic nature of microtubules allows them to explore the intracellular space and mediate the transport of cargoes from the nucleus to the outer edges of the cell and back. They provide physical strength and, through dynamic interactions with subcellular structures, such as kinetochores, cell adhesion sites, or the actin cortex, orchestrate multiple cell biological processes, including chromosome segregation and directional cell migration. Modern methods make it possible to ask for mechanistic principles underlying microtubule dynamics regulation. The main purpose of this book is to provide an up-to-date collection of methods and approaches that are used to investigate microtubule dynamics in vitro and in cells. While it was impossible to cover all available techniques to study microtubule dynamics, I hope this book provides detailed protocols how to perform a broad range of well-established and newly emerging techniques, contributed by experts in the field.

This book was put together with two groups of readers in mind: First, students and postdoctoral researchers starting work in a microtubule laboratory. Second, established researchers in the microtubule field who require a resource for established and new methodologies. The book starts off with a chapter reflecting on how to analyze microtubule dynamics (Chapter 1), followed by detailed descriptions on how to isolate tubulin from different sources and with different posttranslational modifications (Chapters 2–5), methods to study microtubule dynamics and microtubule interactions in vitro (Chapters 6–13), techniques to investigate the ultrastructure of microtubules and associated proteins (Chapters 14 and 15), and assays to study microtubule nucleation, turnover, and force production in cells (Chapters 16–19). Finally, we discover approaches to isolate novel microtubule-associated proteins and their interacting proteins (Chapters 20 and 21). I hope that the combination of in vitro reconstitution experiments, cellular assays, and structural analysis described in this volume will help to build well-informed mechanistic models of dynamic instability and further our understanding of microtubule-mediated processes.

I would like to thank all authors for their enthusiasm and effort in contributing to this collection of methods, as well as Tim Mitchison and Ted Salmon for providing freely accessible online resources of microtubule methods on their lab Web sites (http://mitchison.med.harvard.edu/protocols.html and http://www.bio.unc.edu/faculty/salmon/lab/salmonprotocols.html), which were an inspiration for many of the chapters and cover additional methods not included in this book. Finally, I wish to thank all those who have suffered from a lack of attention while I was reading and editing the chapters in this book for their patience.

Anne Straube

Contents

Contributors

ANNA AKHMANOVA • *Department of Cell Biology, Erasmus Medical Center, Rotterdam, The Netherlands*

ROBIN ANTROBUS • *Cambridge Institute for Medical Research, Wellcome Trust/MRC Building, Addenbrooke's Hospital, Cambridge, UK*

ISABELLE ARNAL • *CNRS, UMR 6026 Interactions Cellulaires et Moléculaires, IFR 140 Génomique Fonctionnelle Agronomie et Santé, Université de Rennes 1, Rennes, France*

CLIVE R. BAGSHAW • *Department of Chemistry & Biochemistry, University of California at Santa Cruz, Santa Cruz, CA, USA*

SUSANNE BECHSTEDT • *Department of Biology, McGill University, Montréal, QC, Canada*

PETER BIELING • *Department of Cellular and Molecular Pharmacology, University of California, San Francisco, CA, USA*

SOPHIE BLESTEL • *CNRS, UMR 6026 Interactions Cellulaires et Moléculaires, IFR 140 Génomique Fonctionnelle Agronomie et Santé, Université de Rennes 1, Rennes, France; Inria, Centre de Rennes Bretagne Atlantique, Inria, Rennes, France*

BARBARA BORGONOVO • *Max Planck Institute of Molecular Cell Biology and Genetics, Dresden, Germany*

DIANE BRAGUER • *INSERM UMR 911, Centre de Recherche en Oncologie biologique et Oncopharmacologie, Université de la Méditerranée, Marseille, France*

GARY J. BROUHARD • *Department of Biology, McGill University, Montréal, QC, Canada*

JAIME N. CAMPBELL • *Department of Biochemistry and Biophysics, University of North Carolina, Chapel Hill, NC, USA; Molecular and Cellular Biophysics Program, University of North Carolina, Chapel Hill, NC, USA*

DENIS CHRÉTIEN • *CNRS, UMR 6026 Interactions Cellulaires et Moléculaires, IFR 140 Génomique Fonctionnelle Agronomie et Santé, Université de Rennes 1, Rennes, France*

FRÉDÉRIC M. COQUELLE • *CNRS, UMR 6026 Interactions Cellulaires et Moléculaires, IFR 140 Génomique Fonctionnelle Agronomie et Santé, Université de Rennes 1, Rennes, France*

ROBERT A. CROSS • *Centre for Mechanochemical Cell Biology, Warwick Medical School, University of Warwick, Coventry, UK*

JAN DE MEY • *Laboratoire de Biophotonique et Pharmacologie, Faculté de Pharmacie, UMR 7213 CNRS, Université de Strasbourg, Illkirch, France*

MARILEEN DOGTEROM • *FOM Institute for Atomic and Molecular Physics (AMOLF), Amsterdam, The Netherlands*

DAVID N. DRECHSEL • *Max Planck Institute of Molecular Cell Biology and Genetics, Dresden, Germany*

Douglas R. Drummond • *Centre for Mechanochemical Cell Biology, Warwick Medical School, University of Warwick, Coventry, UK*
Claire T. Friel • *School of Biomedical Sciences, University of Nottingham, Nottingham, UK*
Felicia Gaskin • *Department of Psychiatry and Neurobehavioral Sciences, University of Virginia School of Medicine, Charlottesville, VA, USA*
Christopher Gell • *Max Planck Institute of Molecular Cell Biology and Genetics, Dresden, Germany*
Claire Heichette • *CNRS, UMR 6026 Interactions Cellulaires et Moléculaires, IFR 140 Génomique Fonctionnelle Agronomie et Santé, Université de Rennes 1, Rennes, France*
Christina Hoey • *Institute of Structural Molecular Biology, Birkbeck College, London, UK*
Stéphane Honore • *INSERM UMR 911, Centre de Recherche en Oncologie biologique et Oncopharmacologie, Université de la Méditerranée, Marseille, France*
Casper C. Hoogenraad • *Department of Neuroscience, Erasmus Medical Center, Rotterdam, The Netherlands*
Jonathon Howard • *Max Planck Institute of Molecular Cell Biology and Genetics, Dresden, Germany*
Anthony A. Hyman • *Max Planck Institute of Molecular Cell Biology and Genetics, Dresden, Germany*
Carsten Janke • *Department of Signalling, Neurobiology and Cancer, Bat 110 – Centre Universitaire, Institut Curie, CNRS UMR 3306/INSERM U1005, Orsay Cedex, France*
Didier Job • *Institut National de la Santé et de la Recherche Medicale Unité 836, Institut des Neurosciences de Grenoble, Grenoble, France; Groupe Physiopathologie du Cytosquelette, Institut de Recherches en Technologies et Sciences pour le Vivant, Direction des Sciences du Vivant, Commissariat à l'Énergie Atomique, Grenoble, France; Université Joseph Fourier, Grenoble, France*
Susan Kain • *Lister Hospital, Stevenage, UK*
Svenja-Marei Kalisch • *FOM Institute for Atomic and Molecular Physics (AMOLF), Amsterdam, The Netherlands*
Miho Katsuki • *Centre for Mechanochemical Cell Biology, Warwick Medical School, University of Warwick, Coventry, UK*
Irina Kaverina • *Department of Cell and Developmental Biology, Vanderbilt University Medical Center, Nashville, TN, USA*
Nanda Keijzer • *Department of Neuroscience, Erasmus Medical Center, Rotterdam, The Netherlands*
Charles Kervrann • *Inria, Centre de Rennes Bretagne Atlantique, Inria, Rennes, France*
Liedewij Laan • *FOM Institute for Atomic and Molecular Physics (AMOLF), Amsterdam, The Netherlands; Department of Molecular and Cellular Biology, Harvard University, Cambridge, MA, USA*
Benjamin Lacroix • *CRBM, Université Montpellier 2 and 1, CNRS, Montpellier, France*

Chapter 1

How to Measure Microtubule Dynamics?

Anne Straube

Abstract

Microtubules are one of the most spectacular features in the cell: long, fairly rigid tubules that provide physical strength while at the same time serving as tracks of the intracellular transport network. In addition, they are the main constituents of the cell division machinery, and guide axonal growth and the direction of cell migration. To be able to fulfil such diverse functions, microtubules have to be arranged into suitable patterns and remodelled according to extra- and intracellular cues. Moreover, the delicate regulation of microtubule dynamics and the dynamic interactions with subcellular structures, such as kinetochores or cell adhesion sites, appear to be of crucial importance to microtubule functions. It is, therefore, important to understand microtubule dynamics and its spatiotemporal regulation at the molecular level.

In this chapter, I introduce the concept of microtubule dynamics and discuss the techniques that can be employed to study microtubule dynamics in vitro and in cells, for many of which detailed protocols can be found in this volume. Microtubule dynamics is traditionally assessed by the four parameters of dynamic instability: growth and shrinkage rates, rescue and catastrophe frequencies, sometimes supplemented by pause duration. I discuss emerging issues with and alternatives to this parameter description of microtubule dynamics.

Key words: Microtubule, Dynamic instability, Microtubule assembly, Catastrophe, Rescue, GTP cap, GFP-Tubulin, EB3, CLIP-170

1. The Discovery of Microtubule Dynamics and Current Understanding of Dynamic Instability

Since the discovery that the spindle fibres correlate with microtubules, their unstable, dynamic nature was noted; using birefringence and correlated electron microscopy to image microtubules in the mitotic spindle, Inoue and Sato described the unusually labile properties of microtubules that underwent cyclic breakdown and reformation and were readily disassembled by lowering the temperature (1). They realised that low concentrations of

Anne Straube (ed.), *Microtubule Dynamics: Methods and Protocols*, Methods in Molecular Biology, vol. 777,
DOI 10.1007/978-1-61779-252-6_1,

colchicine allow the filaments to shorten and perform work, thus explaining how microtubules can move chromosomes during anaphase.

In order to really understand their mechanochemical properties, microtubules needed to be studied in isolation outside of the cell. Tubulin, the material microtubules are built of, was isolated as the colchicine-binding protein by Borisy, Shelanski, and Taylor from various microtubule-rich sources, such as dividing cells, brain tissue, and sperm tails, in 1967 (2, 3) and was given its name by Mohri the year after (4). But it took another 5 years before the first success in reforming microtubules in vitro could be reported. Weisenberg (5) identified the basic requirements to polymerise rat brain tubulin from solution: a calcium chelator (EGTA), magnesium ions, ATP or GTP, and heat (37°C). Pulse-chase experiments with tritium-labelled GTP had suggested an opposite-end assembly and disassembly treadmilling mechanism for tubulin as occurs in actin (6). However, Mitchison and Kirschner recognised that the range of polymer lengths observed in (glutaraldehyde-fixed) microtubule populations can only be explained if some microtubules were growing while others were shrinking. They proposed a mechanism that involved stochastic switching between phases of growth and shrinkage and called this "dynamic instability" (7). A few years later, such behaviour could indeed be observed directly in microtubules assembled both in vitro (8, 9) and in cells (10, 11).

The molecular basis for dynamic instability is thought to be the GTP cap (7, 12): GTP-tubulin is incorporated at the growing end. Addition of an overlying subunit will complete the active site and render buried subunits competent for GTP hydrolysis. Thus, the majority of microtubule lattice consists of GDP-tubulin. GDP-tubulin adopts a curved conformation when not buried in the microtubule lattice. Assembly is disfavoured for this curved conformation, and exposure of GDP-tubulin at the microtubule end is thought to result in outward curling of protofilaments and rapid microtubule disassembly (13–15). A cap of "straight" GTP-tubulin protects growing microtubule ends; its loss results in catastrophe while it is reformed during rescue events. It is thought that this cap contains between 10 and 100 GTP-tubulin molecules (16–18). If uncoupled "stochastic" hydrolysis at a constant rate is assumed, GTP-tubulin would decay exponentially from the tip and the cap size would increase with the growth speed. Both these characteristics resemble the comets observed by EB proteins (19, 20), and nucleotide-state dependence of EB binding to microtubules (21) suggests that EB proteins might be markers for the GTP cap.

Microtubule dynamics is usually assessed by the determination of the four parameters of dynamic instability: growth speed, shrinkage speed, catastrophe frequency, and rescue frequency (9). However, microtubules can also spend considerable time in "pause", a state with very little apparent length changes that is observed in vitro as well as in cells (9, 20, 22, 23). Definitions for such pause events vary in the literature and depend on the spatiotemporal resolution of imaging and image analysis (17, 22, 24–26). Pauses have been suggested to represent unproductive elongation of a few protofilaments, with growth being reinstated once these break off (20). Thermodynamic modelling of microtubule assembly suggested that a microtubule can exist in a state during which some of its protofilaments grow while others shrink, thus no length change is visible with light microscopic techniques (27). Recent measurements of microtubule length fluctuations at 3.5 nm and 10 Hz resolution reveal that pauses are composed of repeated switching between growth and shortening excursions of up to five tubulin layers, which are below the resolution of light microscopy (17). In many cells, pauses are reported to be very frequent and comprise a range of different states. Microtubules captured at the cell cortex in mammalian cells and switching rapidly between growth and shrinkage phases (28, 29) can appear paused if observed at low spatial or temporal resolution (Fig. 1). Similar short episodes of growth and shrinkage are probably occurring at the kinetochore and may drive chromosome oscillations (30). Thus, such "pauses" are in truth very dynamic length oscillations controlled by an army of +TIPs, cortical capture factors, and kinetochore components (31, 32). These events exceed the range of nanoscale fluctuations, and are thus mechanistically different from pauses that occur stochastically during growth phases (26).

A relatively simple thermodynamical model of dynamic instability (i.e., involving stochastic switching between the two states, growth and shrinkage, driven by the protection/loss of the GTP cap) very well describes nanoscale length fluctuations during microtubule assembly (17, 27). Recent measurements of single tubulin incorporation times in spindle microtubules suggest that a random walk model describes the observed microtubule lifetime distributions more accurately than the existing two-state dynamic instability models (33). While we currently lack the molecular understanding to build a mechanistic model that reflects all aspects of dynamic microtubule behaviour observed in vivo and in vitro, recent technical advances allow the measurement of microtubule dynamics at single tubulin-layer resolution, enabling discrete modelling of tubulin addition, loss, GTP hydrolysis, and conformational changes that underlie the remarkable phenomenon of microtubule dynamics.

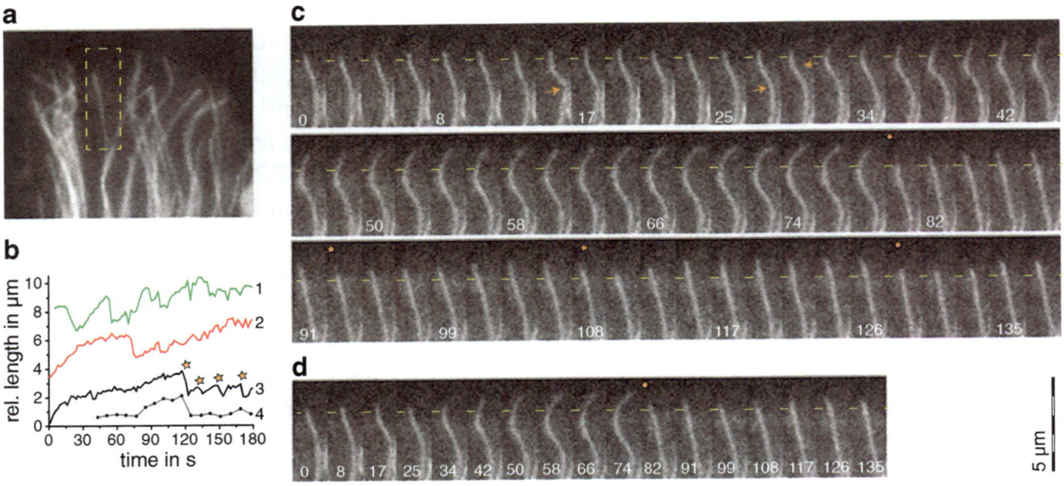

Fig. 1. Fast transitions at the cell cortex. (**a**) Microtubules "captured" at the cell cortex in differentiating C2C12 cells (see (28) for details and experimental procedures). Stills from a time-lapse movie of the microtubule in the boxed area are shown in (**c**) and (**d**) at different sampling rates. (**b**) Life history of three different microtubules (curves 1–3) within the area shown in (**a**). Microtubule length was measured with segmented lines to account for bending under compressive load. Please note that bending outside the area in (**a**) could not be measured due to high microtubule density. Curve 4 is showing length fluctuations for the microtubule in curve 3 at a lower sampling rate. All curves have been shifted along the *y*-axis for clarity. Transition events of catastrophe followed by immediate rescue for the microtubule in curves 3 and 4 are highlighted by stars. (**c**) Stills from a time-lapse movie of a "captured" microtubule showing growth against compressive load which leads to the introduction of bends into the polymer (*arrows*). Stars highlight transition events of catastrophes followed by immediate rescues. These events correspond to those highlighted on curve 3 in (**b**). (**d**) Same microtubule as in (**c**) at a four times lower sampling rate. Note that short-scale fluctuations are not visible under these conditions.

2. Imaging Dynamics of Individual Microtubules In Vitro

To image the assembly and disassembly of individual microtubules, high temporal and spatial resolution is necessary. Nucleation from isolated axonemes, centrosomes, or short stabilised microtubule seeds allows control of the number of filaments formed (so as to minimise the depletion of free tubulin by microtubule formation), observation of assembly over a relatively broad range of tubulin concentrations (9), control of protofilament number, and readout of polarity (34).

The first direct observation of dynamic instability was performed using darkfield microscopy to image microtubule assembly in vitro (8). Darkfield is a contrast enhancement technique, whereby most of the illumination light is excluded from the image. Only light scattered by the specimen is collected to form the image, thus creating a dark background with bright objects. This allows imaging or direct visualisation of single microtubules with a contrast that is comparable to a fluorescent image (Fig. 2a) without the need and risks of introducing a fluorophore into the

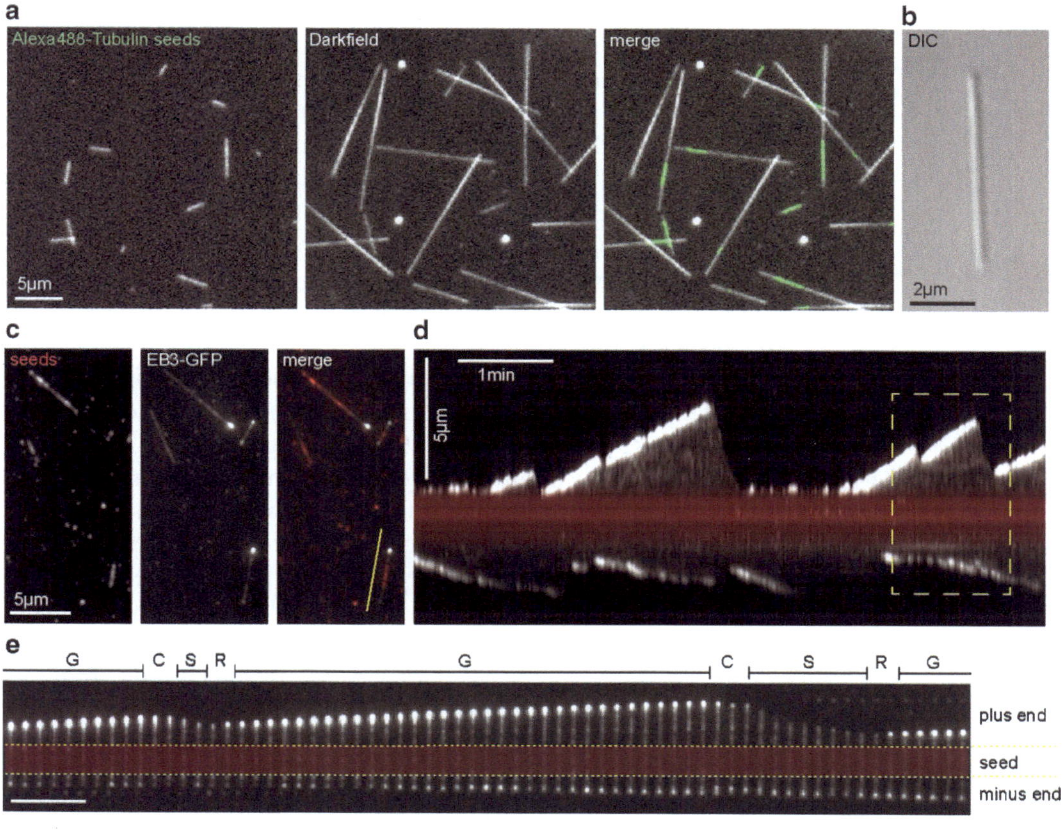

Fig. 2. Visualisation techniques to study the dynamics of individual microtubules in vitro. (**a**) Assembly of unlabelled tubulin on the ends of Alexa488-labelled microtubule seeds visualised by darkfield microscopy (see (34) for experimental procedures). Images in this panel were kindly provided by Frauke Hussmann, CMCB Warwick. (**b**) A single microtubule imaged by VE-DIC. This image was kindly provided by Mishan Britto, CMCB Warwick. (**c**) TIRF imaging of EB3-GFP tracking the growing ends of (unlabelled) microtubules assembling on GMPCPP-stabilised, Rhodamine-labelled seeds (see (39) for experimental procedures). (**d**) Kymograph (or time-space plot) of the microtubule next to the *line* in (**c**) shows phases of and transitions between growth and shrinkage. The seed is overlaid in *red*. Note that the EB3-GFP signal at the microtubule plus end is saturated to allow visualisation of the lattice. (**e**) Stills from the microtubule for time points in the boxed area in (**d**). The seed is highlighted with *dotted lines*. Phases of growth (G) and shrinkage (S) as well as transition events, catastrophe (C), and rescue (R) at the plus end are indicated above. Note that the amount of EB protein bound to the microtubule tip depends on the state: shrinking ends bind no EB3-GFP, growing ends bind EB3 proportional to the growth speed. Time difference between frames: 1 s. *Scale bar*: 5 μm.

tubulin (35). The high image contrast allows the robust application of automatic image analysis tools (34). The disadvantages of darkfield microscopy are the requirement for extremely clean surfaces, as any raised feature will create a signal and the need for very strong illumination, which has the potential to cause damage to the sample.

Another technique to visualise single, unlabelled microtubules in vitro (9, 36) and also in some (very flat) cells (10, 37) is video-enhanced differential interference contrast (VE-DIC) microscopy. DIC relies on the differential effect two components have on the optical path of a split beam; interference during the recombination of the split beams (which have a small offset) specifically enhances the differences in an image, and thus the contrast at the edges of objects. This allows visualisation of objects, such as microtubules, which have little absorbance (Fig. 2b). DIC of individual microtubules requires a quite complicated setup (38) and results in relatively low-contrast images that cannot be directly observed visually. However, kymograph analysis of DIC-imaged dynamic microtubules is possible (39), and recent advances in image filtering and analysis algorithms allow the automatic estimation of dynamic instability parameters from such kymographs (40).

An increasingly popular way to visualise the dynamics of individual microtubules in vitro is total internal reflection fluorescence (TIRF) microscopy (19, 41) (Fig. 2c). In TIRF, the excitation light is totally reflected at the glass/water interface and generates an electromagnetic field (evanescent wave) that selectively excites fluorescent molecules close to the coverslip surface (42). This limited fluorescence excitation has two great benefits: it reduces photodamage induced in the bulk sample and results in a superior signal-to-noise ratio as only molecules in a ~100-nm deep layer are excited. But of course, this technique requires labelling with a fluorophore either on tubulin or a tubulin-binding protein. Microtubules that contain tubulin coupled to a fluorophore are more susceptible to photodamage and one could speculate that this applies to other proteins too (35). Thus, care has to be taken to limit exposure. Recently, the Akhmanova lab has shown that dynamics data obtained with unlabelled tubulin in VE-DIC and TIRF (in the presence of mCherry-EB3, which labels microtubule ends in TIRF) are virtually identical (41). As EB3 alters microtubule dynamics on its own and binds to a variety of microtubule dynamics regulators (39, 41), its use as a marker in microtubule dynamics assays should ideally be limited to physiological concentrations (about 50 nM in undifferentiated mammalian cells, own unpublished data).

All methods described above are limited by the resolution of light microscopy (about 200–300 nm). Attempts have been made to study tubulin assembly at single heterodimer resolution using optical trapping setups (17, 43). Bead displacements in the optical trap can be determined using a quadrant detector with an accuracy of few nanometres sufficient to allow visualisation of, for example, individual steps taken by a walking kinesin molecule (44). Kinesin's 8 nm steps correspond to the length of a tubulin heterodimer, thus such a setup should in principle allow detection of addition or loss of a single layer of tubulin to the end of a microtubule. To do this, microtubules attached to beads are grown against a microfabricated barrier and are pushed back as they elongate. The positional change of the bead attached to the microtubule (or the axoneme it is growing from) is determined with high spatial and temporal accuracy. While two studies using this technique interpret the data differently and therefore come to different conclusions, the power to resolve tubulin assembly at below 8 nm resolution has been demonstrated (17) and will help to gain detailed molecular insight into the mechanisms of microtubule instability regulators in the future.

The interpretation of such precise microtubule length measurements requires knowledge of the structure of microtubule ends. Also, as these measurements are done under a (albeit small) load, the flexibility of microtubule end structures needs to be considered. The flexibility of microtubules can be measured in different ways: as the shape change due to thermal fluctuations; as the bending in hydrodynamic flow or under compressive load; and as the behaviour during relaxation after application of a drag force in the optical trap (45–48). While the values for flexural rigidity of microtubules obtained by different labs and using different techniques vary by a factor of >10 (see (49) for a comparison), microtubules can be regarded as relatively stiff compared to other cytoskeletal filaments. Whether this applies to all structures that have been observed at the ends of microtubules in the electron microscope is questionable. Microtubule ends in cells are rarely blunt ended (50, 51), and sheets, flared, or tapered ends with one or few protofilaments leading are commonly observed on elongating microtubules (50–54). Thus, measured length fluctuations will be the sum of tubulin addition/loss and structural changes, such as sheet closure (53). It is likely that microtubule growth under compressive load will affect the length or number of leading protofilaments (17), which could have a significant effect on growth behaviour. There are indications that the end structures change when microtubules grow in the presence of +TIPs, such as EB1 (55) or XMAP250 (43), which will make the interpretation of nanoscale length measurements even more complicated, and the correlation of dynamic measurements under load and without load with electron microscopy of end structures will be important to gain mechanistic information about the regulation of microtubule dynamics.

3. Imaging Microtubule Dynamic Instability in Cells

Dynamic instability has been visualised in cells using VE-DIC (10). However, this technique is restricted to very thin parts of cells. The introduction of labelled tubulin or MAPs is necessary to study microtubule dynamics in most cell types. Tubulin labelling can be done chemically, and the labelled tubulin is then injected into cells (56, 57). Alternatively, DNA encoding fusions of fluorescent proteins (FPs) with tubulin can either be transiently transfected/injected or stably integrated into the genome to label microtubules. Depending on the amount of label used, microtubules can be labelled uniformly or "speckled" (24, 58, 59).

There is an immense variety of FPs available, but eGFP remains the fluorophore with the best overall characteristics regarding brightness, photostability, and utility in fusions (60), and eGFP-α-tubulin is widely used as a microtubule marker. GFP-Tub3 (one of the two α-tubulin genes in budding yeast) can complement the *tub3* null mutant (59). However, GFP-tubulin cannot fully replace endogenous tubulin in yeast and cell viability is impaired at high expression levels of GFP-tubulin fusion proteins (61, 62), so a low dosage of excitation light should be used to obtain meaningful dynamics data. Generally, FPs with a tendency to oligomerise do not perform well as tubulin fusions; in some cases, a well-designed linker or "GFPised" C-terminus can produce a fusion that is functionally more normal (mCherry-Tubulin: (63); GFP-EB3: (39); EB1-GFP: (64)).

Beyond tubulin itself, FPs can be fused to lattice-decorating MAPs (65, 66) or +TIPs, such as EB family proteins and Clip170 (67, 68). As there is a viable GFP-Clip170 knock-in mouse (69) and dynamics measurements from cells using Cy3-labelled tubulin or YFP-Clip170 result in very similar microtubule growth characteristics (39), fluorescent Clip170 appears to be a marker for plus ends that is safe to use. Using a +TIP rather than uniformly labelled microtubules will not allow the measurement of shrinkage speed or rescue rates as these markers are lost from the microtubule end at catastrophe. However, in crowded cell areas, such as in neurites, at the front of migrating epithelial cells or in the mitotic spindle where tracking of individual microtubule ends is otherwise challenging, the use of +TIPs allows an estimate of changes in microtubule dynamics (70–72). +TIPs also allow determination of the rate of microtubule nucleation at different sites in the cell as a new EB comet is "born" when a microtubule is nucleated (73–75). Of course, new EB comets also appear during rescue events, when the new plus end grows after microtubule severing and when a growing microtubule enters the focal plane if imaging was restricted to a single focal plane (which is often necessary to image fast enough and limit photo-damage to the cell). The frequency of these events is strongly dependent on the sample (cell type, cell cycle/differentiation stage, thickness) and assessments have to be made for each case using uniformly labelled microtubules to determine the frequency of rescues, severing, and microtubules leaving/entering the focal plane.

It is also possible to assess microtubule dynamics without directly observing the microtubule end. A convenient way to estimate tubulin turnover in the spindle makes use of photoactivatable FPs. This technique was pioneered by Mitchison who synthesised a caged fluorescein and used this as a photoactivatable dye to label tubulin (76). He observed microtubule flux in mitotic spindles, but also noted very fast turnover of (probably) non-kinetochore MTs and slow turnover of (probably) kinetochore MTs. Today, the researcher has a wide choice of photoactivatable or photoconvertible FPs (77–80). Fluorescence decay after photoactivation allows determination of the half-life of tubulin molecules within the microtubule polymer, and at the same time the translocation of the photoactivated mark provides readout for microtubule flux (30). When applied with precision, the spatial regulation of microtubule dynamics can be probed using this method. Recently, an exciting new technique to determine the lifetime of tubulin subunits in the microtubule polymer has been applied to spindle microtubules: Needleman and colleagues took speckle microscopy a step further and imaged single tubulin molecules in *Xenopus* meiotic spindles (33). As free tubulin dimers move very fast, only those dimers that get incorporated into the microtubule polymer are imaged as discrete points. This allows determination of the distribution of lifetimes of all microtubules in the field of view and (if combined with markers for kinetochores, K-fibres or antiparallel microtubule overlaps) has the power to detect different dynamic properties of microtubule subpopulations within the mitotic spindle.

4. Analysis of Microtubule Dynamics in Cells

While determining the four (or five, if pause duration is included) parameters of dynamic instability will provide a good description of a microtubule dynamics phenotype in vitro, the spatiotemporal regulation of microtubule dynamics in cells makes this a more complex affair. Microtubules generally exhibit persistent growth in the cytoplasm and high transition frequencies at cell boundaries (29). The asymmetry in catastrophe frequencies ensures that at a given time, the majority of microtubule plus ends are close to the cell cortex (29, 81), and thus microtubule-based intracellular transport is most efficient. During mitosis, growth rates seem to increase during prophase and growth persistence is higher for spindle than astral microtubules (71). Thus, in order to get a complete understanding of how microtubule dynamics is regulated, microtubule subpopulations have to be studied separately, during different cell cycle and differentiation stages, free in the cytoplasm, and in proximity to cell boundaries or other subcellular interaction sites, such as kinetochores or focal adhesions.

Why might obtaining the mean values for the four parameters of dynamic instability not be the best way to describe microtubule dynamics in cells?

Growth speeds are usually a Gaussian distribution, sometimes with a positive skew. The latter might be caused by dynein-driven microtubule sliding (74). Rates during a growth phase can be quite variable (and disrupted by short pauses or even loss of several tubulin layers) (17, 26, 82). Thus, different values will be obtained if speed is measured as the average over a complete growth phase as compared to the measurement of instantaneous speeds between successive frames. Both approaches have merit, and additional insight might be gained by determining distributions for both instantaneous and overall growth speeds. There is an indication of spatial control of microtubule growth speed in cells, with slower growth being observed near the cell edges (70).

Shrinkage speeds are usually a continuous distribution with a positive skew (29, 62, 70, 83). Therefore, giving a mean might not be the best way to report these. Accurate measurement of shrinkage speeds depends a lot on the sampling rate as fast shrinkage events (speeds above 1.5 μm/s) have been reported in different systems (29, 62, 82) and are easily missed. The large range of shrinkage speeds measured suggests that there are different modes of shrinkage, which might explain why different microtubule end conformations have been observed in electron microscopy studies (14, 50, 54). Outward curling of protofilaments probably leads to very rapid catastrophes while shrinkage with blunt ends could be relatively slow. Indications for spatial control of shrinkage speeds have been reported: catastrophes induced at the cell cortex lead to faster shrinkage compared to spontaneous catastrophes occurring in the cytoplasm (28). Also microtubules have been observed to shrink considerably faster when contained within bundles than when separate (62).

Catastrophe frequencies are position-dependent in most cell types: the majority of catastrophes occur at cell boundaries (29, 84). Also the value obtained for catastrophe frequency depends largely on whether small-scale fluctuations are considered or designated as "pauses" (Fig 1).

Rescue frequency: Unless the small-scale transitions of captured microtubule ends are considered (Fig. 1), rescues are rare and hard to observe events in most cell types. This makes it difficult to obtain statistically robust data on cytoplasmic rescue events.

The emerging picture of microtubule dynamic behaviour suggests that the two-state model with stochastic switching between relatively constant periods of growth and shrinkage has deficiencies and is in need of a revision. Likewise, the measurement of the four parameters of dynamic instability might not be the most suitable way to describe microtubule dynamics. So, how can we build better-informed quantitative models of microtubule dynamics and find robust ways to describe the effect of microtubule regulators? Ideally, one would record complete microtubule life histories from nucleation to complete disassembly, correlating phases of the life history with bundling state and tip position/interactions (free in cytoplasm/close to boundary/pulling ER tubule along (85)/at focal adhesion (84) etc.). Parameters that could be readily compared between different datasets would be: (1) the distribution of speeds during growth and shrinkage phases; (2) the distribution of length or duration of growth or shrinkage phases (including short pauses and reversals); (3) the autocorrelation coefficient as a measure of how persistent such length excursions are; (4) dwell times in proximity to certain subcellular structures (e.g., cell membrane, adhesion sites, kinetochores) (28, 81, 86); and (5) the distribution of microtubule lifetimes. Measurements quantitatively describing length excursions and interaction modes should help robustly to describe both microtubule dynamics in wild-type cells and alterations in dynamics produced in mutants, disease states, or following drug treatment.

Acknowledgements

I would like to thank Frauke Hussmann and Mishan Britto for contributing images and Douglas Drummond and Rob Cross for critical comments on the manuscript. I am grateful to Anna Akhmanova for hosting me in her lab to learn how to perform in vitro microtubule plus-end-tracking assays. This work was funded by a Marie Curie Cancer Care programme grant to A.S.

References

1. Inoué, S. and Sato, H. (1967), Cell motility by labile association of molecules. The nature of mitotic spindle fibers and their role in chromosome movement. *J Gen Physiol* **50**: Suppl:259–92.
2. Borisy, G. G. and Taylor, E. W. (1967), The mechanism of action of colchicine. Binding of colchincine-3H to cellular protein. *The Journal of Cell Biology* **34**: 525–33.
3. Shelanski, M. L. and Taylor, E. W. (1967), Isolation of a protein subunit from microtubules. *The Journal of Cell Biology* **34**: 549–54.
4. Mohri, H. (1968), Amino-acid composition of "Tubulin" constituting microtubules of sperm flagella. *Nature* **217**: 1053–4.
5. Weisenberg, R. C. (1972), Microtubule formation in vitro in solutions containing low calcium concentrations. *Science* **177**: 1104–5.
6. Margolis, R. L. and Wilson, L. (1981), Microtubule treadmills--possible molecular machinery. *Nature* **293**: 705–11.
7. Mitchison, T. and Kirschner, M. (1984), Dynamic instability of microtubule growth. *Nature* **312**: 237–42.
8. Horio, T. and Hotani, H. (1986), Visualization of the dynamic instability of individual microtubules by dark-field microscopy. *Nature* **321**: 605–7.
9. Walker, R., O'brien, E., *et al.* (1988), Dynamic instability of individual microtubules analyzed by video light microscopy: rate constants and transition frequencies. *The Journal of Cell Biology* **107**: 1437.
10. Cassimeris, L., Pryer, N. K., and Salmon, E. D. (1988), Real-time observations of microtubule dynamic instability in living cells. *J Cell Biol* **107**: 2223–31.

11. Sammak, P. J. and Borisy, G. G. (1988), Direct observation of microtubule dynamics in living cells. *Nature* **332**: 724–6.
12. Howard, J. and Hyman, A. A. (2009), Growth, fluctuation and switching at microtubule plus ends. *Nat Rev Mol Cell Biol* **10**: 569–74.
13. Hyman, A. A., Chrétien, D., *et al.* (1995), Structural changes accompanying GTP hydrolysis in microtubules: information from a slowly hydrolyzable analogue guanylyl-(alpha,beta)-methylene-diphosphonate. *The Journal of Cell Biology* **128**: 117–25.
14. Müller-Reichert, T., Chrétien, D., *et al.* (1998), Structural changes at microtubule ends accompanying GTP hydrolysis: information from a slowly hydrolyzable analogue of GTP, guanylyl (alpha,beta)methylenediphosphonate. *Proc Natl Acad Sci USA* **95**: 3661–6.
15. Mandelkow, E. M., Mandelkow, E., and Milligan, R. A. (1991), Microtubule dynamics and microtubule caps: a time-resolved cryo-electron microscopy study. *J Cell Biol* **114**: 977–91.
16. Drechsel, D. N. and Kirschner, M. W. (1994), The minimum GTP cap required to stabilize microtubules. *Curr Biol* **4**: 1053–61.
17. Schek, H. T., 3rd, Gardner, M. K., *et al.* (2007), Microtubule assembly dynamics at the nanoscale. *Curr Biol* **17**: 1445–55.
18. Walker, R. A., Pryer, N. K., and Salmon, E. D. (1991), Dilution of individual microtubules observed in real time in vitro: evidence that cap size is small and independent of elongation rate. *J Cell Biol* **114**: 73–81.
19. Bieling, P., Laan, L., *et al.* (2007), Reconstitution of a microtubule plus-end tracking system in vitro. *Nature* **450**: 1100–5.
20. Tirnauer, J. S., Grego, S., *et al.* (2002), EB1-microtubule interactions in Xenopus egg extracts: role of EB1 in microtubule stabilization and mechanisms of targeting to microtubules. *Mol Biol Cell* **13**: 3614–26.
21. Zanic, M., Stear, J. H., *et al.* (2009), EB1 recognizes the nucleotide state of tubulin in the microtubule lattice. *PLoS One* **4**: e7585.
22. Grego, S., Cantillana, V., and Salmon, E. D. (2001), Microtubule treadmilling in vitro investigated by fluorescence speckle and confocal microscopy. *Biophys J* **81**: 66–78.
23. Rusan, N. M., Fagerstrom, C. J., *et al.* (2001), Cell cycle-dependent changes in microtubule dynamics in living cells expressing green fluorescent protein-alpha tubulin. *Mol Biol Cell* **12**: 971–80.
24. Shelden, E. and Wadsworth, P. (1993), Observation and quantification of individual microtubule behavior in vivo: microtubule dynamics are cell-type specific. *J Cell Biol* **120**: 935–45.
25. Waterman-Storer, C. M. and Salmon, E. D. (1997), Actomyosin-based retrograde flow of microtubules in the lamella of migrating epithelial cells influences microtubule dynamic instability and turnover and is associated with microtubule breakage and treadmilling. *J Cell Biol* **139**: 417–34.
26. Keller, P. J., Pampaloni, F., and Stelzer, E. H. K. (2007), Three-dimensional preparation and imaging reveal intrinsic microtubule properties. *Nature Methods* **4**: 843–6.
27. VanBuren, V., Cassimeris, L., and Odde, D. J. (2005), Mechanochemical model of microtubule structure and self-assembly kinetics. *Biophys J* **89**: 2911–26.
28. Straube, A. and Merdes, A. (2007), EB3 regulates microtubule dynamics at the cell cortex and is required for myoblast elongation and fusion. *Curr Biol* **17**: 1318–25.
29. Komarova, Y. A., Vorobjev, I. A., and Borisy, G. G. (2002), Life cycle of MTs: persistent growth in the cell interior, asymmetric transition frequencies and effects of the cell boundary. *J Cell Sci* **115**: 3527–39.
30. Amaro, A. C., Samora, C. P., *et al.* (2010), Molecular control of kinetochore-microtubule dynamics and chromosome oscillations. *Nat Cell Biol* **12**: 319–29.
31. van der Vaart, B., Akhmanova, A., and Straube, A. (2009), Regulation of microtubule dynamic instability. *Biochem Soc Trans* **37**: 1007–13.
32. Maiato, H., DeLuca, J., *et al.* (2004), The dynamic kinetochore-microtubule interface. *J Cell Sci* **117**: 5461–77.
33. Needleman, D. J., Groen, A., *et al.* (2010), Fast microtubule dynamics in meiotic spindles measured by single molecule imaging: evidence that the spindle environment does not stabilize microtubules. *Mol Biol Cell* **21**: 323–33.
34. Katsuki, M., Drummond, D. R., *et al.* (2009), Mal3 masks catastrophe events in Schizosaccharomyces pombe microtubules by inhibiting shrinkage and promoting rescue. *J Biol Chem* **284**: 29246–50.
35. Vigers, G. P., Coue, M., and McIntosh, J. R. (1988), Fluorescent microtubules break up under illumination. *The Journal of Cell Biology* **107**: 1011–24.
36. Bormuth, V., Howard, J., and Schäffer, E. (2007), LED illumination for video-enhanced

DIC imaging of single microtubules. *J Microsc* **226**: 1–5.
37. Allen, R. D., Allen, N. S., and Travis, J. L. (1981), Video-enhanced contrast, differential interference contrast (AVEC-DIC) microscopy: a new method capable of analyzing microtubule-related motility in the reticulopodial network of Allogromia laticollaris. *Cell Motil* **1**: 291–302.
38. Salmon, E. D. and Tran, P. (2007), High-resolution video-enhanced differential interference contrast light microscopy. *Methods Cell Biol* **81**: 335–64.
39. Komarova, Y., De Groot, C. O., *et al.* (2009), Mammalian end binding proteins control persistent microtubule growth. *J Cell Biol* **184**: 691–706.
40. Smal, I., Grigoriev, I., *et al.* (2010), Microtubule Dynamics Analysis Using Kymographs And Variable-Rate Particle Filters. *IEEE Trans Image Process*
41. Montenegro Gouveia, S., Leslie, K., *et al.* (2010), In Vitro Reconstitution of the Functional Interplay between MCAK and EB3 at Microtubule Plus Ends. *Curr Biol* **20**: 1717–22.
42. Axelrod, D., Thompson, N. L., and Burghardt, T. P. (1983), Total internal inflection fluorescent microscopy. *J Microsc* **129**: 19–28.
43. Kerssemakers, J. W. J., Munteanu, E. L., *et al.* (2006), Assembly dynamics of microtubules at molecular resolution. *Nature* **442**: 709–12.
44. Carter, N. J. and Cross, R. A. (2005), Mechanics of the kinesin step. *Nature* **435**: 308–12.
45. Felgner, H., Frank, R., and Schliwa, M. (1996), Flexural rigidity of microtubules measured with the use of optical tweezers. *J Cell Sci* **109 (Pt 2)**: 509–16.
46. Venier, P., Maggs, A. C., *et al.* (1994), Analysis of microtubule rigidity using hydrodynamic flow and thermal fluctuations. *J Biol Chem* **269**: 13353–60.
47. Gittes, F., Mickey, B., *et al.* (1993), Flexural rigidity of microtubules and actin filaments measured from thermal fluctuations in shape. *J Cell Biol* **120**: 923–34.
48. Mickey, B. and Howard, J. (1995), Rigidity of microtubules is increased by stabilizing agents. *J Cell Biol* **130**: 909–17.
49. Kikumoto, M., Kurachi, M., *et al.* (2006), Flexural rigidity of individual microtubules measured by a buckling force with optical traps. *Biophys J* **90**: 1687–96.
50. Zovko, S., Abrahams, J. P., *et al.* (2008), Microtubule plus-end conformations and dynamics in the periphery of interphase mouse fibroblasts. *Mol Biol Cell* **19**: 3138–46.
51. Hoog, J. L., Schwartz, C., *et al.* (2007), Organization of interphase microtubules in fission yeast analyzed by electron tomography. *Dev Cell* **12**: 349–61.
52. Arnal, I., Karsenti, E., and Hyman, A. A. (2000), Structural transitions at microtubule ends correlate with their dynamic properties in Xenopus egg extracts. *J Cell Biol* **149**: 767–74.
53. Chrétien, D., Fuller, S. D., and Karsenti, E. (1995), Structure of growing microtubule ends: two-dimensional sheets close into tubes at variable rates. *The Journal of Cell Biology* **129**: 1311–28.
54. Simon, J. R. and Salmon, E. D. (1990), The structure of microtubule ends during the elongation and shortening phases of dynamic instability examined by negative-stain electron microscopy. *J Cell Sci* **96 (Pt 4)**: 571–82.
55. Vitre, B., Coquelle, F. M., *et al.* (2008), EB1 regulates microtubule dynamics and tubulin sheet closure in vitro. *Nat Cell Biol* **10**: 415–21.
56. Keith, C. H., Feramisco, J. R., and Shelanski, M. (1981), Direct visualization of fluorescein-labeled microtubules in vitro and in microinjected fibroblasts. *The Journal of Cell Biology* **88**: 234–40.
57. Saxton, W. M., Stemple, D. L., *et al.* (1984), Tubulin dynamics in cultured mammalian cells. *J Cell Biol* **99**: 2175–86.
58. Waterman-Storer, C. M., Desai, A., *et al.* (1998), Fluorescent speckle microscopy, a method to visualize the dynamics of protein assemblies in living cells. *Curr Biol* **8**: 1227–30.
59. Carminati, J. L. and Stearns, T. (1997), Microtubules orient the mitotic spindle in yeast through dynein-dependent interactions with the cell cortex. *The Journal of Cell Biology* **138**: 629–41.
60. Davidson, M. W. and Campbell, R. E. (2009), Engineered fluorescent proteins: innovations and applications. *Nature Methods* **6**: 713–7.
61. Ding, D. Q., Chikashige, Y., *et al.* (1998), Oscillatory nuclear movement in fission yeast meiotic prophase is driven by astral microtubules, as revealed by continuous observation of chromosomes and microtubules in living cells. *J Cell Sci* **111 (Pt 6)**: 701–12.
62. Steinberg, G., Wedlich-Söldner, R., *et al.* (2001), Microtubules in the fungal pathogen Ustilago maydis are highly dynamic and determine cell polarity. *J Cell Sci* **114**: 609–22.
63. Shaner, N. C., Campbell, R. E., *et al.* (2004), Improved monomeric red, orange and yellow

fluorescent proteins derived from Discosoma sp. red fluorescent protein. *Nat Biotechnol* **22**: 1567–72.

64. Jankovics, F. and Brunner, D. (2006), Transiently reorganized microtubules are essential for zippering during dorsal closure in Drosophila melanogaster. *Dev Cell* **11**: 375–85.
65. Faire, K., Waterman-Storer, C. M., *et al.* (1999), E-MAP-115 (ensconsin) associates dynamically with microtubules in vivo and is not a physiological modulator of microtubule dynamics. *J Cell Sci* **112 (Pt 23)**: 4243–55.
66. Olson, K. R. and Olmsted, J. B. (1999), Analysis of microtubule organization and dynamics in living cells using green fluorescent protein-microtubule-associated protein 4 chimeras. *Methods Enzymol* **302**: 103–20.
67. Perez, F., Diamantopoulos, G. S., *et al.* (1999), CLIP-170 highlights growing microtubule ends in vivo. *Cell* **96**: 517–27.
68. Stepanova, T., Slemmer, J., *et al.* (2003), Visualization of microtubule growth in cultured neurons via the use of EB3-GFP (end-binding protein 3-green fluorescent protein). *J Neurosci* **23**: 2655–64.
69. Akhmanova, A., Mausset-Bonnefont, A.-L., *et al.* (2005), The microtubule plus-end-tracking protein CLIP-170 associates with the spermatid manchette and is essential for spermatogenesis. *Genes Dev* **19**: 2501–15.
70. Matov, A., Applegate, K., *et al.* (2010), Analysis of microtubule dynamic instability using a plus-end growth marker. *Nature Methods* **7**: 761–8.
71. Piehl, M. and Cassimeris, L. (2003), Organization and dynamics of growing microtubule plus ends during early mitosis. *Mol Biol Cell* **14**: 916–25.
72. Stepanova, T., Smal, I., *et al.* (2010), History-dependent catastrophes regulate axonal microtubule behavior. *Curr Biol* **20**: 1023–8.
73. Efimov, A., Kharitonov, A., *et al.* (2007), Asymmetric CLASP-dependent nucleation of noncentrosomal microtubules at the trans-Golgi network. *Dev Cell* **12**: 917–30.
74. Srayko, M., Kaya, A., *et al.* (2005), Identification and characterization of factors required for microtubule growth and nucleation in the early C. elegans embryo. *Dev Cell* **9**: 223–36.
75. Straube, A., Brill, M., *et al.* (2003), Microtubule organization requires cell cycle-dependent nucleation at dispersed cytoplasmic sites: polar and perinuclear microtubule organizing centers in the plant pathogen Ustilago maydis. *Mol Biol Cell* **14**: 642–57.
76. Mitchison, T. J. (1989), Polewards microtubule flux in the mitotic spindle: evidence from photoactivation of fluorescence. *J Cell Biol* **109**: 637–52.
77. McKinney, S., Murphy, C., *et al.* (2009), A bright and photostable photoconvertible fluorescent protein. *Nature Methods* **6**: 131–3.
78. Patterson, G. H. and Lippincott-Schwartz, J. (2002), A photoactivatable GFP for selective photolabeling of proteins and cells. *Science* **297**: 1873–7.
79. Subach, F. V., Patterson, G. H., *et al.* (2009), Photoactivatable mCherry for high-resolution two-color fluorescence microscopy. *Nature Methods* **6**: 153–9.
80. Subach, F. V., Patterson, G. H., *et al.* (2010), Bright monomeric photoactivatable red fluorescent protein for two-color super-resolution sptPALM of live cells. *J Am Chem Soc* **132**: 6481–91.
81. Drummond, D. R. and Cross, R. A. (2000), Dynamics of interphase microtubules in Schizosaccharomyces pombe. *Curr Biol* **10**: 766–75.
82. Gildersleeve, R. F., Cross, A. R., *et al.* (1992), Microtubules grow and shorten at intrinsically variable rates. *J Biol Chem* **267**: 7995–8006.
83. Yvon, A. M. and Wadsworth, P. (1997), Non-centrosomal microtubule formation and measurement of minus end microtubule dynamics in A498 cells. *J Cell Sci* **110 (Pt 19)**: 2391–401.
84. Efimov, A., Schiefermeier, N., *et al.* (2008), Paxillin-dependent stimulation of microtubule catastrophes at focal adhesion sites. *Journal of Cell Science* **121**: 196–204.
85. Grigoriev, I., Gouveia, S. M., *et al.* (2008), STIM1 is a MT-plus-end-tracking protein involved in remodeling of the ER. *Curr Biol* **18**: 177–82.
86. Mimori-Kiyosue, Y., Grigoriev, I., *et al.* (2005), CLASP1 and CLASP2 bind to EB1 and regulate microtubule plus-end dynamics at the cell cortex. *J Cell Biol* **168**: 141–53.

Chapter 2

Purification of Tubulin from Porcine Brain

Christopher Gell, Claire T. Friel, Barbara Borgonovo, David N. Drechsel, Anthony A. Hyman, and Jonathon Howard

Abstract

Microtubules, polymers of the heterodimeric protein αβ-tubulin, give shape to cells and are the tracks for vesicle transport and chromosome segregation. In vitro assays to study microtubule functions and their regulation by microtubule-associated proteins require the availability of purified αβ-tubulin. In this chapter, we describe the process of purification of heterodimeric αβ-tubulin from porcine brain.

Key words: Tubulin, Purification, Porcine brain, Microtubules, Phosphocellulose

1. Introduction

The study of microtubules, reconstituted in vitro from purified heterodimeric αβ-tubulin, has provided valuable insights into the properties of microtubules as well as the function of many microtubule-associated proteins (MAPs) (1–5). In particular, the use of fluorescently labelled tubulin in single-molecule in vitro assays has developed our understanding of microtubule polymerisation (6), depolymerisation (7), and the role of MAPs in these processes (3, 5, 8). The lack of an efficient system for the over-expression of αβ-tubulin requires that this protein be obtained by purification from animal brains (usually porcine or bovine). The process of purification (outlined in Fig. 1) consists of homogenisation of the brains, followed by cycles of tubulin polymerisation (in the presence of GTP), centrifugation, and depolymerisation (9–11). Thereby, the particular property of αβ-tubulin to form microtubules is used to isolate soluble tubulin from cell membranes and non-MAPs. The resulting solution of soluble tubulin plus MAPs

Anne Straube (ed.), *Microtubule Dynamics: Methods and Protocols*, Methods in Molecular Biology, vol. 777,
DOI 10.1007/978-1-61779-252-6_2,

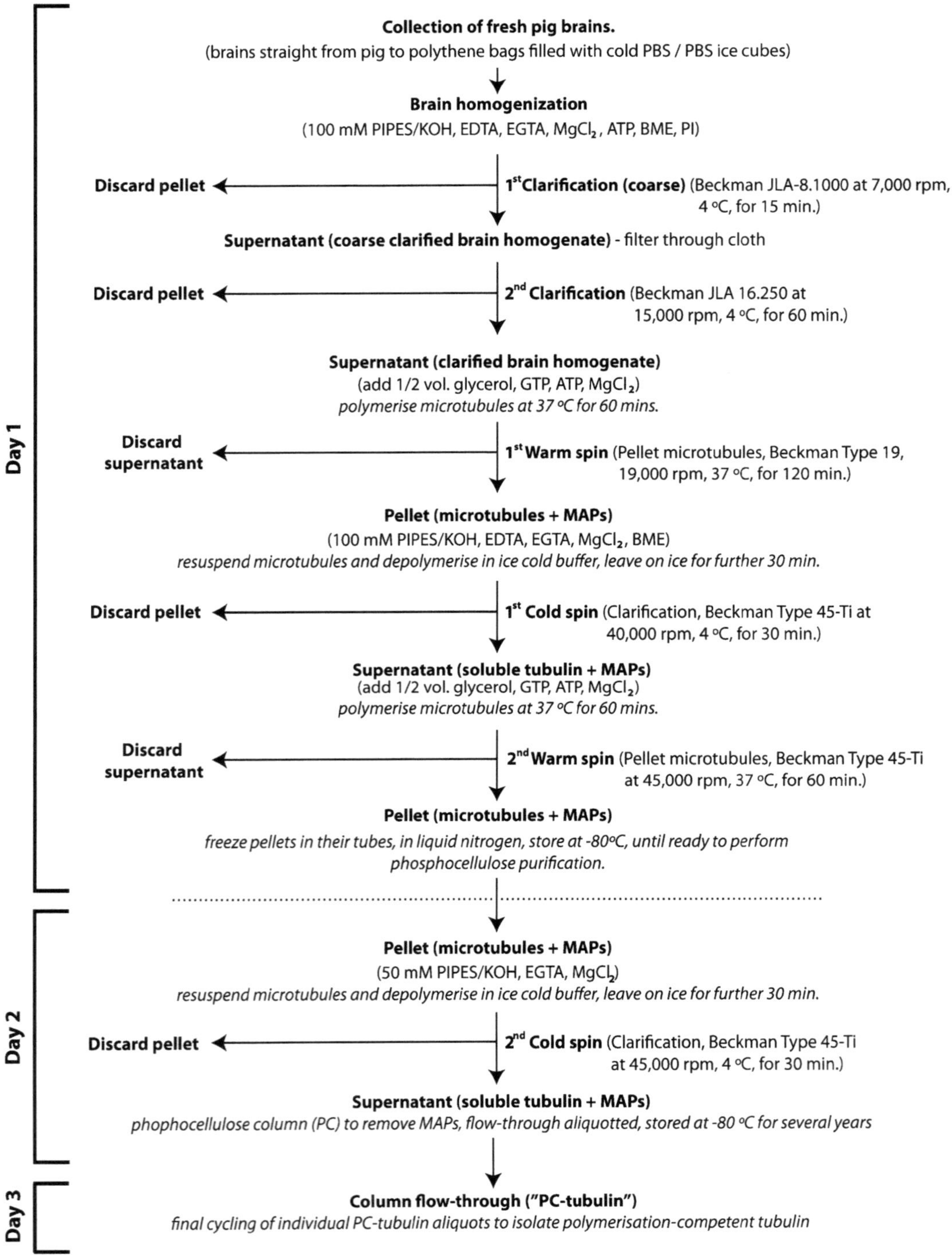

Fig. 1. Outline of the steps in the purification of $\alpha\beta$-tubulin from porcine brains, using alternate steps of polymerisation (in the presence of GTP) and depolymerisation, followed by the removal of MAPs on a phosphocellulose column. Figure style based on ref. 10.

is then passed over a phosphocellulose (PC) column in order to remove the remaining MAPs, which bind to the column while the soluble tubulin flows through. The purified tubulin is generally stored long-term in this state. Prior to use for subsequent in vitro studies, a final cycling (a step of polymerisation–depolymerisation) ensures that only polymerisation-competent tubulin is present. The protocol presented here describes the brain homogenisation, initial cycles of polymerisation/depolymerisation, and phosphocellulose column stages of the purification.

Below is the protocol we use for the purification of porcine brain tubulin on a large scale (typically, starting with ~75 pig brains and resulting in several grams of purified tubulin, see Fig. 1). However, the basic protocol can be scaled down, the most important consideration being the capacity of the available centrifuges; it is essential to consider the reductions and increases in volume expected at each stage and to match this to the available centrifuges to minimise loss. To this end, throughout the protocol, we give an indication of the *typical* volumes expected at each stage (references to Notes). However, significant variability (differing size of the pig brains obtained; amount of supernatant recovered in each step; volume needed for re-suspension of pellets) should be expected. We note that the investment of time necessary for the preparation is somewhat independent of the starting material. Thus, large preparations are an efficient use of time; given the fact that tubulin is stable for years when stored at -80°C.

2. Materials

Unless otherwise noted, the quantities (volumes) given are those that we typically find necessary for a preparation starting with ~75 pig brains (~10 kg). (Items 1–17 and 18–30 detail the requirements for the initial tubulin preparation and phosphocellulose column, respectively).

1. Phosphate-buffered saline (PBS) to make ice cubes (~7.5 L).
2. PBS (~20 L, chilled to 4°C).
3. Large, strong plastic bags, each sufficient to hold 2.5 L PBS ice cubes, 3 L cold PBS, and ~25 pig brains.
4. Sufficient glassware (beakers, measuring cylinders, etc.) pre-cooled or pre-warmed as necessary.
5. Large flasks with magnetic stirrers for tubulin polymerisation; pre-warm flasks to 37°C (e.g. Wheaton Magna Flex Spinner Flasks, 3 L).
6. ~75 pig brains (typically, 10 kg, final yield ~1–3 g of purified tubulin). Brains should be as fresh as possible. It is important

to get the brains into the mixture of chilled/ice PBS as quickly as possible. We achieve a delay of only a few minutes from pig to PBS. The time that the brains spend chilled in this way should also be kept to a minimum. However, in our particular case, transport times dictate a delay of over an hour at this stage and high yields of tubulin are nonetheless obtained.

7. Glycerol (~6 L, pre-warmed to 37°C).
8. Buffer P (polymerisation buffer): 100 mM PIPES/KOH pH 6.85, 0.1 mM EDTA, 2 mM EGTA, 0.5 mM $MgCl_2$ (final concentrations). Prepare 3 L of 5× Buffer P.
9. Protease inhibitors (PI) 1,000× stock solutions: 1 mg/ml aprotinin in water, 1 mg/ml TAME in water, 10 mg/ml trypsin inhibitor in water, 1 mg/ml pepstatin in ethanol, 1 mg/ml TPCK in ethanol, 5 mg/ml leupeptin in ethanol, 100 mM PMSF in ethanol. PI stock solutions can be prepared prior to the purification and stored at −20°C. We typically prepare 10 mL of *each* PI stock solution.
10. 100 mM Mg-ATP (~260 mL, ATP is dissolved in a 100 mM $MgCl_2$ solution and the pH adjusted to 6.8 using KOH). The concentration should be checked by absorption at 260 nm ($\varepsilon = 15{,}400\ M^{-1}\ cm^{-1}$) (12), and solution should be stored at −20°C.
11. 200 mM GTP (~60 mL, GTP is dissolved in distilled water, and the pH adjusted to 6.8 with KOH). The concentration should be checked by absorption at 260 nm ($\varepsilon = 11{,}700\ M^{-1}\ cm^{-1}$) (12], and solution should be stored at −20°C.
12. β-mercaptoethanol (BME).
13. Buffer P+ (prepared on the morning of the preparation): Buffer P, 1× of *each* PI (see Item 9), 1 mM ATP (see Item 10), 0.1% BME.
14. Blender for brain homogenisation (e.g. a 4-L Waring Blender, 38BL30).
15. Dounce homogenisers (Kontes, Vineland, NJ, USA; 885300-0040, 885301-0040, 885302-0040).
16. Appropriate centrifuges, rotors, and centrifuge tubes, pre-warmed or pre-cooled as necessary throughout the preparation.
17. Three large heated water baths for polymerisation. We use large (~50 L) plastic storage boxes, filled with water, each fitted with a heating recirculating water bath head (e.g. Thermo Haake, 003-2859). One bath set to 55°C, and two set to 37°C. Baths must be large enough to accommodate polymerisation flasks (see Item 5).
18. 10× column buffer (CB): 500 mM PIPES/KOH pH 6.85, 10 mM EGTA, 2 mM $MgCl_2$. (Prepare 5 L).

19. CB: 50 mM PIPES/KOH pH 6.85, 1 mM EGTA, 0.2 mM $MgCl_2$ (20 L); prepared from 10× CB.
20. CB + 1 M KCl (12 L); prepared using 10× CB and KCl.
21. Whatman P11 cellulose phosphate (PC, 300 g Whatman 4071050; ~4 mL wet resin/1 g dry weight).
22. 0.5 M NaOH (5 L).
23. 0.5 M HCl (5 L).
24. 0.5 M K-phosphate pH 6.8 (13 L).
25. ddH_2O (5 L).
26. BSA (Fraction V, protease-free; SERVA, 11926).
27. 5 × 2-L beakers, 1 × 4-L beaker, 2× stirring rods.
28. Aspirator pump (Vacuubrand ME16C) with 10 L trap.
29. Peristaltic pump; tubing.
30. Housings for column and pre-column (XK 50/20, XK 50/100 column housings; GE Healthcare 18-8753-01, 18-1000-71).

3. Methods

3.1. Several Weeks Before the Preparation

3.1.1. Equipment and Resource Arrangements

1. Arrange for the collection of pig brains as early as possible on the day of preparation. The use of fresh pig brains is essential.
2. Order sufficient quantities of all reagents.
3. Book all equipment (centrifuges, rotors, warm/cold room time, balances, douncers, blenders, etc. as necessary in your lab; details of the centrifuges, rotors, and speeds we use are stated in the text and notes).
4. Arrange for sufficient manpower for the preparation, ensuring that each person is aware of his or her duties. For a 75 brain prep, we find that the involvement of a minimum of four people is necessary.

3.1.2. Phosphocellulose Column Preparation

Column preparation (Items 18–30) largely follows that described by the Mitchison Lab, with modifications (13). The dry resin is acid/base-treated, defined, and then packed into the column housing. After packing, the resin beds are washed with high-salt, then low-salt buffer, treated with BSA to block non-specific binding to the phosphocellulose, and finally equilibrated with buffer.

Due to the high viscosity of the crude microtubule protein loaded on the PC column, a smaller guard column containing 100 mL of the same cycled resin is connected upstream of the main 1 L column.

1. Base treatment of PC: Add 50 g of dry PC to each of 5 × 2-L beakers containing 1 L of 0.5 M NaOH. Stir gently until resin is wetted and re-suspended into a uniform slurry. Let the resin settle for 5 min and then aspirate the supernatant, including any fines. Quickly add 1 L 0.5 M K-phosphate to the settled resin, stir, check that the pH is near neutral, and then let stand for 5 min. Again aspirate the supernatant, add 1 L of H_2O, re-suspend the resin by stirring, and then allow to settle for 5 min prior to aspirating the supernatant.
2. Acid treatment of PC: Add 1 L 0.5 M HCl to the settled resin, re-suspend, and allow to settle for 5 min. Aspirate the supernatant, add 1 L 0.5 M K-phosphate, stir, check the pH is near neutral, and let settle again for 5 min prior to aspirating the supernatant. Then, combine the resin in all the beakers into a single 4-L beaker and add 3 L 0.5 M K-phosphate. Allow to settle, and aspirate the supernatant. Add 1 L CB + 1 M KCl (see Item 20), stir the resin, allow to settle, and then aspirate. Repeat this wash step with 1 L CB + 1 M KCl twice more.
3. Packing the columns: Pour the resin slurry into the column housings in the cold room and then pack the bed at ~15 mL/min using a peristaltic pump attached to the column outlet to control the flow rate. Attach a flow adapter to the top of the column housing and run 7 L of CB + 1 M KCl through at a flow rate of 5–10 mL/min.
4. Washing, blocking, and equilibrating the column: Continue washing with 10 L of CB monitoring the conductivity to ensure that KCl is fully eluted. Adjust the level of the flow adapter to allow an adequate reserve of buffer at the top of the bed, as the resin expands when going from high to low salt. Load 300 mL of 30 mg/mL BSA in CB, chase with 700 mL of CB, and then stop the flow for 2 h to allow adsorption of the BSA. This step blocks tubulin from binding irreversibly to the fresh resin and needs to be performed only once prior to fractionating crude microtubule protein on the PC column. Finally, wash out the BSA with 2 L CB + 1 M KCl and then equilibrate the column with 10 L of CB. *This last CB wash should be done in the evening before the preparation.*

3.2. Several Days Before the Preparation

1. Freeze 7.5 L PBS in plastic bags, ~1 L per bag, break into rough ice cubes.
2. Chill a further 20 L of PBS.
3. Prepare all stock solutions: buffers, nucleotides, and protease inhibitors; ensure that an excess of these solutions is available, see Materials for details.
4. Pack the PC column, block with BSA, and wash with high salt as described in Subheading 3.1.2, steps 3 and 4.

3.3. Evening Before the Preparation

1. Collect all centrifuge rotors to be used in the preparation and pre-warm/cool as necessary.
2. Check the operation of all centrifuges and other essential equipment (e.g. blenders).
3. Place glycerol at 37°C.
4. Equilibrate PC column with 1× CB as described in Subheading 3.1.2, step 4.

3.4. Morning of the Preparation

Early on the morning of the preparation, one team can collect the brains (see Item 6) while a second makes the following preparations:

1. Pre-cool centrifuges necessary for the first two coarse clarification spins to 4°C.
2. If additional centrifuges are available, pre-warm to 37°C ready for subsequent warm spin (microtubule pelleting).
3. Thaw any frozen stock solutions.
4. Prepare cold and warm rooms with all necessary equipment.
5. Prepare Buffer P+ (see Item 13).

3.5. Purification of Tubulin and MAPs from Brains

3.5.1. Brain Homogenisation

1. Store pig brains in PBS ice/chilled PBS-filled bags until ready for use.
2. Add a mixture of brains (see Note 1) and Buffer P+ to the blender using a ratio of 1 L of Buffer P+ to 1 kg of brains; this typically results in ~18 L final volume for ~75 brains.
3. Homogenise the brain/buffer mixture using short pulses of the blender, separated by pauses in order to avoid heating the homogenate.
4. The final homogenate should have the colour and consistency of a good strawberry milkshake.

3.5.2. Coarse Clarification of Homogenate

1. Clarify the brain homogenate (~18 L total volume) by centrifugation at ~7,000 rpm (~12,000 × *g*, Beckman JLA-8.1000 rotor, see Note 2) for 15 min at 4°C. (Also see Note 3 for important tips related to this and all subsequent centrifugation steps).
2. Recover the supernatant from the centrifuge bottles by pouring carefully into chilled glass beakers. During this step, we filter the supernatant through coarse cotton as it is poured (we use cloth diapers) to prevent any of the loose pellet being collected (see Note 4).
3. Perform a second clarification step of the recovered supernatant (clarified brain homogenate, total volume now ~7 L) by centrifugation at ~15,000 rpm (~33,000 × *g*, Beckman JLA 16.250 rotor) for 60 min at 4°C (see Note 5).

4. Recover the supernatant which contains the soluble protein (typically, ~6 L of supernatant is recovered). This is used for the first tubulin polymerisation step.
5. Check the temperature of the centrifuges required for the next warm spin and ensure that they are at 37°C. Clean and pre-warm centrifuge tubes/rotors as necessary.

3.5.3. Microtubule Polymerisation/ Depolymerisation Cycles

1. To the chilled supernatant, add (final concentrations) 1 mM GTP, 1.5 mM ATP, and 4 mM $MgCl_2$.
2. Add a half volume of pre-warmed glycerol (33% v/v final).
3. Transfer the mixture to a suitable stirring flask (see Note 6).
4. Warm the mixture as quickly as possible to 30°C; we do this by placing the flask in a water bath at 55°C. The polymerisation mixture should be continuously stirred and the temperature monitored closely (see Note 7).
5. Once the solution has reached 30°C (and no higher), transfer the flask to a water bath at 37°C. Incubate for 1 h at 37°C, stirring occasionally (see Notes 7 and 8).
6. After incubation, transfer the viscous solution (now a total volume of ~9 L) to suitable centrifuge tubes, and pellet the microtubules (and MAPs) by centrifugation at ~19,000 rpm (~53,000 × *g*, Beckman Type 19 rotor) for 120 min at 37°C (see Note 9).
7. After centrifugation, ensure that centrifuges and rotors reused in the subsequent step are pre-cooled to 4°C.
8. Discard the supernatant (see Note 10).
9. Store the pellets on ice and re-suspend in ice-cold Buffer P with additionally 0.1% (v/v) BME; typically, ~10 mL per pellet is sufficient for re-suspension (see Note 11).
10. Dounce the re-suspended microtubule/MAPs solution (working in a cold room) until a homogeneous solution is obtained; thorough re-suspension of the pellets is essential (see Note 11).
11. If the pellets are re-suspended separately or in batches (see Note 11), pool the entire solution (keeping on ice), mix well, and determine the approximate concentration using a Bradford assay (see Note 12). If necessary, lower the concentration to ~25 mg/ml (see Note 13) with the addition of cold Buffer P.
12. Incubate the solution for a further 30 min on ice to ensure complete depolymerisation of the microtubules. The total volume is now typically ~840 mL.
13. Clarify the soluble tubulin/MAPs solution by centrifugation at 40,000 rpm (~186,000 × *g*, Beckman Type 45-Ti rotor) for 30 min at 4°C (see Note 14 and 15). After this spin, set the centrifuges to 37°C as required for the next warm spin.

14. Recover the supernatant for the second round of tubulin polymerisation (the reduced volume of the clarified solution is typically now ~560 mL from 840 mL); repeat Subheading 3.5.3, Steps 1–5.
15. After incubation, transfer the viscous solution (typically, ~840 mL) to suitable centrifuge tubes and pellet the microtubules (and MAPs) by centrifugation at ~45,000 rpm (~235,000 × g, Beckman Type 45-Ti rotor) for 60 min at 37°C (see Notes 15 and 16).
16. Discard the supernatant: At this point, the purification can be paused. We do this by freezing the pellets still in their centrifuge tubes. First, cool the pellets and tubes by placing them on ice for 5 min. Then, snap-freeze in liquid nitrogen and store at −80°C. We have successfully paused the protocol at this point for several months with no apparent loss in final tubulin yield.
17. Re-suspend half (six) of the pellets in buffer 1× CB and using ~20 mL/pellet. (This typically results in ~200 mL total volume, see Note 17.) The remaining frozen pellets can be re-suspended and purified at a later date in a second run over the PC column after cleaning and re-equilibration of the resin.
18. Dounce the re-suspended microtubule/MAPs solution (working in a cold room) until a homogeneous solution is obtained.
19. Pool the solution (keeping on ice), mix well, and allow the microtubules in the solution to depolymerise for a further 30 min on ice (see Note 17).
20. Clarify the soluble tubulin/MAPs solution one final time at 45,000 rpm (~235,000 × g, Beckman Type 45-Ti rotor) for 30 min at 4°C (see Note 18).
21. Recover the supernatant, taking care not to disturb the unstable pellet, into a pre-cooled glass bottle (recovered volume ~150 mL) kept on ice discarding the pellets. Check the protein concentration by Bradford assay (see Note 12) and if necessary, adjust to 20 mg/mL with 1× CB. Solution (soluble tubulin/MAPs) is now ready for phosphocellulose column chromatography.

3.5.4. Purification of Tubulin on a Phosphocellulose Column

1. Introductory note: Due to the high viscosity of the tubulin sample and the compressibility of the PC column bed, a pre-column, one tenth the volume of the main column, is used. The pre-column binds much of the pink, smelly, sticky stuff that normally clogs the column and leads to low flow rates and resin compression. The pre-column can either be run alone or in-line with the main column. If run alone, the flow through is pooled and loaded immediately onto the main PC

column. For a 100 mL pre-column, tubulin begins to elute ~70 mL after loading.

2. The sample is then applied at 5 mL/min to the large 1 L column and then, after washing with 150 mL of 1× CB, fraction collection can be started (8 mL/fraction). The flow rate may need to be lowered while loading or running the column in order to keep the back pressure below the limit for the column housing. Generally, a flow rate of at least 3 mL/min can be maintained.
3. The tubulin elutes as a clear, slightly yellow solution. Check the protein concentration by Bradford Assay using BSA as the standard (see Note 12) and pool the most concentrated fractions (typically, ~160 mL total) as well as side fractions (~50 mL) with less tubulin. Aliquots are fast-frozen in cryotubes (4 mL) by immersion in liquid N_2 and stored at −80°C.
4. Analysis of the fractions eluted from the PC column by SDS-PAGE shows that the bulk of the MAPs and other contaminants are bound by the PC column, leaving a highly purified tubulin preparation (Fig. 2).

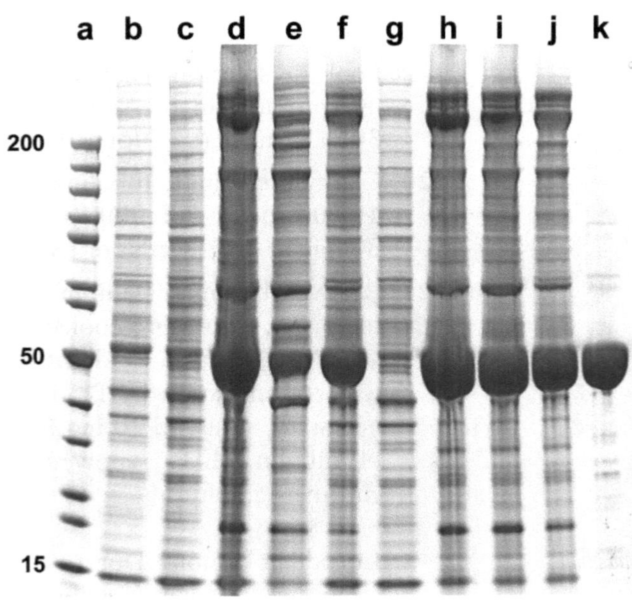

Fig. 2. SDS-PAGE gel showing protein content at each stage in a typical tubulin purification. Lanes: (**a**) Fermentas protein ladder 10–200 kDa; (**b**) clarified brain homogenate; (**c**) first warm spin supernatant; (**d**) first warm spin pellet; (**e**) first cold spin pellet; (**f**) first cold spin supernatant; (**g**) second warm spin supernatant; (**h**) second warm spin pellet; (**i**) second cold spin pellet; (**j**) second cold spin supernatant (PC column load); (**k**) PC column flow through ("PC tubulin"). The tubulin band at ~50 kD is overloaded in order to monitor the level of contaminating MAPs that are largely removed after chromatography on PC.

5. To regenerate the column, wash with 3 CV of 1 M KCl in 1× CB at 3 mL/min; this high salt wash contains the MAPs that can be collected if desired. Re-equilibrate the resin by washing with 3–5 CV of 1× CB and monitor the conductivity to ensure that the KCl is removed. Finally, wash with 5 CV of 1× CB containing 0.1% NaN_3 for long-term storage at 4°C.

4. Notes

1. Some protocols suggest dissection of the brains and removal of the meninges and blood clots. However, using pig brains and our particular blender, we have found this to be an unnecessary delay. Such steps may be necessary when using bovine brain or different equipment.
2. We have three Beckman Avanti J-20 centrifuges available with JLA-8.1000 rotors, giving a capacity of 18 × 1 L, usually sufficient for the coarse clarification of the homogenate from ~75 brains. Note that this first clarification is not strictly necessary; large centrifuges, like the J-20s, may not be available. However, this step does result in a significant initial reduction of volume (~18 to 7 L), resulting in an increased final tubulin yield using a smaller number of centrifuges in the subsequent steps.
3. In this and all subsequent centrifugation steps, we state the rpm, the relative centrifugal force, the rotor used, and the total spin time in our preparation; of course, any suitable rotor/rpm/time combination can be substituted. A number of checks should be carried out prior to each spin: all centrifuge tubes should be balanced to at least 1 mg; rotors should be checked for damage, particularly to their over-speed discs; o-rings should be checked for wear, re-greased, and replaced as necessary; all threads should be treated with spincote; each centrifuge tube should be checked for wear (including o-rings, if present); and typically, centrifuge tubes should always be filled completely to reduce the chance of collapse and if insufficient solution is present, then the tubes should be topped up with the appropriate buffer.
4. This first coarse clarification results in a small dense pellet with a much larger loose foam-like pellet. Discard both the pellet and the foam.
5. We use five Beckman Avanti J-25 centrifuges with JLA 16.250 rotors, giving a maximum capacity of 30 × 250 mL. After this second clarification, small pellets are obtained and the volume of recovered supernatant is typically ~6 L.

6. See Item 5. However, any container is suitable as long as it is possible to continuously stir the solution and monitor the temperature (see Note 7).
7. It is essential to bring the tubulin solution above 30°C quickly; slow polymerisation affects the final tubulin yield. We have had good results by ensuring that the glycerol, added to the chilled supernatant, is at 37°C and the polymerisation solution is added to a pre-warmed (37°C) flask. The flask is then transferred to a large water bath set at 55°C. This raises the temperature of the solution quickly (~5 min). It is essential to ensure that the solution is stirred continuously during this time and that the temperature inside the flask is closely monitored. The flask must be removed from the 55°C bath when the temperature reaches 30°C, or there is a risk that the high temperature will denature the tubulin close to the glass; the flask is then transferred to a separate water bath set at 37°C.
8. As polymerisation proceeds, the solution becomes more viscous and undergoes a slight colour/clarity change.
9. The total volume of the polymerisation mixture is ~9 L, requiring 36×250-mL centrifuge bottles. We use four Beckman Optima LE-80K centrifuges with Type 19 rotors and two Beckman Avanti J-25s with JLA-16.250 rotors (resulting in a reduced *g*-force of 33,000×*g* for the JLA-16.250). Note that Type 19 centrifuge tubes will likely collapse slightly at 37°C in this spin (even when filled) and must not be reused. Also note that Beckman Avanti J-25 centrifuges have no facility for heating. However, using a pre-warmed rotor and pre-warmed tubes (37°C) and the centrifuge at room temperature, there is minimal loss of tubulin.
10. It is sensible to retain a small volume of this supernatant and all subsequent supernatants (and re-suspended pellet solutions) to perform gel electrophoresis to monitor the progress of the purification (Fig. 2).
11. It is essential to re-suspend the pellets completely. Similarly, it is desirable to re-suspend in as low a volume that allows complete re-suspension. Typically, we begin by attempting re-suspension using 5 mL of Buffer P for each pellet and use a soft spatula to scrape and break the pellet from the walls of the centrifuge tube. Then, homogenise the solution using a dounce homogeniser; use the dounce slowly to avoid pulling a vacuum. Additional dilution may then be necessary, particularly if the re-suspension is "holding" bubbles. This dilution, along with additional "washing" of the tubes and dounces, typically results in using the equivalent of ~10 mL per pellet. Conveniently (for the next centrifugation step), the total volume (including the additional volume of the pellets) is typically 800–900 mL. Several steps of douncing/incubation

on ice may be necessary to maximise re-suspension and depolymerisation of the microtubule pellet.

12. The total protein concentration (consisting of tubulin and any MAPs) can be determined using a standard Bradford assay (14).
13. Lowering the concentration assists depolymerisation, as well as diluting out glycerol and other components prior to clarification; nevertheless, the concentration should not be lowered below 25 mg/mL; otherwise, the efficiency of polymerisation will be affected in the next steps.
14. We now typically use two Beckman LE-80K centrifuges with Type 45-Ti rotors (giving a total capacity of 12 × 70 mL = 840 mL). The recovered volume of the supernatant is typically ~560 mL.
15. In the Type 45-Ti rotor, at the required centrifugation speeds, it is *essential* to fill the centrifuge tubes completely; otherwise, collapse *will* occur. If necessary, tubes can be topped up with *cold* or *warm* buffer P, as appropriate.
16. The volume of the polymerisation solution is now ~840 mL due to the added glycerol. We use two Beckman LE-80K centrifuges with Type 45-Ti rotors (giving a total capacity of 12 × 70 mL = 840 mL).
17. The same advice applies to the re-suspension as before (see Note 11); note, however, the different buffer and final volumes. The increased re-suspension volume (20 mL/pellet compared to 10 mL/pellet) is to ensure depolymerisation, but also to ensure that the solution is not too viscous, improving the effectiveness of the final clarification spin. It is particularly important to re-suspend the pellets well at this stage in order to minimise the viscosity of the solution loaded onto the PC column.
18. Re-suspension of half of the pellets (six) with around 20 mL of buffer results in a total volume of ~200 mL. Thus, the clarification can easily be carried out in a single Beckman LE-80K centrifuge with a Type 45-Ti rotor. If necessary, to ensure that three tubes are filled completely (3 × 70 mL = 210 mL), the volume of the re-suspension should be adjusted to 210 mL. Significant pellets are obtained after this clarification, and the volume of supernatant recovered is typically ~150 mL.

Acknowledgements

The authors acknowledge the input of many people in the development of these protocols. Protocols published on the Web, especially by the Mitchison lab, have proved invaluable in the development of this protocol. In more recent years, many members, past and present, of the Howard, Diez, and Hyman labs at the MPI-CBG have been involved in performing this purification. In particular, we thank Regine Hartmann, Corina Braeuer, Marija Zanic, Gero Fink, Tony Ashford, and Andy Hunter. Finally, C.G. and C.T.F. thank Gary Brouhard, both for sharing the misery of getting this protocol up and running (again) and for promising (after "the incident") never to run another PC column.

References

1. Horio, T. and Hotani, H. (1986), Visualization of the dynamic instability of individual microtubules by dark-field microscopy. *Nature* **321**: 605–7.
2. Mitchison, T. and Kirschner, M. (1984), Dynamic instability of microtubule growth. *Nature* **312**: 237–42.
3. Howard, J. and Hyman, A. A. (2009), Growth, fluctuation and switching at microtubule plus ends. *Nat Rev Mol Cell Bio* **10**: 569–74.
4. Desai, A. and Mitchison, T. J. (1997), Microtubule polymerization dynamics. *Annu Rev Cell Dev Biol* **13**: 83–117.
5. Howard, J. and Hyman, A. A. (2003), Dynamics and mechanics of the microtubule plus end. *Nature* **422**: 753–8.
6. Brouhard, G. J., Stear, J. H., *et al.* (2008), XMAP215 is a processive microtubule polymerase. *Cell* **132**: 79–88.
7. Helenius, J., Brouhard, G., *et al.* (2006), The depolymerizing kinesin MCAK uses lattice diffusion to rapidly target microtubule ends. *Nature* **441**: 115–9.
8. Akhmanova, A. and Steinmetz, M. O. (2008), Tracking the ends: a dynamic protein network controls the fate of microtubule tips. *Nat Rev Mol Cell Biol* **9**: 309–22.
9. Borisy, G. G., Marcum, J. M., *et al.* (1975), Purification of tubulin and associated high molecular weight proteins from porcine brain and characterization of microtubule assembly in vitro. *Ann N Y Acad Sci* **253**: 107–32.
10. Castoldi, M. and Popov, A. V. (2003), Purification of brain tubulin through two cycles of polymerization-depolymerization in a high-molarity buffer. *Protein Expr Purif* **32**: 83–8.
11. Shelanski, M. L., Gaskin, F., and Cantor, C. R. (1973), Microtubule assembly in the absence of added nucleotides. *Proc Natl Acad Sci USA* **70**: 765–8.
12. Burton, K. (1969), Spectral data and pK values for purines, pyrimidines, nucleosides, and nucleotides., in: Data for Biochemical Research, R.M.C. Dawson, et al., Editors., Oxford University Press: Oxford and New York.
13. Mitchison Lab: "Large Scale Tubulin Preparation". *(http://mitchison.med.harvard.edu/protocols/tubprep.html).*
14. Bradford, M. M. (1976), A rapid and sensitive method for the quantitation of microgram quantities of protein utilizing the principle of protein-dye binding. *Anal Biochem* **72**: 248–54.

Chapter 3

Purification of Tubulin from the Fission Yeast *Schizosaccharomyces pombe*

Douglas R. Drummond, Susan Kain, Anthony Newcombe, Christina Hoey, Miho Katsuki, and Robert A. Cross

Abstract

The fission yeast *Schizosaccharomyces pombe* is an attractive source of tubulin for biochemical experiments as it contains few tubulin isoforms and is amenable to genetic manipulation. We describe the preparation of milligram quantities of highly purified native tubulin from *S. pombe* suitable for use in microtubule dynamics assays as well as structural and other biochemical studies.

S. pombe cells are grown in bulk in a fermenter and then lysed using a bead mill. The soluble protein fraction is bound to anion-exchange chromatography resin by batch binding, packed in a chromatography column and eluted by a salt gradient. The tubulin-containing fraction is ammonium sulphate precipitated to further concentrate and purify the protein. A round of high-resolution anion-exchange chromatography is carried out before a cycle of polymerisation and depolymerisation to select functional tubulin. Gel filtration is used to remove residual contaminants before a final desalting step. The purified tubulin is concentrated, and then frozen and stored in liquid nitrogen.

Key words: *S. pombe*, Yeast, Tubulin, Purification, Microtubule dynamics

1. Introduction

Mammalian brain is commonly used as a source of tubulin for biochemical assays, such as microtubule dynamics experiments. Tubulin forms a high percentage of the soluble protein fraction in brain considerably simplifying subsequent purification (1, 2). However, mammalian cells typically have multiple tubulin isoforms present with a diverse range of post-translational modifications (3). This diversity is a potential complication for any biochemical analysis as the behaviour of individual isoforms may differ (4–6).

Anne Straube (ed.), *Microtubule Dynamics: Methods and Protocols*, Methods in Molecular Biology, vol. 777,
DOI 10.1007/978-1-61779-252-6_3, © Springer Science+Business Media, LLC 2011

Further limitations of the mammalian brain tubulin are the difficulties inherent in creating mutations in the tubulin which would necessitate production of transgenic animals or the use of in vitro cell culture, with the further technical complications of separating the mutant from the non-mutant tubulins and the avoidance of mixed heterodimers of alpha and beta tubulins.

Many of these difficulties have been overcome for other proteins by using prokaryotic recombinant expression systems. However, commonly used prokaryotic systems, such as *Escherichia coli*, lack the chaperones required for successful folding of eukaryotic tubulins (7) so that the tubulin is produced in an unfolded form. Refolding the tubulin by incubation in eukaryotic cell extracts (8) gives limited yields. There is one report of successful refolding of a plant tubulin from urea (9). The bacterial tubulin BtubA/B from prosthecobacter is expressed in *E. coli* in a folded functional form (10, 11), but has not been observed to form dynamic microtubules.

To overcome these limitations, other eukaryotic sources of tubulin have been sought. *Schizosaccharomyces cerevisiae* has few isoforms and is genetically tractable. The purification of both native (12–17) and tagged tubulins (4, 18, 19) has been reported with successful expression and purification of mutant tubulins in sufficient quantities for microtubule dynamics assays. Uchimura et al. have also developed a method for the purification of lethal mutants without requiring tagging of the tubulin (20, 21).

The fission yeast *S. pombe*, like budding yeast, has two alpha (*nda2*, *atb2*) (22) and one beta (*nda3*) (23) tubulin genes. From the genomic sequence of *S. pombe* (24), these have predicted molecular weights of 51.1, 50.5, and 49.4 kDa for $\alpha 1$, $\alpha 2$, and β tubulins (25), respectively. Western blots of total protein extracts from *S. pombe* separated by 2-D gel electrophoresis identified three polypeptides corresponding to the two alpha and one beta tubulin present in the genome (26). Charge or mass variations resulting from post-translational modifications were not detected. Alfa et al. (27) reported a partial purification of *S. pombe* tubulin in limited quantities using ion-exchange chromatography. This study demonstrated that all of the tubulin present within *S. pombe* was tyrosinated at the C-terminal position of the alpha tubulin, again suggesting that *S. pombe* would be a good source of homogeneous tubulin ideal for functional biochemical analysis. *S. pombe* is a model organism for the study of microtubules and microtubule-binding proteins and their functions in vivo (28, 29) in relation to the cell cycle and cellular morphogenesis, and these protein activities are now being reconstituted in microtubule dynamics assays in vitro (30–32). However, these studies have utilised tubulin from other species to form microtubules due to the lack of purified *S. pombe* tubulin.

In order to study the biochemical properties and microtubule dynamics of *S. pombe* tubulin and the effects of its associated proteins in a faithfully reconstituted system, we developed a purification procedure for untagged native *S. pombe* tubulin that yields about 10 mg of tubulin from an 80 l cell culture with a purity of ~99%. Smaller scale purifications are also possible (see Note 1). The procedure is essentially a modification of the method developed by Barnes et al. and Davis et al. for the purification of *S. cerevisiae* tubulin (12, 14). The protocol works both with wild-type strains of *S. pombe* yielding a preparation with a mix of α1/β and α2/β tubulin heterodimers, and with strains where the non-essential α2 gene (33) has been replaced with a second copy of the α1 gene, yielding a pure single isoform preparation of α1/β tubulin (34) with an increased yield.

An 80 l culture of *S. pombe* is grown in a fermenter and collected by centrifugation. The pellet of *S. pombe* cells is re-suspended and the cells disrupted using a bead mill to break open the cell wall (see Fig. 1a). The soluble extract of *S. pombe* proteins is bound to a DE52 (DEAE) weak anion-exchange resin suspended in solution. The ion-exchange resin is packed in a chromatography column and the tubulin eluted by a salt gradient (see Fig. 2).

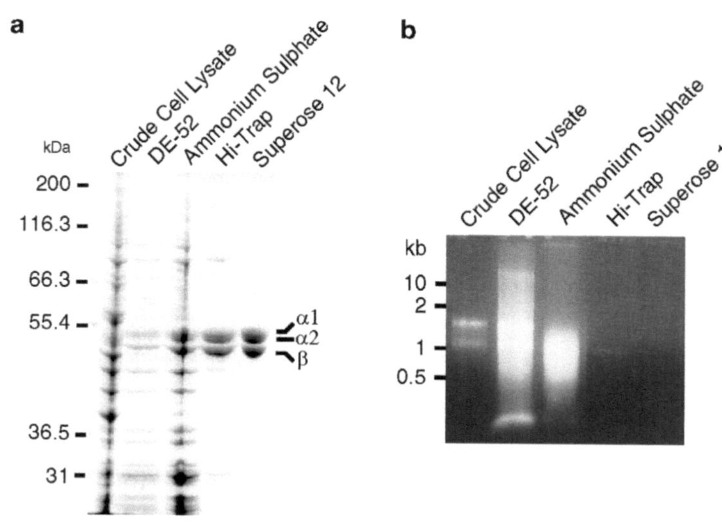

Fig. 1. (**a**) Protein samples taken after each major step during the purification of tubulin from wild-type *S. pombe* cells containing α1, α2, and β tubulin isoforms were separated by SDS-PAGE. The 12.5% polyacrylamide gel was stained with Coomassie blue. Each step of the procedure removes contaminating proteins from the tubulin. (**b**) Protein samples after each step of the purification procedure were separated on a 4% agarose gel in TBE buffer. The gel was stained with ethidium bromide. Nucleic acid from the cell lysate is concentrated on the DE52 column and co-purifies with the tubulin fractions. Each step of the procedure reduces the nucleic acid contamination level; however, only gel filtration chromatography on Superdex 200 removes the final traces of nucleic acid.

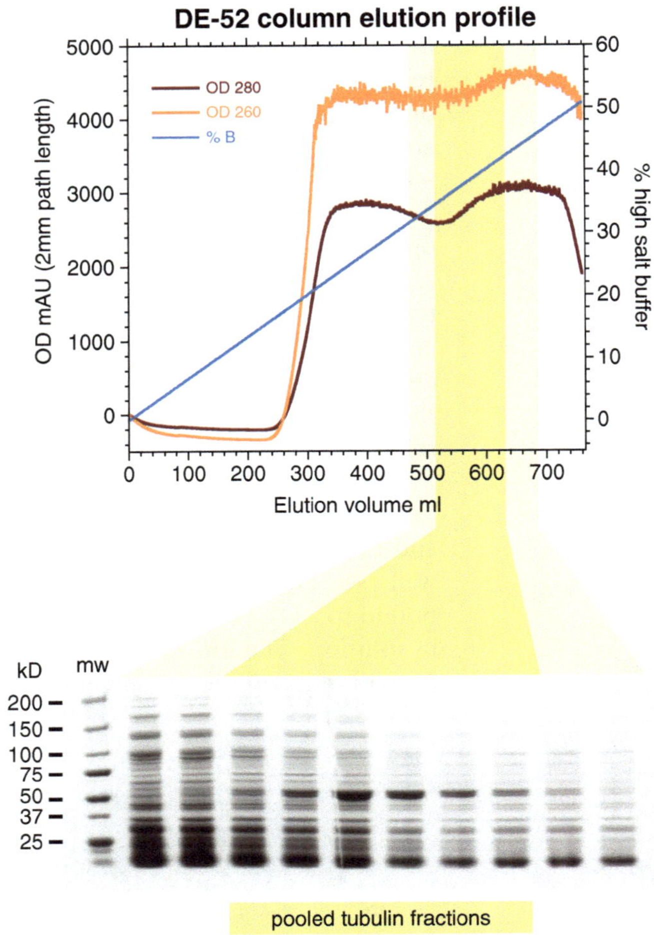

Fig. 2. Elution profile of DE52 anion-exchange chromatography column monitored at 260 and 280 nm with the high-salt buffer gradient. The shaded area shows the fractions analysed by SDS-PAGE. After staining with Simply Blue, the tubulin-containing fractions were selected and pooled. The OD_{260} is higher than OD_{280} because of nucleic acids present in all fractions. Tubulin elutes at about 34% of the high-salt buffer corresponding to 0.34 M NaCl in the 100 mM PIPES buffer.

The tubulin-containing fractions are precipitated using ammonium sulphate to concentrate and further purify the protein. A major contaminant during the purification is nucleic acid (see Fig. 1b), and most of the purification steps reduce both protein and nucleic acid contaminants in the tubulin fraction. A further round of ion-exchange chromatography is carried out using higher resolution HiTrap Q strong anion-exchange resin (see Fig. 3). A cycle of tubulin polymerisation, pelleting of microtubules, and depolymerisation is then used to enrich functional tubulin. The disassembled tubulin is further purified on a gel filtration column

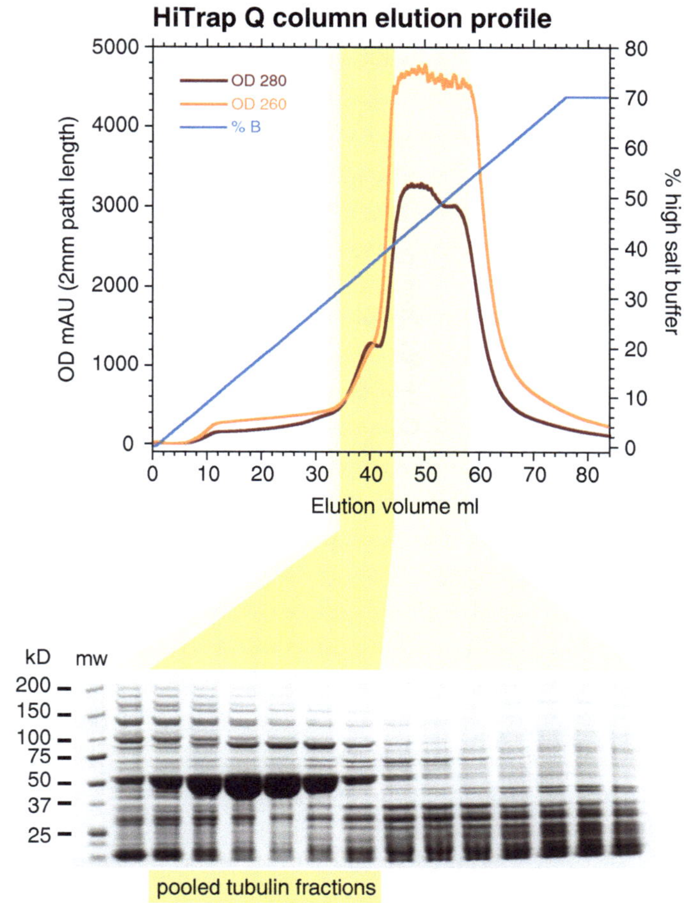

Fig. 3. Elution profile of HiTrap Q HP anion-exchange chromatography column monitored at 260 and 280 nm. The gradient of high-salt buffer is shown. The shaded area shows the fractions analysed by SDS-PAGE. After staining with Simply Blue (Invitrogen), the tubulin-containing fractions were selected and pooled. The OD_{260} is higher than OD_{280} because of nucleic acids present in most of the fractions eluted from the column. Only tubulin-containing fractions have $OD_{280} > OD_{260}$. Tubulin elutes at about 32% of the high-salt buffer corresponding to about 0.32 M of NaCl in the 100 mM PIPES buffer.

(see Fig. 4) to remove residual proteins, nucleic acids, and the GMPCPP used for polymerisation before a final desalting step. The purified tubulin is concentrated in a spin concentrator, and then frozen and stored in liquid nitrogen. This method yields about 10 mg of tubulin from an 80 l culture with a typical purity of 98–99% (see Fig. 5). Tubulin prepared by this method has been successfully used in studies of *S. pombe* kinesin motor protein Klp2 (34), structural studies of the +TIP binding protein Mal3 (the *S. pombe* homologue of EB1) (35), and functional studies on the effect of Mal3 on the dynamics of microtubules formed from *S. pombe* tubulin (36).

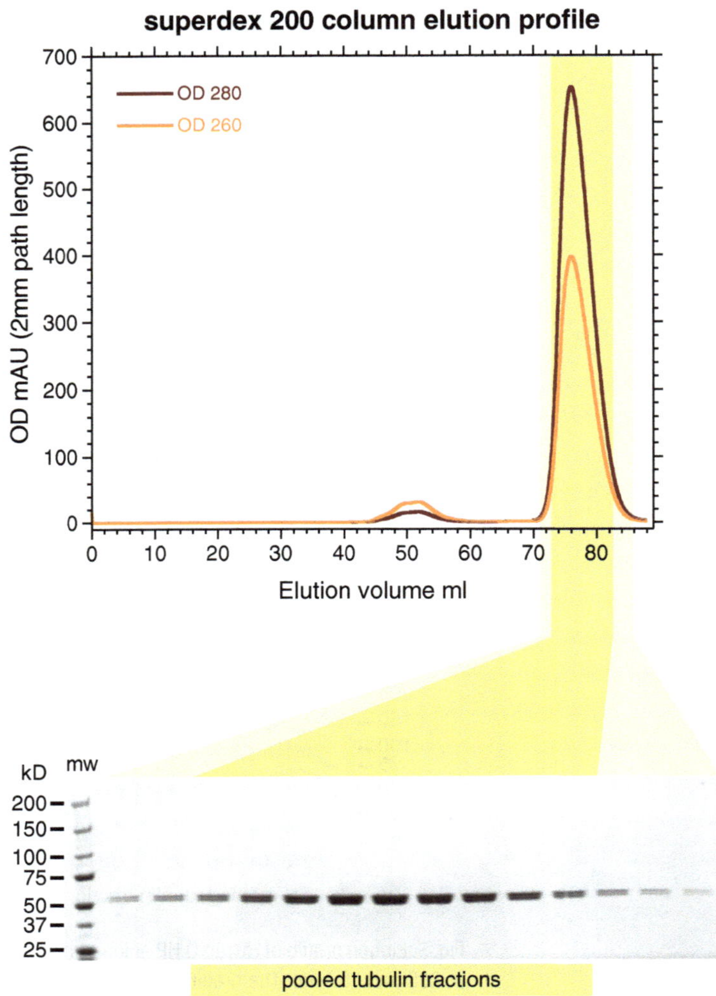

Fig. 4. Elution profile of Superdex 200 XK16/60 gel filtration chromatography column monitored at 260 and 280 nm. The shaded area shows the fractions analysed by SDS-PAGE. After staining with Simply Blue, the tubulin-containing fractions were selected and pooled. Nucleic acid contaminants elute first at 40 ml ($OD_{260} > OD_{280}$) and the tubulin protein as a single peak at 70 ml ($OD_{280} > OD_{260}$). Smaller buffer components, such as GMPCPP, elute later in a volume equivalent to the total column volume of about 120 ml.

2. Materials

2.1. Yeast Media

1. YPD: 0.5% (w/v) yeast extract, 1% (w/v) peptone, 1% (w/v) glucose.
2. YPD agar: 2% (w/v) agar, 0.5% (w/v) yeast extract, 1% (w/v) peptone, 1% (w/v) glucose.

2.2. Stock Solutions

The volume (given in brackets behind each solution name) is the quantity required for one purification (see Notes 1–3).

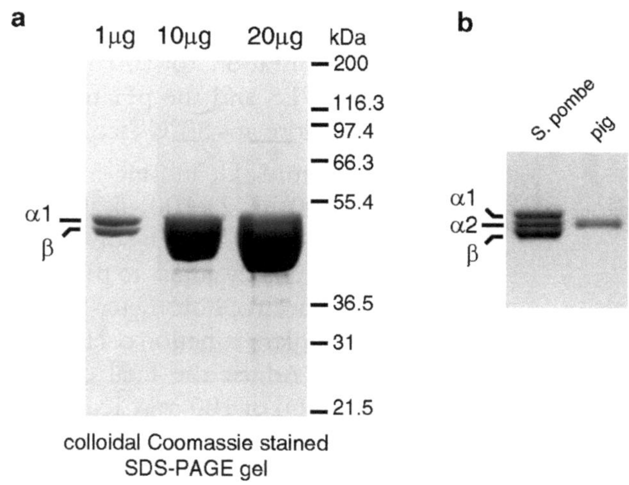

Fig. 5. (**a**) Samples of tubulin purified from a strain containing only the α1 and β tubulin isoforms were separated on a 10% Bis-Tris NuPAGE gel in MOPs buffer. The gel was stained with Simply Blue, a sensitive form of colloidal Coomassie stain. Tubulin purity estimated from densitometry measurements of these gels is typically 98–99%. These low levels of contaminating proteins are detectable only when the gel is heavily loaded with tubulin protein. (**b**) Samples of *S. pombe* tubulin from a wild-type strain containing α1, α2, and β isoforms and pig brain tubulin were separated by SDS-PAGE on a 12.5% gel. The purified pig brain tubulin co-migrates with *S. pombe* tubulin α2 isoform and can be used as a marker to identify the *S. pombe* tubulin bands on gels during the purification.

1. 10 M KOH (≥88% purity) (see Note 4).
2. 10 M NaOH (≥98% purity) (see Note 4).
3. 1 M K-PIPES, pH 6.9 (2.0 l): PIPES free acid (≥97% purity) is re-suspended in H_2O and adjusted to pH 6.9 using KOH. Filter and store at 4°C (see Notes 5–8).
4. 1 M $MgSO_4$ (130 ml): (≥99% purity), filter and store at 4°C.
5. 1 M $MgSO_4$ (5 ml): 1.00 ± 0.04 M solution (Sigma M3409), store at 4°C.
6. 0.5 M Ethylene glycol-bis(2-aminoethylether)-*N,N,N′,N′*-tetraacetic acid (EGTA) (80 ml): (≥97% purity, Sigma E-4378). Adjust to pH 7.0 with KOH, filter and store at 4°C.
7. 4 M NaCl (700 ml): (>99.9% purity), filter and store at room temperature.
8. 2 M KCl (130 ml): (>99.0% purity), filter and store at room temperature.
9. 100 mM Guanosine 5′-triphosphate (GTP), (45 ml as 4 × 10 ml, 5 × 1 ml aliquots): (Sodium salt, 95% purity, Jena Bioscience GMBH). Dissolve the powder in cold 100 mM K-PIPES pH 6.9. Immediately adjust to pH 7.0 with KOH. Measure the GTP OD_{260} in a spectrophotometer and calculate the concentration from the molar extinction coefficient

of GTP at pH 7.0, E = 11,700 M^{-1} cm^{-1} (37). Adjust the final concentration to 100 mM GTP by addition of 100 mM K-PIPES and the pH to 7.0 by addition of KOH. Aliquot and store at −80°C (see Notes 9–11).

10. 100 mM Guanosine-5′ diphosphate (GDP), (600 μl as 2 × 200 μl, 2 × 100 μl aliquots): (Sodium salt, ≥97% purity, Sigma G-7127). Dissolve in cold 100 mM K-PIPES pH 6.9. Immediately adjust to pH 7.0 with KOH. Measure the OD_{260} in a spectrophotometer and calculate the concentration from the molar extinction coefficient at pH 7.0, E = 11,800 M^{-1} cm^{-1} (37). Adjust the final concentration to 100 mM GDP by addition of 100 mM K-PIPES and the pH to 7.0 by addition of KOH. Aliquot and store at −80°C.
11. 200 mM Guanosine-5′-[(α,β)-methyleno]triphosphate (GMPCPP), (30 μl): (Sodium Salt, >95% purity, Jena Bioscience, Germany). Purchased as a solution from the manufacturer and then stored as 5-μl aliquots at −80°C.
12. 1 M Dithiothreitol (DTT), (40 ml as 4 × 10-ml aliquots): (>99% purity). Dissolve in cold H_2O; filter and store at −20°C (see Note 12).
13. 5× PEM (800 ml): 500 mM K-PIPES pH 6.9, 5 mM $MgSO_4$, 10 mM EGTA. Adjust to pH 6.9 with KOH; filter and store at 4°C (see Note 13).
14. 8 M Guanidinium chloride (1 ml): (Guanidine HCl >99% purity, Melford MB2002). Dissolve and filter. Store at room temperature.
15. Ultrapure Glycerol (800 ml): (99.5% purity), MP Biomedicals Inc, cat no 800689.
16. Ammonium sulphate: As powder (>99.5% purity, Aristar grade BDH).

2.3. Protease Inhibitors (see Note 14)

1. 100 mM Nα-p-Tosyl-L-arginine methyl ester hydrochloride (TAME), (40 ml as 4 × 10-ml aliquots): (≥98% purity, Sigma T-4626). Dissolve in H_2O. Store at −80°C.
2. 500× AAE (18 ml as 2 × 8-ml and 1 × 2-ml aliquots): 0.5 mg/ml aprotinin (Peptides International), 0.5 mg/ml 4-(2-aminoethyl) benzenesulfonyl fluoride hydrochloride (AEBSF), (≥97% purity, Sigma A-8456); 0.5 mg/ml E-64 (Peptides International). Dissolve in H_2O and store at −80°C.
3. 500× PC (18 ml as 2 × 8-ml and 1 × 2-ml aliquots): 0.5 mg/ml Pepstatin A (Peptides International), 0.5 mg/ml Chymostatin (Peptides International). Dissolve in dimethyl sulfoxide (DMSO; Sigma Aldrich 27,685-5) and store at −80°C.
4. 500× Antipain (18 ml as 2 × 8-ml and 1 × 2-ml aliquots): 0.5 mg/ml Antipain (Peptides International). Dissolve in H_2O and store at −80°C.

5. 250× Leupeptin (38 ml as 4×8-ml and 3×2-ml aliquots): 0.5 mg/ml Leupeptin (Peptides International). Dissolve in H_2O and store at −80°C (see Note 15).

2.4. Buffers Required for One Purification

1. 0.2 M NaOH (1 l): Filter and store at 4°C.
2. 20% (v/v) ethanol (3 l): (≥99% purity). Filter and store at 4°C.
3. 200 mM K-PIPES pH 6.9 for DE52 equilibration (1.1 l): Dilute 1 M K-PIPES pH 6.9 stock, adjust to pH 6.9 with KOH, filter and store at room temperature.
4. DE52 high-salt equilibration buffer (1.1 l): 100 mM K-PIPES, 10 mM $MgSO_4$, 2 mM EGTA, 1 M NaCl, 10% (v/v) glycerol, adjust to pH 6.9 with KOH, filter and store at room temperature.
5. DE52 low-salt equilibration buffer (3 l): 100 mM K-PIPES, 10 mM $MgSO_4$, 2 mM EGTA, 50 mM NaCl, 10% (v/v) glycerol, adjust to pH 6.9 with KOH, filter and store at room temperature.
6. Lysis buffer (4 l): 100 mM K-PIPES pH 6.9, 10 mM $MgSO_4$, 2 mM EGTA, 50 mM NaCl. Adjust to pH 6.9 with KOH and 3.86 l with H_2O, filter and store at 4°C. Immediately before use, add the labile components to make up to 4 l: 40 ml of 100 mM GTP, 20 ml of 1 M DTT, 40 ml of 100 mM TAME, 8 ml of 500× AAE, 8 ml of 500× Antipain, 8 ml of 500× PC, 16 ml of 250× Leupeptin.
7. Low-salt DE52 buffer (3.0 l): 100 mM K-PIPES pH 6.9, 10 mM $MgSO_4$, 2 mM EGTA, 50 mM NaCl, 10% (v/v) glycerol. Adjust to pH 6.9 with KOH, make up to 2.952 l volume with H_2O, filter and store at 4°C. Immediately before use, add the labile components: 3 ml 100 mM GTP (0.1 mM final concentration), 15 ml 1 M DTT, 6 ml of 500× AAE, 6 ml of 500× PC, 6 ml of 500× Antipain, 12 ml of 250× Leupeptin.
8. DE52 high-salt buffer (500 ml): 100 mM K-PIPES pH 6.9, 10 mM $MgSO_4$, 2 mM EGTA, 1 M NaCl, 10% (v/v) glycerol. Adjust to pH 6.9 with KOH, make up to 491.5 ml with H_2O, filter and store at 4°C. Immediately before use, add the labile components: 0.5 ml 100 mM GTP (0.1 mM final concentration), 3 ml 1 M DTT, 1 ml of 500× AAE, 1 ml of 500× PC, 1 ml of 500× Antipain, 2 ml of 250× Leupeptin.
9. Re-suspension buffer (1 l): 1× PEM, 50 mM KCl. Adjust to pH 6.9 with KOH and make up to 791.2 ml, filter and store at 4°C. Immediately before use, add:

   1 ml of 100 mM GTP (0.1 mM final concentration), 2 ml of 500× AAE, 2 ml of 500× PC, 2 ml of 500× Antipain, 4 ml of 250× Leupeptin.

10. HiTrap Q low-salt buffer (400 ml): 1× PEM, 50 mM KCl, adjust to pH 6.9 with KOH and make up to 400 ml, filter and store at 4°C. Immediately before use, add 200 μl of 100 mM GDP (50 μM final concentration).
11. HiTrap Q high-salt buffer (200 ml): 1× PEM, 1 M KCl, adjust to pH 6.9 with KOH and make up to 200 ml. Filter and store at 4°C. Immediately before use, add 100 μl of 100 mM GDP (50 μM final concentration).
12. Superdex 200 equilibration buffer (300 ml): 0.5× PEM, filter and store at 4°C.
13. Superdex 200 gel filtration buffer (1 l): 1× PEM, 150 mM NaCl. Adjust to pH 6.9 with KOH and make up to 999.5 ml, filter and store at 4°C. Immediately before use, add 500 μl of 100 mM GTP (50 μM final concentration).
14. Depolymerisation buffer (5 ml): 20 mM PIPES, 1 mM $MgSO_4$, 5 mM $CaCl_2$. Adjust to pH 6.9 using NaOH. Immediately before use, add 50 μl of 100 mM GDP (final concentration 1 mM); (see Note 16).
15. 1× PEM (1 l): Dilute 5× PEM stock with H_2O (1× PEM is 100 mM K-PIPES, 1 mM $MgSO_4$, 2 mM EGTA pH 6.9). Adjust to pH 6.9 with KOH, filter and store at 4°C.
16. Pellet wash buffer (5 ml): 1× PEM (from 1 l 1× PEM stock). Immediately before use, add 0.5 μl of 200 mM GMPCPP (20 μM final concentration).
17. Tubulin storage buffer (750 ml): 750 ml of 1× PEM (from 1.1 l 1× PEM stock). Immediately before use, add 150 μl of 100 mM GDP (20 μM final concentration).

2.5. Chromatography Media and Columns

1. DE52 DEAE cellulose (Whatman).
2. XK50/30 chromatography column with AK50 FF fast-flow adaptors (GE Healthcare).
3. HiTrap Q HP column, 5 ml volume (GE Healthcare).
4. Two HiPrep 26/10 rapid desalting columns (GE Healthcare).
5. HiLoad Superdex 200 prep grade XK16/60 gel filtration column (GE Healthcare).

2.6. Consumables

1. 500 μm glass beads (425–600 μm acid-washed glass beads, Sigma G8772).
2. 0.2-μm pore vacuum filters (Millipore Express plus, PES membrane).
3. Spin concentrators (Vivaspin 15R 10,000 mwco HY, Vivascience).
4. 500-ml polycarbonate centrifuge bottles (Sorvall, 03944).
5. 14-ml polypropylene tubes (BD Falcon 14 ml 17 × 100 mm, polypropylene 352059).

6. 50-ml polycarbonate Oakridge tubes (Nalgene 3138).
7. 2-ml microcentrifuge tubes (Eppendorf).
8. 50-ml polypropylene tube (BD Falcon, 352098).
9. 5-ml polypropylene tubes (BD Falcon 5-ml polypropylene tubes 12 × 75 mm part no 352063).
10. 1.5-ml polypropylene microcentrifuge tubes with separate lid (Sarstedt 72.692).
11. 3-ml polyallomer ultracentrifuge tubes (Beckman Coulter 349623).
12. 500-μl screw cap tubes with O ring for liquid nitrogen (PGC Scientifics, 16-8115-04).
13. Small-volume disposable cuvette (Uvette, Eppendorf).
14. Precast protein gels (Novex 10% Bis-Tris gels run in MOPs buffer, Invitrogen).
15. Rapid, sensitive protein gel stain (Simply Blue SafeStain, Invitrogen).

2.7. Specialist Equipment

1. 80 l fermenter (New Brunswick).
2. Continuous-flow centrifuge (CEPA).
3. Conductivity meter (Orion).
4. Handheld homogeniser (Braun).
5. Bead mill (Dyno-Mill KDL-A, Willy A Bachofen).
6. Chiller for bead mill (Tricool Minac 10).
7. High-capacity peristaltic pump (504S, Watson Marlow).
8. Haemocytometer (Improved Neubauer, Hawksley).
9. Low-speed centrifuge (Evolution, Sorvall).
10. Low-speed rotor for 500-ml bottles (SLA3000, Sorvall).
11. Low-speed rotor of 50-ml tubes (SS-34, Sorvall).
12. Refrigerated microcentrifuge (5417R, Eppendorf).
13. Overhead stirrer with 150-mm Teflon mixer (Velp Scientifica type PW, with 150-mm plain-end Teflon mixer).
14. Chromatography system (Akta purifier 10 with 2 mm flow cell, 1 × 10 ml and 2 × 50 ml Superloops).
15. Peristaltic pump (P1, GE healthcare).
16. Ultracentrifuge and rotor for 3-ml sample tubes (TLA100.3 rotor, Optima Max-XP centrifuge, Beckman Coulter).
17. Low-speed refrigerated centrifuge with swing-out rotor for 50-ml tubes (Centra MP4R, IEC).
18. Spectrophotometer for measuring small-volume cuvettes (Cary 50, Varian).

3. Methods

3.1. Overview of S. pombe Cell Growth

1. Day 1: Prepare yeast media (Subheading 2.1).
2. Day 1: Streak cells on plate (Subheading 3.3, step 1).
3. Day 3: Inoculate small starter culture (Subheading 3.3, step 2).
4. Day 4: Inoculate large starter culture (Subheading 3.3, step 3).
5. Day 5: Inoculate large culture in fermenter (Subheading 3.3, step 4).
6. Day 6: Harvest cells and freeze cell pellet (Subheading 3.3, steps 5–7).

3.2. Overview of S. pombe Tubulin Purification (see Fig. 1)

1. Day 1: Prepare stock solutions (Subheading 2.2 to 2.3).
2. Day 1: Prepare buffers for day 2 (Subheading 2.4, items 1–9).
3. Day 1: Equilibrate DE52 resin (Subheading 3.4).
4. Day 2: Lyse cells in bead mill (Subheading 3.5, steps 1–7).
5. Day 2: Remove insoluble cell debris by centrifugation (Subheading 3.5, step 8).
6. Day 2: Batch bind soluble fraction to DE52 anion-exchange resin (Subheading 3.6, steps 1–5).
7. Day 2: Pack DE52 resin into a chromatography column and elute tubulin using a salt gradient (Subheading 3.6, steps 6–9, see Fig. 2).
8. Day 2: Ammonium sulphate precipitation of the tubulin (Subheading 3.7).
9. Day 3: Prepare buffers for day 4 (Subheading 2.4, items 10–18).
10. Day 3: Equilibrate chromatography columns for day 4 (Subheading 3.8).
11. Day4: Pellet ammonium sulphate precipitate, re-suspend and desalt using HiPrep 26/10 desalting column (Subheading 3.9).
12. Day 4: Separate tubulin using higher resolution anion-exchange chromatography on HiTrap Q column (Subheading 3.10, see Fig. 3).
13. Day 4: Select functional tubulin by a cycle of microtubule polymerisation and depolymerisation (Subheading 3.11).
14. Day 4: Separate tubulin by size using Superdex 200 gel filtration chromatography (Subheading 3.12, see Fig. 4).
15. Day 4: Desalt tubulin into PEM buffer using a HiPrep 26/10 desalting column (Subheading 3.13, step 1).

16. Day 4: Concentrate tubulin using a Vivascience Vivaspin 15R concentrator 10,000 mwco HY (Subheading 3.13, steps 2–6).
17. Day 4: Freeze and store tubulin in small aliquots in liquid nitrogen (Subheading 3.13, step 7).
18. Day 5: Analyse tubulin purity and concentration (Subheading 3.14, see Fig. 5).
19. Day 5: Wash chromatography columns (Subheading 3.15).

3.3. Growth of S. pombe Yeast (see Note 17)

1. Streak out *S. pombe* cells on agar plates and incubate at 32°C for 2–3 days.
2. Pick a single colony from the plate and inoculate a 50 ml starter culture of YPD in a flask and grow overnight at 30°C with shaking.
3. Use the 50 ml starter culture to inoculate a 1 l YPD culture and grow overnight at 30°C with shaking to a cell density of about 5×10^7 cells/ml.
4. Use the 1 l culture to inoculate 80 l of YPD, pre-warmed to 30°C in a fermenter at a density of $4–5 \times 10^5$ cells/ml (see Note 18).
5. Grow the cells in the fermenter at 30°C with aeration (30 cu. ft/min) and stirring (100–500 rpm) for about 16–20 h until the cells reach a density of $3–3.5 \times 10^7$ cells/ml (see Note 19).
6. Reduce the fermenter temperature to 20°C and harvest the cells using a continuous-flow centrifuge at 6,000 rpm ($6,084 \times g$).
7. Remove the cell paste from the centrifuge, place in a shallow tray to form a thin layer (<5 mm thick), and rapidly freeze using liquid nitrogen. Weigh and then store the cell pellet at −80°C (see Note 20). Usually, 500–600 g of cell pellet is produced.

3.4. Equilibrate DE52 Chromatography Resin

1. At room temperature in a 2-l beaker, wash 180 g of DE52 chromatography resin with 1.1 l of 200 mM K-PIPES pH6.9 by very gentle stirring using a magnetic stirrer. Measure the pH and adjust if necessary with HCl or KOH to pH 6.9 (see Note 21).
2. Stop stirring and allow the resin to settle for 15 min before decanting off the layer of buffer above the DE52 resin which contains the fines.
3. Measure the conductivity of the high-salt DE52 equilibration buffer (usually about 108 mS/cm).
4. Add 1.1 l of DE52 high-salt equilibration buffer to the DE52 resin, mix, and then allow to settle for 30 min.

5. Check the conductivity of the buffer above the DE52 resin which should be similar to the conductivity of the starting buffer. If not, repeat this equilibration step.
6. Carefully decant off the layer of buffer above the DE52 resin.
7. Measure the conductivity of the DE52 low-salt equilibration buffer stock (usually about 17 mS/cm).
8. Add 1 l of DE52 low-salt equilibration buffer to the DE52 resin, mix, and then allow it to settle for 30 min before decanting off the top layer of buffer.
9. Repeat the low-salt wash twice and on the final wash check the conductivity of the supernatant, which should be similar to the starting buffer. If the conductivity is different by more than a few mS/cm, then repeat the low-salt washing step.
10. After removing the final supernatant, cover the beaker containing the DE52 resin with cling film and store at 4°C.

3.5. Lysis of *S. pombe* Cells

1. Place 500–600 g of frozen *S. pombe* cell pellet in a 5-l beaker. Add lysis buffer to the cell pellet to a total volume of ~1.8 l. Gently stir using a magnetic stirrer until the larger lumps of cells have dispersed. Use a handheld homogeniser to speed the process and ensure that no large lumps of cells remain (see Note 22).
2. While the cell pellet is defrosting, prepare the bead mill (see Note 23). Flasks and beakers containing the liquids for the bead mill and for collecting the processed fluid are kept on ice during processing. Flush through the bead mill with 2 l of pre-cooled H_2O.
3. Wash through the bead mill chamber with 1 l of lysis buffer.
4. Remove 0.5 ml of the cell suspension as the 0% lysis sample for step 7.
5. Pump the cell suspension through the bead mill at 100 ml/min using a high-capacity peristaltic pump and collect the processed fluid in a 5-l beaker on ice. Remove a sample of the processed cells to check for lysis.
6. Once the cell suspension has almost completely passed through the bead mill, quickly transfer the processed suspension back to the inlet beaker and process again. Usually, processing requires four passes through the mill at a flow rate of about 100 ml/min.
7. Check cell lysis after each pass. Make 1/100 dilution of lysate samples in H_2O, and count the number of intact cells using a haemocytometer. Continue processing the cells for up to four passes until >80–90% lysis is achieved (see Note 24).
8. Following the last pass of the cell lysate, rinse through the grinding chamber with about 600 ml of lysis buffer until the

eluate runs clear. Combine this rinse buffer with the cell lysate (see Note 25).

9. Transfer the cell lysate to 500-ml polycarbonate centrifuge bottles and centrifuge in an SLA-3000 rotor at 12,000 rpm (24,336 × *g*) for 30 min at 4°C.

3.6. Chromatography on DE52

1. Transfer the supernatant, containing the soluble cell lysate, to a 5-l beaker containing the equilibrated DE52 resin.
2. In a cold room (4°C), gently stir the DE52 for 30 min to keep it in suspension in cell lysate using an overhead stirrer with a large 150-mm Teflon mixer.
3. Switch off stirrer and leave the DE52 resin to settle for 30 min.
4. Carefully decant excess lysis buffer from the settled DE52 resin.
5. Wash the DE52 resin by adding 450 ml of low-salt DE52 buffer, mix gently, allow to settle for 50 min, and then pour or pipette off the supernatant.
6. Pack the DE52 resin into an X50/30 column fitted with fast-flow adaptors and connect the column to an Akta purifier chromatography system. The pressure limit for the column is 0.3 MPa (see Note 26).
7. Wash the column at 6 ml/min with DE52 low-salt buffer until a stable baseline is reached. This usually takes 750–1,000 ml (3–4-column volumes for a 240-ml bed) (see Note 27).
8. Elute the tubulin using a gradient of 0–60% of the high-salt DE52 buffer in 900 ml at 5 ml/min (180 min, about 4 CV assuming a 240-ml resin bed) and collect 10 ml fractions in 14-ml polypropylene tubes (see Note 28).
9. The tubulin should elute at 30–40% of the DE52 high-salt buffer. Analyse samples of alternate fractions by SDS-PAGE to identify the fractions containing tubulin. Usually, the tubulin elutes in about ten fractions. Discard the low-concentration fractions at the start and end that have higher levels of contaminants (see Fig. 2 and Note 29).

3.7. Ammonium Sulphate Precipitation

1. Measure the volume of the pooled tubulin containing fractions and place them in a container, such as a glass bottle, with a cap. Add a magnetic stirring bar and place on a magnetic stirrer in a cold room.
2. Precipitate the tubulin by slowly adding 375 mg of ammonium sulphate for each 1 ml of solution while gently stirring. Once all of the ammonium sulphate has dissolved, store at 4°C (see Notes 30 and 31).

3.8. Equilibration of Chromatography Columns

1. Equilibrate a HiPrep Superdex 200 (prep grade) 16/60 column using 2 CV (248 ml) H_2O at a flow rate of <1 ml/min (0.3 MPa pressure limit), 1 CV (124 ml) of equilibration buffer at 1 ml/min and overnight with 2× CV (248 ml) of Superdex gel filtration buffer.
2. Connect two HiPrep 26/10 desalting columns together in series to a peristaltic pump. Equilibrate with H_2O overnight at a low flow rate (0.5 ml/min). Connected in series, the total CV is 106 ml (maximum pressure is 0.15 MPa per column or 0.3 MPa for two). Wash with at least 2 CV (212 ml).

3.9. Recover and Desalt the Ammonium Sulphate Precipitate

1. Gently re-suspend the ammonium sulphate precipitate and transfer to 50-ml polycarbonate Oakridge tubes. Pellet the ammonium sulphate precipitate using an SS-34 rotor at 9,143 rpm (10,000 ×*g*) in an Evolution centrifuge for 30 min at 4°C.
2. During the centrifuge run of step 1, equilibrate a HiPrep 26/10 desalting column with 265 ml (5 CV) of re-suspension buffer at 8 ml/min.
3. A large cloudy pellet should be visible after step 1. In a cold room, carefully pour off the supernatant and leave the tube inverted to drain for a few seconds on a paper towel. Re-suspend all of the pellets in a *total* volume of 12 ml of re-suspension buffer and dissolve by rotating the sealed tube on a roller mixer in a cold room for 30–45 min.
4. Transfer the re-suspended pellets to 2-ml microcentrifuge tubes and centrifuge at 13,000 rpm (18,000 ×*g*) for 2 min at 4°C in a refrigerated microcentrifuge.
5. Pool the supernatants in a 50-ml polypropylene tube. Use a peristaltic pump to load the pooled supernatants onto a 50 ml Superloop (see Note 32).
6. Desalt the sample into re-suspension buffer on the HiPrep 26/10 desalting column from step 2 at 4 ml/min (maximum loading volume is 15 ml and pressure limit is 0.15 MPa). Collect 2.5 ml fractions in 5-ml polypropylene tubes. The protein elutes as a single peak. Pool all the fractions from the peak.

3.10. Chromatography on HiTrap Q

1. Equilibrate a 5-ml HiTrap Q HP column using 25 ml of HiTrap Q low-salt buffer (5 CV), 25 ml of HiTrap Q high-salt buffer (5 CV), and 50 ml of HiTrap Q low-salt buffer (10 CV) at 5 ml/min. The pressure limit for the HiTrap column is 0.3 MPa.
2. Use a peristaltic pump to directly load the pooled fractions onto the HiTrap Q column at 5 ml/min. Wash the unbound protein from the column with 25 ml of low-salt HiTrap Q buffer at 5 ml/min (5 CV).

3. Elute the tubulin with a 75 ml gradient of 0–70% HiTrap Q high-salt buffer at 5 ml/min (15 CV) and collect 1 ml fractions in 1.5-ml tubes.
4. During step 3, switch on Optima MAX-XP ultracentrifuge to pre-warm to 32°C and pre-warm a TLA 100.3 rotor (see Note 33).
5. Check fractions from step 3 for tubulin by analysing samples by SDS-PAGE. Tubulin starts eluting at 30–31% high-salt buffer (see Fig. 3). Pool fractions containing tubulin, avoiding any that contain high proportions of contaminating proteins, and keep on ice.

3.11. Tubulin Cycling

1. Use a pipette to accurately measure the volume of the pooled tubulin fractions, which is usually about 10 ml. Polymerise the tubulin by adding GMPCPP from a 200 mM stock to give a final concentration of 0.6 mM. Add 1/10 volume of DMSO and 1/10 volume of glycerol.
2. Divide the sample among 4–6 3-ml polyallomer ultracentrifuge tubes, cover with parafilm, and incubate at 32°C for 1 h (see Note 34).
3. Pre-warm pellet wash buffer and depolymerisation buffer at 32°C.
4. Pellet the polymerised microtubules by centrifuging for 10 min at 68,000 rpm (124,405 × *g*) at 32°C in a TLA-100.3 rotor in an Optima Max-XP ultracentrifuge.
5. Carefully remove the supernatant. Gently pipette 200 μl of pre-warmed pellet wash buffer over each pellet and remove the buffer at once to remove residual polymerisation mix solution.
6. Re-suspend the tubulin pellets in 5 ml total volume of warm depolymerisation buffer by adding about 0.9 ml of buffer to each pellet and mixing at room temperature for 10 min by gentle occasional pipetting. Incubate on ice for 1 h with gentle mixing by occasional pipetting (see Note 35).
7. While the pellets are re-suspending, pre-cool the Optima MAX-XP ultracentrifuge and TLA-100.3 rotor to 4°C (see Note 36).
8. While pellets are re-suspending, equilibrate a HiPrep 26/10 rapid desalting column with 265 ml of tubulin storage buffer at 10 ml/min (5 CV).
9. Combine the depolymerised tubulin from Step 6 into two of the 3-ml ultracentrifuge tubes. Centrifuge at 68,000 rpm, (124,405 × *g*) for 10 min at 4°C in a TLA-100.3 rotor in an Optima Max-XP ultracentrifuge.
10. Pool the supernatants containing the tubulin in a 50-ml polypropylene tube.

3.12. Gel Filtration Chromatography

1. Use a peristaltic pump to load the depolymerised tubulin onto a 10 ml Superloop. The maximum sample volume for the Superdex 200 XK16/60 column is 5 ml and column pressure limit is 0.3 MPa.
2. Run the Superdex 200 XK16/60 column at 1 ml/min with Superdex buffer, inject the sample onto the column, and collect 1 ml fractions in 1.5-ml tubes. Tubulin elutes at about 64–70 ml as a single peak with a total volume of about 9 ml (see Fig. 4). Pool the tubulin fractions (see Note 37).

3.13. Desalt, Concentrate, and Freeze Tubulin for Storage

1. Use a peristaltic pump to load the tubulin fractions on a 50 ml Superloop. Desalt the tubulin into Tubulin storage buffer at 5 ml/min using the HiPrep 26/10 rapid desalting column from Subheading 3.11, step 8 (maximum loading volume is 15 ml and pressure limit is 0.15 MPa). Collect 1 ml fractions in 1.5-ml tubes. The tubulin elutes as a single peak in about 15 ml total volume. Pool and mix the fractions in a 50-ml polypropylene tube on ice.
2. Use a spectrophotometer to measure OD_{280} of the tubulin sample and calculate its concentration from the molar extinction coefficient $E_{280} = 108{,}390\ cm^{-1}\ M^{-1}$. Calculate the reduction in volume required to concentrate the tubulin to 40–50 μM. With a yield of 8–10 mg tubulin, the final volume required should be about 1.5–2.0 ml (see Notes 38 and 39).
3. Pre-rinse a Vivaspin 15R 10,000 mwco spin concentrator with 15 ml tubulin storage buffer at 4,100 rpm (2,675 × *g*) for 10 min at 4°C in a Centra MP4R refrigerated centrifuge with a swing-out rotor (see Note 40).
4. Discard the rinse buffer from the spin concentrator and add the protein sample. Centrifuge at 4,100 rpm (2,675 × *g*) in the Centra MP4R centrifuge with swing-out rotor for 20 min at 4°C. Check the volume remaining and estimate the time required to reach the desired volume (see Note 41).
5. Transfer the concentrated tubulin solution from the spin concentrator into a 50-ml polypropylene tube.
6. Rinse the concentrator membrane with 0.5 ml of cold tubulin storage buffer. Centrifuge the concentrator at 4,100 rpm (2,675 × *g*) for 2 min at 4°C in the Centra MP4R with swing-out rotor to collect the wash buffer. Combine the wash buffer with the concentrated tubulin solution from step 5 and ensure that it is fully mixed by gentle pipetting, without causing air bubbles (see Note 42).
7. In a cold room, use a pipette with pre-cooled disposable tips to dispense 5–20-μl aliquots of the concentrated tubulin into pre-cooled 500-μl screw cap tubes. Immerse the tubes in liquid nitrogen to rapidly freeze and store in liquid nitrogen (see Note 43).

3.14. Analyse Tubulin Yield and Purity

1. Measure the concentration of 3–5 tubulin aliquots. Mix 5 μl of the tubulin aliquot with 15 μl H_2O and 60 μl of 8 M guanidinium chloride by vortexing. Each tubulin aliquot sample should be diluted and measured independently. Use a sample containing 5 μl of the tubulin storage buffer in place of the tubulin aliquot to take a zero absorbance reading. Measure OD_{280} in a small-volume cuvette in a spectrophotometer. Calculate the mean tubulin concentration using the molar extinction coefficient $E_{280} = 108{,}390\ M^{-1}\ cm^{-1}$ (see Note 44).
2. Measure an absorbance scan in the range 220–400 nm. There should be a clear peak at about 280 nm with $OD_{280} > OD_{260}$ (see Note 45).
3. Analyse tubulin samples by SDS-PAGE and stain with a high-sensitivity protein stain. Use several loadings up to at least 20 μg tubulin per lane to assess the tubulin purity (see Fig. 5). A final yield of about 10 mg of tubulin at a concentration of 40 μM and purity 98–99% should be obtained from 500 to 600 g of cell pellet (see Notes 46 and 47).

3.15. Wash Chromatography Columns

1. Connect the two HiPrep 26/10 rapid desalting columns in series (0.3 Mpa total pressure limit) and wash using 53 ml (0.5 CV) of 0.2 M NaOH, 530 ml (5 CV) of H_2O, and 530 ml (5 CV) of 20% (v/v) ethanol at 5 ml/min before storage at 4°C (see Note 48).
2. Wash the Superdex 200 XK16/60 column with 25 ml (0.2 CV) 0.2 M NaOH at 1 ml/min, 500 ml (4 CV) H_2O at 1 ml/min, and 500 ml (4 CV) 20% (v/v) ethanol at 0.5 ml/min. Store the columns at 4°C (see Notes 48 and 49).

4. Notes

1. All solutions are made using ultra-pure water (>18 Mohm) as produced by reverse osmosis-based purification units (Elga or Millipore). Volumes are given for one purification from an 80 l culture of cells with 500–600 g of cell pellet. However, smaller scale preparations are possible down to 20 l of culture with a proportional reduction in buffer volumes and column sizes for all steps, apart from the HiPrep desalting and Superdex 200 columns. Below this scale, further reduction of volume becomes impractical because of the bead mill chamber size and greater non-specific losses causing a disproportionate reduction in yield.
2. All solutions are vacuum filtered through 0.2-μm pore filters.

3. The pH of all solutions is measured at 25°C at their final concentration.
4. Store the KOH and NaOH stock solutions in plastic (high-density polyethylene or similar) containers.
5. PIPES is also named as Piperazine-*N,N′*-bis(2-ethanesulfonic acid); 1,4-piperazinediethanesulfonic acid; or Piperazine-1,4-bis(2-ethanesulfonic acid).
6. The PIPES powder does not dissolve fully until the pH is raised by addition of KOH. Immersion of a pH probe into the undissolved PIPES suspension can affect the functioning of the pH probe; therefore, it is best to add KOH until the PIPES dissolves before starting to monitor the pH.
7. Since heat is often generated with the addition of KOH, it is important to use a pH probe and meter with temperature measurement and compensation. This corrects the pH meter reading for any rise in temperature; however, it does not compensate for the temperature dependence of the PIPES buffer. This must be calculated using the measured temperature of the solution and the temperature coefficient of PIPES buffer which is −0.085/10°C (38, 39). Thus, the pH of a PIPES solution at 35°C is 0.085 pH units lower than the same solution measured at 25°C.
8. KOH is used to adjust pH as the slowly hydrolysed GTP analogue GMPCPP used to stabilise microtubules is reported, in tests using brain tubulin, to undergo slower hydrolysis in solutions containing K^+ compared to Na^+ ions (40).
9. GTP is labile. Store the lyophilised powder at −80°C.
10. GTP hydrolyses under acidic conditions. Therefore, make the stock solution using cooled buffer solution and work quickly to adjust the pH.
11. With labile components, such as GTP, that are added to buffers just before use, it is convenient to make appropriate-sized aliquots for use during the purification to avoid cycles of freeze thawing the stock solution.
12. DTT is labile at high temperatures. Store the powder at 4°C. Thaw aliquots and store on ice before adding to pre-cooled buffers just before use.
13. PEM (100 mM K-PIPES, 1 mM $MgSO_4$, 2 mM EGTA pH 6.9 (41)) is conveniently prepared as a 5× stock for the preparation of other buffers by dilution of the stock. Since microtubule dynamics are sensitive to the concentration of Mg^{2+} ions, the $MgSO_4$ is added from a solution with a precisely determined concentration (Sigma 1.00 ± 0.04 M, stock solution 5). This can be difficult to achieve when making the

solution with normal lab "dry" chemical stocks, which may have absorbed moisture on storage.

14. Store powder stocks of peptide inhibitors at −20°C, preferably with the bottles in a sealed container containing a desiccant, such as silica gel. Once in solution, most peptide inhibitors are unstable even when frozen at −20°C. Store aliquots of solutions at −80°C. We have found that most inhibitor stock solutions, apart from Leupeptin, can be stored for at least 6 months with no discernable effect upon the purification.
15. Leupeptin is labile and is best made fresh for each purification. Store at −80°C for a few days.
16. NaOH is used as Na^+ ions accelerate hydrolysis of GMPCPP, at least in brain tubulin (40).
17. *S. pombe* cells can also be grown using standard methods in any rich or defined minimal yeast media, such as YES or EMM (42).
18. Shake flasks can also be used to grow the cells; however, a fermenter and a continuous-flow centrifuge are more convenient for handling the large culture volumes. We use an off-site fermenter facility (NIMR, UK) for bulk cell preparations and have the cell pellets shipped to us on dry ice.
19. It is critical not to exceed a cell density of 5×10^7 cells/ml before harvesting the cells; otherwise, tubulin degradation products will be present in the preparation. All measurements of culture density are made using a haemocytometer to directly count the cell number.
20. The cell pellet is stored at −80°C and can be stored for several years with no loss of tubulin yield. Cell pellets can also be safely shipped on dry ice.
21. If mixing is too vigorous, too many fines are generated. Removing fines is important as they impede the buffer flow through the column, increasing the backpressure and requiring a reduction in the flow rate.
22. The defrosting step can take up to 60 min with no detrimental effect on the purification. Remove the magnetic stirrer before using the homogeniser, and take care not to generate frothing or foam that might denature the proteins. It is important to disperse any lumps of cells that might block the flow through the bead mill. As soon as all frozen lumps have dispersed, transfer the beaker onto ice to keep the buffer cold.
23. The bead mill is a Dyno-Mill KDL-A with a 300 ml tungsten carbide lined chamber and polyurethane agitator discs. A stainless steel chamber would also be suitable, but the tungsten carbide liner allows the use of denser beads, such as

zirconium, though it has the disadvantage of being more sensitive to cracking if the chamber contents freeze. We have found no advantage to using denser beads when processing *S. pombe* cells. The chamber is filled to 80% capacity with 240 ml of 500 μm glass beads. The Mill is operated on its fastest rotation speed giving a circumferential speed of 14 m/s with 63 mm agitators. Cooling is very important to maintain a process temperature of 4°C. Tap water or recirculating water chillers do not have adequate cooling capacity, and we use a Tricool Minac 010 chiller unit with a 1.85 kW cooling capacity at −20°C which uses standard (blue) antifreeze (car engine coolant) (50% v/v) as the coolant liquid. This should not be mixed with newer anti-corrosion types (red) as the mixture can sometimes form a viscous gel. The temperature of the chiller is manually adjusted during the run to maintain the process temperature at 4°C. While pre-cooling the system, disconnect the grinding chamber from the cooling circuit and set the chiller to −10°C. Process fluid (either wash buffers or cell suspension) must be pumped through the Mill chamber using a large-capacity peristaltic pump before starting the Mill; otherwise, the Mill seals will be destroyed. While the Mill is running, the Mill chamber is also connected to the chiller unit cooling circuit. The temperature of the chiller may need to be reduced to −18°C during the processing run, depending on the room temperature. The Mill seals are surrounded by an external lubricating fluid. We normally use a triple seal arrangement with a closed tube containing only H_2O to avoid contamination of the process fluid. The gap on the Mill chamber outflow that retains the glass beads within the Mill is set using a nominal 0.1 mm spacer, which widens slightly at the 4°C operating temperature. The Mill should not be run without cooling, or the gap may close as the Mill warms up causing damage to the Mill.

24. Usually, four passes are sufficient to achieve >80–90% cell lysis. Very low lysis might be caused by too high a flow rate of the process fluid through the chamber, so reducing the flow rate might help. Take care that the temperature of the process fluid does not rise if the flow rate is reduced. A drop in lysis efficiency between purifications could be caused by a change in the density of the cell suspension used. A consistent reduction in lysis efficiency might indicate that either the glass beads in the Mill chamber or the agitator discs have become worn and require replacing.

    In the event of low lysis, more passes of the cells are possible; however, prolonged processing times should be avoided to reduce possible effects on the tubulin functionality. We have used up to five passes and still produced normal levels of functional tubulin.

25. After processing, switch off the DynoMill, then stop the process flow, and set the cooling temperature to 4°C. This avoids damage to the Mill. The Mill and beads are washed by first pumping water through the chamber until it runs clear, followed by about 1 l of 20% (v/v) ethanol. The Mill is stored with 20% ethanol in the chamber between runs to inhibit microbial contamination.
26. We describe the use of columns and conditions for an Akta purifier 10. Previously, we have used an FPLC system (GE Healthcare) and the method could be carried out using any low-pressure chromatography system. The Akta is run in a 4°C cold room to maintain tubulin activity during the many hours of chromatography.
27. It is important to observe the pressure limits of the chromatography resins and columns. If the pressure limit of the system is being exceeded, the flow rate should be reduced; otherwise, the column hardware may fail or the resin may become compressed. The pressure limit for the X50 is 0.3 MPa. To keep backpressure in the system to a minimum, we remove the Akta flow restrictor and connect the outlet from the conductivity cell directly to the fraction collector using 0.75-mm internal diameter (green) tubing. Backpressure before the column is measured by disconnecting the column inlet tubing from the system and using 10% glycerol in water at 6 ml/min (or the appropriate buffer and flow rate for the column). This gives the pre-column pressure which can be added to the column pressure limit to give a total system pressure limit that is safe for the column.
28. Because of high protein concentrations, the standard 10 mm flow cell is replaced by a 2-mm path length flow cell to avoid saturating the absorbance detector. The Akta system is set to monitor OD_{280}, OD_{260}, and conductivity. Nucleic acid is a major contaminant and not readily detectable on the protein gels used to analyse column fractions.
29. Precast gels, such as Novex 10% Bis-Tris gels run in MOPs buffer, and a rapid protein gel stain, such as Simply Blue SafeStain, are suitable as the resolution is good even with the column fractions that contain high concentrations of salt, protein detection is sensitive, and they save time during the purification.
30. It is important to use a high-purity ammonium sulphate with low levels of heavy metal contamination to avoid protein denaturation. Any lumps in the ammonium sulphate should be removed by grinding with a mortar and pestle. The ammonium sulphate is added slowly so that the powder remains in suspension and is allowed to dissolve before more is added. This avoids accumulating undissolved powder at the bottom of the container that would give localised high concentrations

of ammonium sulphate that might cause protein denaturation.

31. The ammonium sulphate precipitate can be stored for several days at 4°C.
32. The Superloop allows large sample volumes to be loaded onto gel filtration columns with no loss or dilution of the sample compared to a conventional sample loop. A maximum volume of 15 ml can be loaded onto the 26/10 fast-desalting column.
33. Both the rotor and centrifuge should be pre-warmed at 32°C. The rotor can be rapidly warmed by placing in a polythene bag and immersing in a water bath.
34. Polyallomer tubes are used, as they are more resistant to DMSO than polycarbonate.
35. Pipette the pellets using the tip of a 1-ml air-displacement pipette. To avoid losses through pellets and tubulin protein sticking to the tip, reuse the same tip for all mixing. Ensure that the pellets are completely re-suspended by the end of the incubation period.
36. The rotor can be rapidly cooled by placing it in a polythene bag in a bucket of iced water.
37. The Superdex 200 chromatography removes residual nucleic acid contamination (Fig 1b). HPLC analysis has confirmed that the GMPCPP from the tubulin polymerisation is removed by this step and that only GTP and GDP remain in the tubulin sample.
38. The tubulin is usually quite dilute with a concentration of about 5 μM. The OD_{280} is measured without dilution using tubulin storage buffer as the absorbance zero. To reduce sample losses, use a spectrophotometer that can measure small volumes, such as Cary50 (Varian), in combination with a small-volume cuvette, such as the disposable Uvette (Eppendorf). This combination can reliably measure the absorbance of protein samples with a volume of 70 μl. Alternatively, use a Nanodrop (Thermo), which can measure 2-μl volume protein samples.
39. The molar extinction coefficient for the tubulin αβ heterodimer is calculated from the protein sequence of the *S. pombe* α and β tubulins (25) using the method of Gill and Von Hippel (43) and adding the extinction coefficient of GTP (37), assuming full occupancy of the α and β tubulin GTP-binding sites. *S. pombe* α1 and α2 tubulins have the same calculated extinction coefficient, so the same molar extinction coefficient is used for both wild-type (α1, α2, β) and single isoform (α1, β) tubulin purifications.
40. The composition of the concentrator filter and design of the concentrator are very important to ensure high protein

recovery. We have successfully concentrated *S. pombe* tubulin to 60 μM using Vivaspin 15R filters.

41. The rate of concentration can vary, so take care not to over-concentrate the tubulin, which can make it difficult to recover and cause losses.

42. Rinsing the filter is important to ensure a high recovery of tubulin. Ensure that the concentrated tubulin is well-mixed as the concentration process, viscosity of the tubulin solution, and flushing of the filter often result in a final solution containing concentration gradients. Avoid causing air bubbles in the sample as this can lead to denaturation of the protein and loss of activity.

43. No glycerol is added before freezing, as this would interfere with subsequent measurements of microtubule dynamics. Aliquots of concentrated tubulin frozen in this way retain >70% activity and can be stored in liquid nitrogen for at least several years with no significant loss of activity. The proportion of active tubulin is at least as high as is obtained by freezing in the presence of glycerol. When recovering tubulin for microtubule dynamics assays, rapidly thaw the tubes by gentle warming and mixing using hand heat and then place on ice just before the last ice melts. Use the tubulin as soon as possible after thawing and do not store for long periods on ice to ensure consistent activity between different assays.

44. The concentration of the individual aliquots should be similar. Large discrepancies could indicate problems with mixing the dilutions, which are viscous and should be vortexed. Genuine differences between aliquots indicate insufficient mixing of the concentrated tubulin before aliquoting.

45. The scan should have a flat baseline. A sloping baseline indicates particulate material in the sample. A high absorbance at 260 nm with no distinct peak at 280 nm ($OD_{260} > OD_{280}$) suggests contamination by nucleic acids.

46. Because of the high protein loadings used to assess tubulin purity, the staining and signal from the tubulin bands can be compressed and non-linear. Therefore, either use a fluorescent gel stain, such as Sypro Red (Invitrogen), and image the gel with a gel-imaging system with linear detection (such as Pharos gel scanner (Bio-Rad)) or use a Colloidal Coomassie stain (Simply Blue, Invitrogen) and carefully calibrate the gel image using an optical density standard (Thor Labs) and image J software (44).

47. We obtain 8–12 mg of tubulin from 500 to 600 g of cell pellet with a typical yield being 10 mg. If the yield is very low or there are problems with the tubulin purity being less than 98%, analyse samples from intermediate stages of the purification by SDS-PAGE to determine where the problem has

occurred. Load an equivalent fraction of each step (1/8,000 of the total volume) so that the relative yield can be easily estimated directly from the gel.

48. Washed and stored in this way, the columns can be reused for many years with no loss of performance and no cross-contamination of subsequent purifications.
49. DE52 resin and the HiTrap Q column are discarded after each purification and not reused.

Acknowledgements

We gratefully acknowledge the Medical Research Council (G0200542), Association for International Cancer Research (09-0221), and Marie Curie Cancer Care for supporting this study.

References

1. Castoldi M, Popov AV. (2003) Purification of brain tubulin through two cycles of polymerization-depolymerization in a high-molarity buffer. Protein Expr Purif **32**:83–88.
2. Mitchison T, Kirschner M. (1984) Microtubule assembly nucleated by isolated centrosomes. Nature **312**:232–237.
3. Luduena RF. (1998) Multiple forms of tubulin: different gene products and covalent modifications. Int Rev Cytol **178**:207–275.
4. Bode CJ, Gupta ML, Suprenant KA, Himes RH. (2003) The two alpha-tubulin isotypes in budding yeast have opposing effects on microtubule dynamics in vitro. EMBO reports **4**:94–99.
5. Newton CN, DeLuca JG, Himes RH, Miller HP, Jordan MA, Wilson L. (2002) Intrinsically slow dynamic instability of HeLa cell microtubules in vitro. J Biol Chem **277**:42456–42462.
6. Detrich HW, 3rd, Parker SK, Williams RC, Jr., Nogales E, Downing KH. (2000) Cold adaptation of microtubule assembly and dynamics. Structural interpretation of primary sequence changes present in the alpha- and beta-tubulins of Antarctic fishes. J Biol Chem **275**: 37038–37047.
7. Lewis SA, Tian G, Cowan NJ. (1997) The alpha- and beta-tubulin folding pathways. Trends Cell Biol **7**:479–484.
8. Shah C, Xu CZ, Vickers J, Williams R. (2001) Properties of microtubules assembled from mammalian tubulin synthesized in Escherichia coli. Biochemistry **40**:4844–4852.
9. Jang MH, Kim J, Kalme S, et al. (2008) Cloning, purification, and polymerization of Capsicum annuum recombinant alpha and beta tubulin. Bioscience, biotechnology, and biochemistry **72**:1048–1055.
10. Sontag CA, Staley JT, Erickson HP. (2005) In vitro assembly and GTP hydrolysis by bacterial tubulins BtubA and BtubB. J Cell Biol **169**:233–238.
11. Schlieper D, Oliva MA, Andreu JM, Lowe J. (2005) Structure of bacterial tubulin BtubA/B: evidence for horizontal gene transfer. Proc Natl Acad Sci USA **102**:9170–9175.
12. Barnes G, Louie KA, Botstein D. (1992) Yeast proteins associated with microtubules in vitro and in vivo. Mol Biol Cell **3**:29–47.
13. Kilmartin JV. (1981) Purification of yeast tubulin by self-assembly in vitro. Biochemistry **20**:3629–3633.
14. Davis A, Sage CR, Wilson L, Farrell KW. (1993) Purification and biochemical characterization of tubulin from the budding yeast Saccharomyces cerevisiae. Biochemistry **32**:8823–8835.
15. Dougherty CA, Sage CR, Davis A, Farrell KW. (2001) Mutation in the beta-tubulin signature motif suppresses microtubule GTPase activity and dynamics, and slows mitosis. Biochemistry **40**:15725–15732.
16. Sage CR, Davis AS, Dougherty CA, Sullivan K, Farrell KW. (1995) beta-Tubulin mutation suppresses microtubule dynamics in vitro and slows mitosis in vivo. Cell Motil Cytoskeleton **30**:285–300.

17. Sage CR, Dougherty CA, Davis AS, Burns RG, Wilson L, Farrell KW. (1995) Site-directed mutagenesis of putative GTP-binding sites of yeast beta- tubulin: evidence that alpha-, beta-, and gamma-tubulins are atypical GTPases [published erratum appears in Biochemistry 1995 Dec 26;34(51):16870]. Biochemistry **34**:7409–7419.
18. Gupta ML, Jr., Bode CJ, Georg GI, Himes RH. (2003) Understanding tubulin-Taxol interactions: mutations that impart Taxol binding to yeast tubulin. Proc Natl Acad Sci USA **100**:6394–6397.
19. Gupta ML, Jr., Bode CJ, Thrower DA, et al. (2002) beta-Tubulin C354 Mutations that Severely Decrease Microtubule Dynamics Do Not Prevent Nuclear Migration in Yeast. Mol Biol Cell **13**:2919–2932.
20. Uchimura S, Oguchi Y, Katsuki M, et al. (2006) Identification of a strong binding site for kinesin on the microtubule using mutant analysis of tubulin. EMBO J **25**:5932–5941.
21. Uchimura S, Oguchi Y, Hachikubo Y, Ishiwata S, Muto E. (2010) Key residues on microtubule responsible for activation of kinesin ATPase. EMBO J **29**:1167–1175.
22. Toda T, Adachi Y, Hiraoka Y, Yanagida M. (1984) Identification of the pleiotropic cell division cycle gene NDA2 as one of two different alpha-tubulin genes in Schizosaccharomyces pombe. Cell **37**:233–242.
23. Hiraoka Y, Toda T, Yanagida M. (1984) The NDA3 gene of fission yeast encodes beta-tubulin: a cold-sensitive nda3 mutation reversibly blocks spindle formation and chromosome movement in mitosis. Cell **39**:349–358.
24. Wood V, Gwilliam R, Rajandream MA, et al. (2002) The genome sequence of Schizosaccharomyces pombe. Nature **415**:871–880.
25. Hertz-Fowler C, Peacock CS, Wood V, et al. (2004) GeneDB: a resource for prokaryotic and eukaryotic organisms. Nucleic Acids Res **32**:D339–343.
26. Hagan IM. (1998) The fission yeast microtubule cytoskeleton. J Cell Sci **111**:1603–1612.
27. Alfa CE, Hyams JS. (1991) Microtubules in the fission yeast Schizosaccharomyces pombe contain only the tyrosinated form of alpha-tubulin. Cell Motil Cytoskeleton **18**:86–93.
28. Sawin KE, Tran PT. (2006) Cytoplasmic microtubule organization in fission yeast. Yeast **23**:1001–1014.
29. Piel M, Tran PT. (2009) Cell shape and cell division in fission yeast. Curr Biol **19**:R823–827.
30. Nakamura M, Zhou XZ, Lu KP. (2001) Critical role for the EB1 and APC interaction in the regulation of microtubule polymerization. Curr Biol **11**:1062–1067.
31. Sandblad L, Busch KE, Tittmann P, Gross H, Brunner D, Hoenger A. (2006) The Schizosaccharomyces pombe EB1 homolog Mal3p binds and stabilizes the microtubule lattice seam. Cell **127**:1415–1424.
32. Bieling P, Laan L, Schek H, et al. (2007) Reconstitution of a microtubule plus-end tracking system in vitro. Nature **450**:1100–1105.
33. Adachi Y, Toda T, Niwa O, Yanagida M. (1986) Differential expressions of essential and nonessential alpha-tubulin genes in Schizosaccharomyces pombe. Mol Cell Biol **6**:2168–2178.
34. Braun M, Drummond DR, Cross RA, McAinsh AD. (2009) The kinesin-14 Klp2 organizes microtubules into parallel bundles by an ATP-dependent sorting mechanism. Nat Cell Biol **11**:724–730.
35. des Georges A, Katsuki M, Drummond DR, Osei M, Cross RA, Amos LA. (2008) Mal3, the Schizosaccharomyces pombe homolog of EB1, changes the microtubule lattice. Nature structural & molecular biology **15**:1102–1108.
36. Katsuki M, Drummond DR, Osei M, Cross RA. (2009) Mal3 masks catastrophe events in Schizosaccharomyces pombe microtubules by inhibiting shrinkage and promoting rescue. J Biol Chem **284**:29246–29250.
37. Rex M. C. Dawson DCE, William H. Elliott, K. M. Jones. (1969) Data for biochemical Research. Oxford University Press, Oxford.
38. Good NE, Winget GD, Winter W, Connolly TN, Izawa S, Singh RM. (1966) Hydrogen ion buffers for biological research. Biochemistry **5**:467–477.
39. Blanchard JS. (1984) Buffers for enzymes. Methods Enzymol **104**:404–414.
40. Hyman AA, Salser S, Drechsel DN, Unwin N, Mitchison TJ. (1992) Role of GTP hydrolysis in microtubule dynamics: information from a slowly hydrolyzable analogue, GMPCPP. Mol Biol Cell **3**:1155–1167.
41. Walker RA, O'Brien ET, Pryer NK, et al. (1988) Dynamic instability of individual microtubules analyzed by video light microscopy: rate constants and transition frequencies. J Cell Biol **107**:1437–1448.
42. Moreno S, Klar A, Nurse P. (1991) Molecular genetic analysis of fission yeast Schizosaccharomyces pombe. Methods Enzymol **194**:795–823.
43. Gill SC, von Hippel PH. (1989) Calculation of protein extinction coefficients from amino acid sequence data. Anal Biochem **182**:319–326.
44. Abramoff MD, Magelhas PJ, Ram SJ. (2004) Image processing with ImageJ. Biophotonics International **11**:36–42.

Chapter 4

Generation of Differentially Polyglutamylated Microtubules

Benjamin Lacroix and Carsten Janke

Abstract

Microtubules are cytoskeletal structures built of alpha- and beta-tubulins. Although tubulins are highly conserved throughout evolution, microtubules can be adapted to a range of different functions. A powerful mechanism that could regulate the functional specialization of microtubules is the posttranslational modification of tubulin molecules. Two tubulin modifications, polyglutamylation and polyglycylation, generate amino acid side chains of different length on tubulin. These modifications are thought to regulate interactions between microtubules and their associated proteins; however, detailed studies of this potential mechanism have not been performed. The investigation of the potential regulatory role of polyglutamylation requires in vitro tools to visualize the molecular events that could be affected by this modification. Classically, in vitro work with microtubules is performed with tubulin from brain tissue; however, this tubulin is highly posttranslationally modified. Here, we describe a method for the purification of tubulin carrying controlled levels of polyglutamylation, which can be used in basic in vitro assays.

Key words: Microtubule, Tubulin, Posttranslational modification, Polyglutamylation

1. Introduction

Polyglutamylation is a posttranslational modification that generates side chains of several glutamate residues on the carboxy-terminal of both alpha- and beta-tubulins (1, 2). These side chains are initiated by the addition of one, the so-called branching point glutamate to the gamma-carboxyl group of a gene-encoded glutamate residue. Side chain elongation leads to the addition of further glutamates that are connected by classical peptide bonds (3). The length of the glutamate side chains can be highly variable.

Polyglutamylation takes place on the carboxy-terminus of tubulin molecules, which is an important interaction site for microtubule-associated proteins (MAPs) and molecular motors (4, 5). Thus, the modification might selectively regulate the interactions

Anne Straube (ed.), *Microtubule Dynamics: Methods and Protocols*, Methods in Molecular Biology, vol. 777,
DOI 10.1007/978-1-61779-252-6_4, © Springer Science+Business Media, LLC 2011

of MAPs and motors with microtubules. This is an exciting hypothesis since it is known that polyglutamylation is present only on specific microtubules, such as the mitotic spindle and midbody during cell division, the ciliary axoneme (6, 7), most neuronal microtubules (1, 8–10) and the centrioles (11), which are the core structures of centrosomes and basal bodies. Thus, polyglutamylation might play an important role in the functional specialization of these microtubules.

To elucidate the role of tubulin polyglutamylation at the molecular level, it is necessary to use in vitro approaches. Most in vitro studies have been performed with tubulin purified from pig or bovine brain using the classical polymerization/depolymerization procedure (12). This method allows for the purification of large amounts of pure tubulin, which is suitable for a range of demanding in vitro methods. However, the disadvantage of brain tubulin is that it contains high levels of posttranslational modifications, including polyglutamylation, and is thus not suited to study the role of this modification. To overcome this limitation, we looked for an alternative tubulin source that allows for the purification of either nonmodified or selectively modified tubulin. Following the observation that cultured cells, such as HeLa cells, contain generally very low levels of polyglutamylation (13), we adapted a protocol for tubulin extraction from yeast (14) to the purification of tubulin from HeLa cells. Tubulin from HeLa cells can be polymerized into microtubules that can then be used for in vitro assays. However, the tubulin yield of this method is much lower than that from classical preparations from brain, and repeated cycles of polymerization/depolymerization cannot be performed. Consequently, the HeLa tubulin is less pure, which can result in shorter microtubules. Nevertheless, this method allows for generating microtubules that carry defined subtypes of polyglutamylation for in vitro comparison. To obtain differentially modified tubulins, we use the recently discovered tubulin polyglutamylases (15, 16), which efficiently generate polyglutamylation when transfected into HeLa cells. By using different types of these polyglutamylases (15), we are also able to define the type of glutamylation generated, and in particular, we can control if short or long glutamate side chains are generated. Here, we describe a rapid method to purify tubulin from HeLa cells (see Note 1) that is easy to set up in a laboratory equipped for basic biochemistry. This method provides selectively polyglutamylated microtubules that can be used in standard in vitro assays.

2. Materials

2.1. Cell Culture and Cell Transfection

1. HeLa S3 cells: ATCC number CCL-2.2. This human cell line is a clonal derivative of the HeLa cell line (CCL2) (17) (see Note 1).

2. Twenty five ∅145-mm (165-cm^2) tissue culture dishes (Greiner Bio-one, Frickenhausen, Germany).
3. Dulbecco's modified Eagle's medium (DMEM; Invitrogen, Paisley, UK) supplemented with 10% fetal bovine serum (FBS Hyclone, Logan, UT) and a mix of antibiotics to a final concentration of 100 Units/ml of Penicillin and 100 μg/ml of Streptomycin (Invitrogen).
4. Phosphate-Buffered Saline (PBS)-EDTA: 0.5 mM sterile EDTA, pH 8.0 in sterile PBS solution (Invitrogen).
5. 150 mM NaCl in ultrapure water (Invitrogen) is filtered through a 0.2-μm sterile filter. Store at 4°C.
6. Transfection reagent: jetPEI (Polyplus, Illkirch, France).
7. Expression plasmids for YFP-tagged polyglutamylase enzymes: A truncated version of murine TTLL4 (amino acids 555–1,193 (15)) for short side chains or a truncated version of murine TTLL6 (amino acids 1–513 (15)) for long side chains.
8. Cell lifter (TPP, Trasadingen, Switzerland).

2.2. Cell Lysis and Tubulin Extraction

1. 15- and 50-ml polypropylene conical tubes (BD Falcon™, Franklin Lakes, NJ, USA).
2. Low-speed centrifuge for sedimentation of cells.
3. Ultrasound cell disruptor.
4. Micro ultracentrifuge and corresponding tubes and rotors, such as OPTIMA™ TL with TLA-100.3 rotor (Beckman, Brea, CA, USA) and polycarbonate centrifuge tubes (13 × 51 mm; Beckman).
5. Resuspension buffer is buffer A: 0.2 M NaCl, 50 mM Mes (2-(*N*-morpholino)ethanesulfonic acid)-NaOH, pH 6.8, 2 mM EGTA, 1 mM $MgCl_2$.
6. Dithiothreitol (DTT) stock solution: 1 M DTT (Sigma, St Louis, MO) in deionized water (see Note 3).
7. NP40 stock solution: 10% NP40 (Sigma) in deionized water (see Note 2).
8. Protease inhibitors (400×): 4 mg/ml aprotinin (Sigma), 4 mg/ml leupeptin (Sigma), and 4 mg/ml 4-(2-aminoethyl)-benzoesulfonyl fluoride (Sigma) in water (see Note 3).
9. GTP solution: Guanosine 5′-triphosphate lithium salt (Sigma) in water or 50 mM Tris–HCl, pH 7 (see Note 3).
10. Taxol (Sigma) stock solutions at 1 mM and 10 mM in dimethylsulfoxide (DMSO; Sigma; see Note 3).

2.3. Chromatography

1. Econo-Pac Columns (Bio-Rad, Hercules, CA, USA).
2. Diethylaminoethyl (DEAE) Sepharose Fast-Flow gel (GE Healthcare, Buckinghamshire, UK).

3. Akta Purifier FPLC system (GE Healthcare).
4. Buffer A: 0.2 M NaCl, 50 mM Mes (2-(*N*-morpholino) ethanesulfonic acid)-NaOH, pH 6.8, 2 mM EGTA, 1 mM $MgCl_2$ filtered through a 0.2-μm polyethersulfone Nalgene filter (Thermo Fisher, Rochester, NY, USA; see Note 4).
5. Buffer B: 1 M NaCl, 50 mM Mes–NaOH, pH 6.8, 2 mM EGTA, and 1 mM $MgCl_2$ (filtered through a 0.2-μm polyethersulfone Nalgene filter; Thermo Fisher; see Note 4).

2.4. SDS-Polyacrylamide Gel Electrophoresis (see Note 5)

1. 5× Laemmli sample buffer: 400 nM Tris–HCl, pH 6.8, 10% SDS, 50% glycerol, 0.01% of bromophenol blue.
2. Electrophoresis cell: Mini-PROTEAN® cell with a PowerPac™ power supply (Bio-Rad).
3. Separating gel buffer (4×): 1.5 M Tris–HCl, pH 9.0, 0.4% SDS.
4. Stacking gel buffer (4×): 0.5 M Tris–HCl, pH 6.8, 0.4% SDS.
5. Acrylamide/bisacrylamide stock solution (74:1): 40% acrylamide solution (Bio-Rad) supplemented with 0.54% bisacrylamide (w/v) powder (Bio-Rad).
6. Ammonium persulfate (APS) stock solution: 10% (w/v) APS (Sigma) in deionized water (see Note 2).
7. Tetramethylethylenediamine (TEMED; Sigma).
8. Running buffer: 50 mM Tris–HCL, 384 mM glycine, 0.1% SDS in deionized water.
9. Coomassie brilliant blue staining solution: 50% ethanol, 40% deionized water, 10% acetic acid, 0.25% Brillant Blue R. Solution is filtered through a paper filter.

2.5. Immuno Blotting to Detect Tubulin Modifications

1. Bjerrum and Schaffer-Nielsen blot transfer buffer: 48 mM Tris –HCl, 39 mM glycin, 0.4% SDS, 20% ethanol (technical grade).
2. Tank blot system: Criterion™ cell with a PowerPac™ power supply (Bio-Rad).
3. Nitrocellulose transfer membrane (Whatman-GE Healthcare).
4. Gel blotting paper 0.8 mm (Whatman-GE Healthcare).
5. Tris-buffered saline (TBS): 20 mM Tris–HCl, pH 7.4, 140 mM NaCl.
6. TBS containing 5% (v/w) fat-free dry milk for blocking and antibody incubation.
7. TBS-T: TBS, 0.1% of Tween 20 (Sigma) for washing.
8. Antibodies: GT335 is a monoclonal antibody that recognizes the branching point of the glutamylation side chain (will soon be commercially available: Inoxxis SA, Epalinges, Switzerland).

12G10 monoclonal anti-alpha-tubulin antibody (Developmental Studies Hybridoma Bank, University of Iowa, Iowa City, IA, USA). 6-11B-1 is a monoclonal antibody specific to acetylated alpha-tubulin (Sigma). PolyE antibody detects long glutamate chains (specific detection of side chains ≥3 E) (15). Anti-mouse or anti-rabbit horseradish peroxidase-coupled antibody (GE Healthcare).

9. ECL Chemo luminescence detection kit (GE Healthcare).

2.6. Tubulin Dialysis

1. NAP5 column (GE Healthcare).
2. Precipitation buffer is buffer A saturated in ammonium sulfate at 4°C.
3. Ultracentrifuge with corresponding tubes and rotors, such as 50.2 Ti (Beckman) with thick-wall polycarbonate centrifuge tubes (25 × 89 mm; Beckman).
4. Bradford Reagent (Sigma).
5. Bovine serum albumin stock solution: 1 mg/ml BSA(Sigma) in deionized water (see Note 3).
6. Spectrophotometer with a wavelength of 595 nm (e.g., V-600; JASCO Corporation, Tokyo Japan).
7. Brinkley BR buffer (BRB80): 80 mM PIPES (piperazine-*N*,*N'*-bis(2-ethanesulfonic acid))-KOH pH 6.8, 1 mM $MgCl_2$, 1 mM EGTA (see Notes 2 and 3).

2.7. Tubulin Polymerization

1. GTP solution: Guanosine 5′-triphosphate lithium salt (Sigma) in water or 50 mM Tris–HCl, pH 7.
2. Taxol (Sigma) stock solutions at 1 mM and 10 mM in DMSO (Sigma); (see Note 3).
3. Micro-ultracentrifuge and corresponding tubes and rotors, such as OPTIMA™ TL with TLA-100.3 rotor (Beckman) with 1.5-ml polyallomer microfuge tubes and tube adaptors (Beckman).
4. Water bath at 37°C.

3. Methods

In this method, we describe the generation of tubulin that carries different subtypes of the posttranslational modification, polyglutamylation. To generate these modifications on the normally nonmodified HeLa tubulin (see Note 1), we transfect the cells with polyglutamylase (TTLL) enzymes, and according to the reaction specificities of these enzymes, specific subtypes of tubulin polyglutamylation are generated in transfected cells (15).

This method is efficient, but it remains important to characterize the obtained tubulin for the precise type of polyglutamylation by immuno blot (an example is shown in Fig. 2b). Different types of polyglutamylation can be generated by transfecting specific polyglutamylases into the cell prior to tubulin purification. Using an enzyme specific to the initiation reaction (e.g., TTLL4 (15)) allows for the generation of tubulin with short side chains while transfection of an elongating enzyme (e.g., TTLL6 (15)) generates tubulin with long side chains. Under the given experimental conditions, both enzymes modify alpha- and beta-tubulins.

Since the amount of tubulin obtained from HeLa cells is insufficient to perform cycles of polymerization/depolymerization, in this method, we separate tubulin (which has an isoelectric point of about 5) on an anion exchange column (DEAE sepharose). By applying a relatively shallow salt gradient during the elution, it is possible to separate tubulin from other acidic proteins and obtain tubulin at relatively high purity (Fig. 1). We use a progressive ammonium sulfate precipitation that allows us to concentrate the tubulin, and at the same time to remove some additional contaminating proteins (Fig. 2a). The precipitated tubulin is desalted with BRB80 buffer and remains active for several months if stored at −80°C.

3.1. Preparation of Cells

HeLa cells (see Note 1) are grown in 15 ml of DMEM with 10% FBS and antibiotics in ∅145-mm (165-cm^2) dishes. Cells are detached from dishes with PBS containing 0.5 mM EDTA and split 1:5 into 145-mm dishes (at least 25 dishes are required for one tubulin purification; see Note 6). To obtain nonmodified tubulin, cells are grown until they reach maximal density.

3.2. Transfection of Cultured Cells

In order to obtain modified tubulin, cells are transfected with a modifying enzyme (TTLL) when they reach around 50% confluence. Transfection is made following the manufacturer's instructions (jetPEI; Polyplus):

1. For the transfection of one ∅145-mm culture dish, 20 μg of DNA is mixed into1 ml of 150 mM NaCl.
2. In a different tube, 40 μl of jetPEI is mixed into 1 ml of 150 mM NaCl.
3. The jetPEI is added to the DNA, and the mix is immediately vortexed and incubated at room temperature for 20 min.
4. The transfection solution is added to 15 ml of fresh DMEM medium with 10% FBS and antibiotics.
5. Old culture medium is discarded from the culture dishes, and replaced by the medium containing the transfection solution.
6. After 2 h, the medium is exchanged for a fresh, regular DMEM medium with 10% FBS and antibiotics.

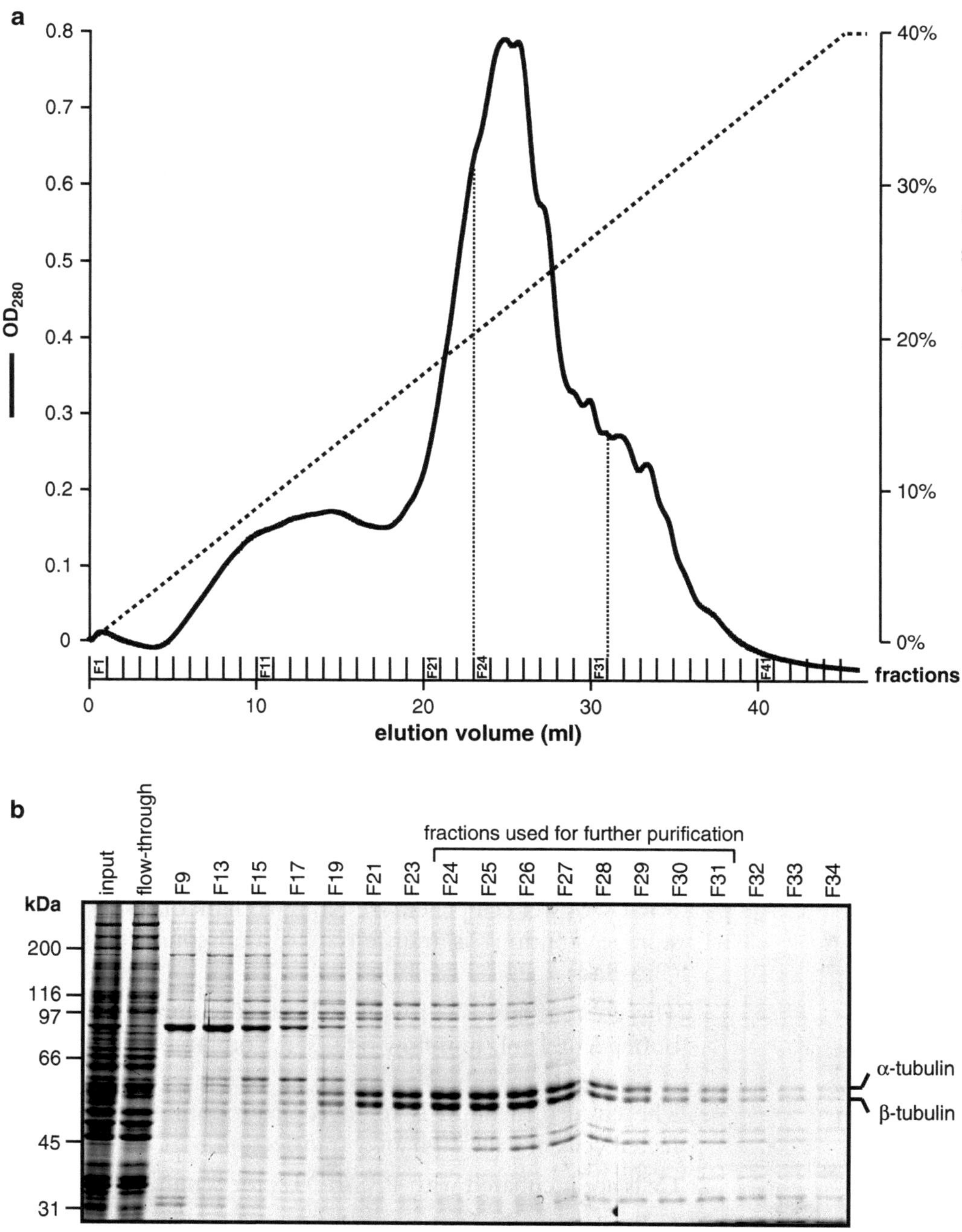

Fig. 1. Purification of tubulin by liquid chromatography. (**a**) Elution profile of the DEAE sepharose column. The protein content was determined by measuring the optical density at 280 nm (OD_{280}). Elution volume and fraction numbers are indicated. The thick dashed line shows the NaCl gradient generated by increasing the percentage of buffer B (right axis). The thin dashed lines label the fractions that have been pooled together for further tubulin purification (see B). (**b**) Samples from the initially loaded cell extract (input), the flow through, and selected elution fractions are analyzed by SDS-PAGE followed by Coomassie brilliant blue staining of the gels. The fractions that have been used for further purification are labeled. Note that using the modified SDS-PAGE conditions described here, alpha- and beta-tubulin can be separated.

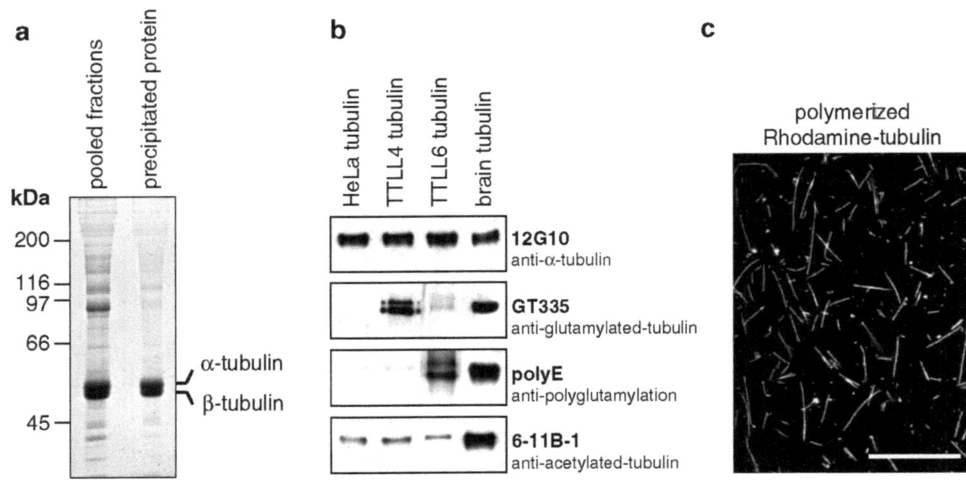

Fig. 2. Final purification steps and analysis of polyglutamylation state. (**a**) The final purification step consists in a precipitation of tubulin with ammonium sulfate. Pooled fractions from the chromatography (Fig. 1) and precipitated proteins are separated on SDS-PAGE and stained with Coomassie brilliant blue. Note that some of the contaminating proteins are absent in the precipitated fraction. (**b**) Immuno blot analysis of differentially modified tubulin. HeLa tubulin is not recognized with the two glutamylation-specific antibodies GT335 and polyE. Combination of strong GT335 labeling and no polyE labeling indicates the presence of short glutamate side chains on TTLL4-modified tubulin whereas polyE labeling of TTLL6-modified tubulin reveals the presence of long side chains. The overall glutamylation level of TTLL6 tubulin is lower as compared to that of TTLL4-modified or brain tubulin (GT335). Acetylation, another tubulin modification, is not affected by the expression of different TTLL enzymes. Anti-alpha-tubulin antibody is used as loading control. (**c**) Microscopy image of HeLa tubulin polymerized in the presence of rhodamine-labeled tubulin. Scale bar is 10 μm.

3.3. Preparation of the Chromatography Column

1. The DEAE sepharose Fast-Flow gel is resuspended by careful shaking.
2. For 1 g of wet cell pellet (see Note 6), 1 ml of resin (without storage solution) is transferred to an Econo-Pac Column (Bio-Rad).
3. The column is washed with at least 10-column volumes of buffer A at 1 ml/min flow rate (see Note 4).

3.4. Preparation of Cell Extracts for the Tubulin Purification

1. The culture medium is removed from confluent nontransfected cells or cells 20 h after transfection.
2. Cells are incubated with PBS-EDTA for 5 min.
3. Cells are detached using a cell lifter and collected in Falcon tubes at room temperature (see Note 7).
4. Cells are centrifuged (200 RCF, 5 min, room temperature) and subsequently washed in 40 ml PBS at room temperature.
5. After sedimentation (200 RCF, 5 min, room temperature), the cells are washed in 40 ml buffer A.
6. The cells are spun down (200 RCF, 5 min, room temperature), the weight of the wet cell pellet is determined, and cells are resuspended in 1 ml of buffer A per 1 g cell pellet.

7. The cell suspension is adjusted to 1 mM DTT, 0.1 mM GTP, 0.2% (v/v) NP40, and 1× protease inhibitors.
8. The cell suspension is transferred onto ice, and cells are disrupted by sonication (2×5 s at 50% intensity using a ∅ 3-mm diameter probe).
9. The cell lysate is incubated on ice for 15 min (see Note 8).
10. The tubulin-containing soluble fraction is obtained by a high-speed centrifugation at 120,000 RCF for 20 min at 4°C.

3.5. Tubulin Purification by Chromatography

1. After centrifugation, the supernatant is loaded on the pre-equilibrated DEAE Fast-Flow column at a flow rate of 1 ml/min. Protein content is detected by measuring the optical density of the flow through at 280 nm (OD_{280}).
2. After completion of the loading, the column is washed with buffer A at a flow rate of 1 ml/min until the OD_{280} has stabilized (no unbound protein is present in the column).
3. Bound proteins are eluted by a salt gradient. This gradient is generated by increasing the concentration of buffer B (containing 1 M NaCl) from 0 to 40% over a period of 40 min at 1 ml/min. Fractions of 1 ml are collected (Fig. 1a).
4. From each fraction, 10 μl are subjected to 3 μl of 5× Laemmli sample buffer, boiled at 95°C for 5 min and analyzed by SDS-PAGE on 10% acrylamide gels (see Subheading 3.6 below for how to prepare gels).
5. The gels are stained for 20 min in Coomassie brilliant blue staining solution and are subsequently transferred to boiling water for destaining.
6. The fractions with the highest tubulin content are pooled together (Fig. 1b).
7. The solution is adjusted to 60% of saturation of ammonium sulfate by adding (drop by drop under constant gentle mixing) 1.5 V of precipitation buffer at 4°C. The solution is incubated for 30 min at 4°C.
8. The precipitated proteins are sedimented by high-speed centrifugation at 120,000 RCF for 45 min at 4°C.
9. The supernatant is discarded, and the pellet is dissolved in 500 μl of BRB80 containing 0.1 mM GTP at 4°C.
10. At 4°C, the solution is loaded onto a NAP5 gel filtration column (GE Healthcare), and the flow through is discarded.
11. The tubulin fraction is eluted from the NAP5 column with 500 μl of BRB80 containing 0.1 mM GTP at 4°C (Fig. 2a).
12. The tubulin concentration is determined using a standard Bradford assay. 1 μl of the protein samples are mixed into 1 ml of a 1:1 dilution of Bradford reagent in deionized water.

For calibration, a standard curve is generated with 0, 0.5, 1, 1.5, and 2 mg/ml BSA, which is equally assayed with the diluted Bradford reagent.

13. For all samples, including the BSA standard, the optical density is determined at 595 nm in a spectrophotometer.
14. The tubulin solution is aliquoted into 20-μl aliquots and frozen in liquid nitrogen. Aliquots are stored at −80°C.

3.6. Analysis of Tubulin Modifications by Immuno Blot

1. To prepare a 10% acrylamide gel, 2 ml of acrylamide/bis solution, 2 ml of separating gel buffer, and 4 ml of deionized water are mixed. Polymerization is started by adding 60 μl of APS stock solution and 8 μl TEMED, and the solution is immediately pipetted into the preassembled gel holder. The gel solution is carefully overlaid with 1 ml of deionized water.
2. After polymerization (20–30 min), the stack gel is prepared by mixing 280 μl of acrylamide/bis solution, 650 μl of stacking gel buffer, and 1.65 ml of deionized water. Polymerization is started by the addition of 10 μl of APS stock solution and 4 μl TEMED, and the solution is immediately transferred onto the top of the polymerized separating gel. A comb is introduced to generate the sample wells, and the gel is allowed to polymerize for at least 30 min.
3. The electrophoresis cell is assembled, combs are removed from the gels, and chambers are filled with running buffer. 2 μl of the purified tubulin solution are adjusted with Laemmli sample buffer, boiled for 5 min (95°C), and separated at 150 V.
4. Proteins are transferred onto a nitrocellulose membrane at 100 V for 60 min in a tank blot system at 4°C.
5. The blot membrane is washed with TBS and incubated with 5% fat-free milk in TBS for 30 min at room temperature.
6. Primary antibodies (GT335, 12G10, polyE) are incubated in TBS-T containing 5% milk for 2 h at room temperature or overnight at 4°C (see Note 9).
7. The membrane is then washed 3 × 3 min in TBS-T under shaking.
8. The secondary antibodies (anti-mouse for GT335 and 12G10; anti-rabbit for polyE) are incubated in TBS-T for 1 h at room temperature. The membranes are washed 5 × 3 min.
9. The antibody signal is detected using the ECL chemiluminescence kit following the manufacturer's instructions.
10. Combination of GT335, polyE, and 12G10 antibody signals allows estimating the extent and subtype of tubulin polyglutamylation (Fig. 2b).

3.7. Microtubule Polymerization

Either freshly prepared tubulin or a thawed aliquot (see Note 10) is used for polymerization.

1. (Optional): The tubulin solution is centrifuged at 4°C for 10 min at 150,000 RCF in order to remove aggregates that can disturb polymerization (this might be important for copolymerization with rhodamine-labeled tubulin, see Note 11).
2. The tubulin solution is adjusted to 1 mM GTP. Taxol can be added to a concentration of 3 μM to help the polymerization and obtain longer microtubules.
3. Microtubule polymerization is induced by incubating the solution at 37°C for 30 min.
4. Taxol is adjusted to a final concentration of 20 μM (1 V of prewarmed BRB80 buffer containing 40 μM taxol is added), and the solution is mixed by careful pipetting. (Optional: To visualize microtubules, rhodamine-labeled tubulin can be incorporated during polymerization. The preparation of rhodamine-labeled tubulin is described in Hyman et al. (18))
5. The polymerized microtubules are spun down at 150,000 RCF for 20 min at 37°C.
6. The supernatant is discarded, and the microtubule pellet is very carefully resuspended in an appropriate volume of prewarmed BRB80 containing 20 μM taxol or another appropriate buffer required for the subsequent assay.
7. Microtubules can now be used for in vitro assays (Fig. 2c).

4. Notes

1. We only tested the method with HeLa or 293 FT cells; however, any cell line could be used. It is critical to test in forehand the endogenous tubulin modification levels and the efficiency of plasmid transfection for the cell line to be used.
2. Stock solutions of APS, NP40, and 5× BRB80 (filtered through a 0.2-μm membrane filter) can be kept at 4°C.
3. Stock solution of BSA (1 mg/ml), DTT (1 M), GTP (0.1 M), taxol (10 μM), protease inhibitors 400×, and BRB80 1× can be aliquoted and kept at −20°C.
4. Neither buffer A nor B contains GTP, and the DEAE column should not be washed with a GTP-containing buffer.
5. The SDS-PAGE used here is a modified version that allows separating alpha- and beta-tubulin (Fig. 1b). Please note that stock solutions are different from standard SDS-PAGE solutions.

6. From our experience, 25 ∅145-mm cell culture dishes confluent with HeLa cells correspond to 2 g of cell pellet.
7. Cells can also be scraped directly into the growth medium with the cell lifter. However, in this case, an additional washing step with PBS is required.
8. After the lysis, cells are kept on ice, which prevents protein degradation, but also induces MT depolymerization and increases the tubulin yield in the purification step.
9. Recommended dilutions for primary antibodies are 1:1,000 for purified polyE, 1:250 for hybridoma supernatant of GT335, and 1:400 for hybridoma supernatant of 12G10.
10. To optimize the quality of microtubules, we recommend thawing the tubulin rapidly (in warm water or by holding the tube in the hand), and to quickly start the polymerization after thawing. Tubulin aliquots should not be refrozen.
11. Some rhodamine-tubulin preparations contain aggregated tubulin that can inhibit microtubule polymerization. Therefore, we strongly recommend spinning down the tubulin solution at 4°C and at high speed before the polymerization step.

Acknowledgments

This work was supported by the CNRS, the Universities of Montpellier 2 and 1, the Association pour la Recherche sur le Cancer (ARC) award 3140, the Agence nationale de la recherche (ANR) awards 05-JCJC-0035 and 08-JCJC-0007, the Fondation pour la Recherche Médicale (FRM) research grant DEQ20081213977, the HFSP program grant RGP 23/2008, and the EMBO Young Investigator Program grant to CJ. BL was supported by a fellowship from the La Ligue contre le Cancer and the EMBO short-term fellowship ASTF 157-2007. We are grateful to M. Donnay, N. Gold, G. Herrada-Aldrian, A. Lawera, J.-C. Mazur, N. Morin, and J. van Dijk (CRBM) as well as T. Surrey (EMBL, Heidelberg, Germany) for technical assistance and providing essential reagents. We also want to thank N. Gold, M.M. Magiera, and K. Rogowski for helpful comments on the manuscript. The monoclonal antibody 12G10 developed by J. Frankel and M. Nelson was obtained from the Developmental Studies Hybridoma Bank developed under the auspices of the NICHD and maintained by the University of Iowa.

References

1. Eddé, B., Rossier, J., Le Caer, J. P., Desbruyeres, E., Gros, F., and Denoulet, P. (1990) Posttranslational glutamylation of alpha-tubulin, *Science 247*, 83–85.
2. Rüdiger, M., Plessman, U., Kloppel, K. D., Wehland, J., and Weber, K. (1992) Class II tubulin, the major brain beta tubulin isotype is polyglutamylated on glutamic acid residue 435, *FEBS Lett 308*, 101–105.
3. Redeker, V., Rusconi, F., Mary, J., Prome, D., and Rossier, J. (1996) Structure of the C-terminal tail of alpha-tubulin: increase of heterogeneity from newborn to adult, *J Neurochem 67*, 2104–2114.
4. Mizuno, N., Toba, S., Edamatsu, M., Watai-Nishii, J., Hirokawa, N., Toyoshima, Y. Y., and Kikkawa, M. (2004) Dynein and kinesin share an overlapping microtubule-binding site, *Embo J 23*, 2459–2467.
5. Skiniotis, G., Cochran, J. C., Muller, J., Mandelkow, E., Gilbert, S. P., and Hoenger, A. (2004) Modulation of kinesin binding by the C-termini of tubulin, *Embo J 23*, 989–999.
6. Bré, M. H., de Nechaud, B., Wolff, A., and Fleury, A. (1994) Glutamylated tubulin probed in ciliates with the monoclonal antibody GT335, *Cell Motil Cytoskeleton 27*, 337–349.
7. Mary, J., Redeker, V., Le Caer, J. P., Rossier, J., and Schmitter, J. M. (1996) Posttranslational modifications in the C-terminal tail of axonemal tubulin from sea urchin sperm, *J Biol Chem 271*, 9928–9933.
8. Wolff, A., de Nechaud, B., Chillet, D., Mazarguil, H., Desbruyeres, E., Audebert, S., Eddé, B., Gros, F., and Denoulet, P. (1992) Distribution of glutamylated alpha and beta-tubulin in mouse tissues using a specific monoclonal antibody, GT335, *Eur J Cell Biol 59*, 425–432.
9. Audebert, S., Desbruyeres, E., Gruszczynski, C., Koulakoff, A., Gros, F., Denoulet, P., and Eddé, B. (1993) Reversible polyglutamylation of alpha- and beta-tubulin and microtubule dynamics in mouse brain neurons, *Mol Biol Cell 4*, 615–626.
10. Audebert, S., Koulakoff, A., Berwald-Netter, Y., Gros, F., Denoulet, P., and Eddé, B. (1994) Developmental regulation of polyglutamylated alpha- and beta-tubulin in mouse brain neurons, *J Cell Sci 107*, 2313–2322.
11. Bobinnec, Y., Moudjou, M., Fouquet, J. P., Desbruyeres, E., Eddé, B., and Bornens, M. (1998) Glutamylation of centriole and cytoplasmic tubulin in proliferating non- neuronal cells, *Cell Motil Cytoskeleton 39*, 223–232.
12. Vallee, R. B. (1986) Reversible assembly purification of microtubules without assembly-promoting agents and further purification of tubulin, microtubule-associated proteins, and MAP fragments, *Methods Enzymol 134*, 89–104.
13. Regnard, C., Desbruyeres, E., Denoulet, P., and Eddé, B. (1999) Tubulin polyglutamylase: isozymic variants and regulation during the cell cycle in HeLa cells, *J Cell Sci 112*, 4281–4289.
14. Davis, A., Sage, C. R., Wilson, L., and Farrell, K. W. (1993) Purification and biochemical characterization of tubulin from the budding yeast Saccharomyces cerevisiae, *Biochemistry 32*, 8823–8835.
15. van Dijk, J., Rogowski, K., Miro, J., Lacroix, B., Eddé, B., and Janke, C. (2007) A targeted multienzyme mechanism for selective microtubule polyglutamylation, *Mol Cell 26*, 437–448.
16. Janke, C., Rogowski, K., Wloga, D., Regnard, C., Kajava, A. V., Strub, J.-M., Temurak, N., van Dijk, J., Boucher, D., van Dorsselaer, A., Suryavanshi, S., Gaertig, J., and Eddé, B. (2005) Tubulin polyglutamylase enzymes are members of the TTL domain protein family, *Science 308*, 1758–1762.
17. Puck, T. T., and Marcus, P. I. (1955) A Rapid Method for Viable Cell Titration and Clone Production with Hela Cells in Tissue Culture: the Use of X-Irradiated Cells to Supply Conditioning Factors, *Proc Natl Acad Sci USA 41*, 432–437.
18. Hyman, A., Drechsel, D., Kellogg, D., Salser, S., Sawin, K., Steffen, P., Wordeman, L., and Mitchison, T. (1991) Preparation of modified tubulins, *Methods Enzymol 196*, 478–485.

Chapter 5

Preparation of Pure Tyrosinated or Detyrosinated Tubulin Isoforms

Laurence Lafanechère and Didier Job

Abstract

Alpha tubulin comprises a C-terminal tyrosine residue, which is subject to cyclic removal from the peptide chain by a still uncharacterized carboxypeptidase and re-addition to the chain by a tubulin tyrosine ligase. We have shown in different animal or human models that the presence or absence of the tyrosine residue had dramatic consequences for both tumor progression and neuronal organization. In cells, tubulin detyrosination impairs the proper localization of CAP-Gly proteins at microtubule + end, compromises the activity of microtubule-depolymerizing motors of the Kinesin 13 family, and favors both spastin microtubule-severing activity and kinesin 1 processivity. The biochemical basis for these cellular effects of tubulin detyrosination can now be investigated in reconstituted systems in vitro using homogeneous solutions of polymerizable tyrosinated or detyrosinated tubulin.

Key words: Tubulin, Posttranslational modifications, Tyrosination cycle, Detyrosination, Nontyrosinatable tubulin, Δ2- tubulin, Tubulin tyrosine ligase

1. Introduction

Tubulin is subject to special posttranslational modifications which principally affect the C-terminal of its alpha subunit (1–3). In the so-called tubulin tyrosination cycle, the C-terminal tyrosine residue of alpha tubulin is cyclically cleaved from and re-added to the peptide chain. The physiological significance of the tubulin tyrosination cycle has been a long-standing enigma. We have successively shown that tubulin detyrosination favored tumor progression in both animal models and human tumors (4, 5), and vitally compromised neuronal organization in whole animals (6). In cells, tubulin detyrosination alters the positioning of mitotic spindle as well as cell morphology and polarity (7, 8). Molecular interactions

Anne Straube (ed.), *Microtubule Dynamics: Methods and Protocols*, Methods in Molecular Biology, vol. 777,
DOI 10.1007/978-1-61779-252-6_5, © Springer Science+Business Media, LLC 2011

modified by tubulin detyrosination in cells have been identified in various laboratories, and concern microtubule + end-binding proteins, such as CAP-Gly proteins (7, 9, 10), depolymerizing motors of the kinesin 13 family (11), conventional kinesin (12, 13), and microtubule-severing protein spastin (14).

Following detyrosination, alpha tubulin can be further truncated for the penultimate glutamate residue, leading to a tubulin form which is resistant to tyrosination, also known as Δ2-tubulin (15–17). The physiological function of Δ2-tubulin, which represents about 40% of brain tubulin (18), is still unknown.

The investigation of the biochemical basis of such modifications benefits from the availability of affinity-purified tubulin preparations comprising polymerizable and homogeneous solutions of tyrosinated or detyrosinated tubulin (18, 19). The preparation of homogeneous solutions of tubulin tyrosination variants requires tubulin tyrosine ligase (TTL) and immunoaffinity matrix as described below.

The following is an update of the original methods. As some of the materials are not easily accessible, alternatives are proposed in Subheading 4.

2. Materials

2.1. Preparation of Immunoaffinity Matrix

1. 2N NaOH.
2. Sepharose 4B.
3. CNBr.
4. $NaHCO_3$/NaCl buffer: 0.1 M $NaHCO_3$, 0.5 M NaCl, pH 8.3.
5. Disposable desalting columns (such as PD-10 desalting Columns sold by GE Heathcare), prepared according to the manufacturer's instructions.
6. 0.5 M Ethanolamine, pH 8.
7. LA/C4 antibody (20, 21).
8. MME: 100 mM MES, 1 mM $MgCl_2$, 1 mM EGTA, pH 6.75.
9. PME: 100 mM PIPES, 1 mM $MgCl_2$, 1 mM EGTA, pH 6.65 (see Note 1).
10. Phosphate-Buffered saline (PBS).
11. Storage buffer: 0.02% NaN_3 in MME PME, or PBS.

2.2. Buffers for TTL Purification

1. 10× Basic Buffer (BB, 500 mL): 250 mM MES, 1 M KCl, 20 mM $MgCl_2$, 10 mM EGTA, pH 6.8 with KOH.
2. Stock solution of 500 mM ATP in MME.

3. Stock solution of 1 M DTT in MME.
4. Stock solution of 20 mM PMSF in ethanol.
5. Buffer with ATP (ATP-B, 500 mL): 1× BB, 1 mM ATP, 0.5 mM DTT, 10% (v/v) Glycerol.
6. Buffer without ATP (B, 500 mL): 1× BB, 0.5 mM DTT, 10% (v/v) Glycerol.
7. Extraction Buffer (EB, 50 mL): 0.02% Triton X100 in ATP-B. Just before use, add 0.02 mM PMSF.
8. Washing buffer (WB, 500 mL): 1× BB, 1 mM ATP, 0.5 mM DTT, 1.5 M NaCl.
9. Elution buffer (EB, 200 mL): 1× BB, 3 M $MgCl_2$, 0.5 mM DTT, 20% (v/v) Glycerol (see Note 2).
10. Dialysis buffer (DB, 3 L): 1× BB, 1 mM ATP, 10% (v/v) Glycerol.

2.3. TTL Kinetics, Assay of TTL Activity

1. 5 mg/mL of phosphocellulose-purified tubulin (PC-tubulin, (22)) in MME or PME.
2. Purified TTL (concentration to be determined, see below).
3. A stock solution of 20 mM tyrosine in DMSO. This solution is light sensitive. It is stable at room temperature for 6 months.
4. ^{3}H-Tyrosine (specific activity circa 50 Ci/mmole). We usually use ^{3}H-Tyrosine from Perkin-Elmer (L-[Ring-3,5-3H]-Tyrosine, >97%, 5mCi (185MBq), NET127005MC).
5. Stock solution of 100 mM DTT in MME.
6. Stock solution of 500 mM ATP in MME. This stock solution can be stored frozen at −20°C.
7. Stock solution of 2.5 M $MgCl_2$ in MME. This stock solution can be stored at −20°C.
8. Stock solution of 2.5 M KCl in MME. This stock solution can be stored at −20°C.
9. 12% TCA solved in water.
10. Pure Ethanol.
11. Whatman GF/C disk filters, 25 mm diameter.

2.4. Separation of Tyrosinated, Detyrosinated, and Δ2-Tubulin

1. Immunoaffinity column, with the rat monoclonal antibody YL½ (see Note 3) coupled to cyanogen bromide (CNBr)-activated sepharose 4B, prepared as described in this chapter. The procedure is described for a 20-mL decanted column. The column contains 5 mg of YL½ antibody per milliliter of activated gel.
2. Disposable desalting columns (such as PD-10 desalting Columns supplied by GE Healthcare, reference 17-0851-01),

prepared according to the manufacturer's instructions, can be used. As an example, we usually prepare at least 20 spin columns for 2–2.5 mL samples, two spin columns for 1 mL samples, and four spin columns for 200 μL samples.

3. Gather enough rapid concentration devices for 15 mL or smaller volumes. We use Centriprep YM-30, from Millipore, reference 4307.
4. ≈ 60 mg of PC-tubulin at a concentration of 10 mg/mL.
5. ≈ 300 μL of pure active TTL (≈ 4 mg/mL).
6. Carboxypeptidase A (CPA) (see Note 4).

2.4.1. Buffers for PC-Tubulin Recycling

1. PME buffer (see Subheading 2.1).
2. Stock solution of 100 mM GTP (in MME).
3. Stock solution of 500 mM $MgCl_2$ (in MME or PME).
4. >99.5% pure anhydrous glycerol.

2.4.2. Buffers for Separation of Tubulin Isoforms

1. PME buffer (see Subheading 2.1): 2.5 L. Most of the PME buffer is used for the preequilibration of spin columns.
2. PME buffer containing 1 M KCl: 250 mL.
3. PME buffer containing 0.1 M KCl: 250 mL. This buffer is obtained by mixing 25 mL of PME buffer containing 1 M KCl with 225 mL of PME buffer.
4. PME buffer containing 0.8 M KCl: 250 mL. This buffer is obtained by mixing 200 mL of PME buffer containing 1 M KCl with 50 mL of PME buffer.
5. D_2O-based PME buffer: 50 mL. The D_2O–PME buffer has the same constitution as PME buffer, except that D_2O is used instead of H_2O. Pure tubulin, which is extremely labile, is very stable in such a buffer.
6. PME buffer containing 6 M Urea: 100 mL.
7. PME buffer with 0.02% NaN_3.
8. 1 M DTT in PME buffer.
9. Reagents and buffers used for TTL tyrosination assay (see Subheading 2.3).

3. Methods

3.1. Preparation of Immunoaffinity Matrix

TTL is purified by immunoaffinity using a monoclonal antibody (LA/C4) able to recognize the native form of TTL. Tubulin isoforms are separated using an immunoaffinity column, with an anti-Tyr tubulin antibody (YL½).

This section describes the procedure used to couple antibodies to sepharose B beads to generate immunoaffinity columns.

Antibodies are coupled to CNBr-activated sepharose 4B according to Cuatrecasas (23).

Caution: This methodology requires the use of CNBr, which is extremely dangerous. Read attentively the full procedure before use. As an alternative, see Note 5.

3.1.1. Preparation of Antibodies

The pH at which the coupling stage is performed determines the degree of coupling since with this procedure it is the unprotonated form of the amino group which is reactive. We obtain a good yield of coupling when IgGs are at pH 8.3 in the $NaHCO_3$/NaCl buffer.

Thus, we recommend to transfer purified IgGs to that buffer before sepharose 4B activation using disposable spin columns. IgGs' concentration should be adjusted to 5 mg/mL. Take a 1-μL aliquot for SDS–PAGE analysis. Keep IgGs at 4°C on ice.

3.1.2. Sepharose 4B Activation

This procedure is for the activation of 10 mL of decanted sepharose (see Note 6).

1. Rinse the sepharose at least three times with 10 mL of distilled water (see Note 7).
2. The following steps are done in a well-ventilated hood, wearing gloves and glasses.
3. Prepare a clean place under a hood with bench paper.
4. Put a pH meter and a magnetic stirrer under the hood.
5. Add 50 mL of 2N NaOH into a 500-mL vacuum flask with a Buchner funnel (coarse disk). The size of the Buchner should be small (30 mL) regarding to the size of the flask. The Buchner and flask, connected to a vacuum device, should be placed under the hood.
6. Prepare a 500-mL beaker containing 200 mL 2N NaOH to rinse instruments that will be in contact with CNBr.
7. Prepare 15 mL of 4°C cold water and keep on ice.
8. Prepare 300 mL of 4°C cold $NaHCO_3$/NaCl buffer and keep on ice.
9. Prepare a timer.
10. Weigh 1 g of CNBr (see Note 8).
11. Suspend the 10 mL sepharose 4B in an equal volume of dH_2O in a beaker with a magnetic stirrer.
12. Plunge the pH electrode into the suspension and start the timer.
13. Slowly add 1 g of CNBr while keeping the pH between 9 and 11 by continuously adding drops of 2N NaOH (see Notes 9, 10,

and 11). The reaction is complete in 10–15 min, as indicated by the almost stable pH.

14. Quickly transfer the suspension to the Buchner funnel and wash under suction first with 15 mL cold water and then with 300 mL cold $NaHCO_3$/NaCL buffer (see Note 12).

3.1.3. Coupling Procedure

1. Take the IgGs to be coupled, already in cold $NaHCO_3$/NaCl buffer. IgGs should be in a volume equal to the volume of the packed, "activated" sepharose (see Note 13).
2. Add the IgGs to the moist, washed sepharose.
3. Immediately mix the suspension (see Note 14). This will lead to the coupling of IgGs to the sepharose beads at a concentration of 5 mg of IgGs per mL of sepharose.
4. Gently stir the suspension either 2 h at room temperature or overnight at 4°C. We usually use a rotating stirrer for tubes.
5. Allow the sepharose to sediment (see Note 7). Keep the supernatant to check its protein content and for SDS–PAGE analysis to have an idea of the coupling yield. Usually, the reaction is very efficient and the coupling is apparently total.
6. Wash with at least 3 volumes of $NaHCO_3$/NaCl buffer (see Note 7).
7. Rinse with at least 3 volumes of 0.5 M Ethanolamine.
8. Saturate by gently stirring the suspension in ethanolamine for 2 h at room temperature.
9. Wash with at least 3 volumes of $NaHCO_3$/NaCl buffer (see Note 7).
10. Equilibrate in storage buffer.
11. Store at 4°C.

3.2. Purification of Tubulin Tyrosine Ligase

Initially, TTL was isolated from porcine brain by immunoaffinity chromatography (20), using a mouse monoclonal antibody specific for the ligase. This procedure was adapted to purify TTL from large-scale culture of insect cells infected with baculovirus and expressing high level of TTL. We describe this adapted procedure (see Note 15).

3.2.1. Purification

Large amount of cells expressing TTL were obtained by growing them in suitable reactors. These cells can then be kept for several years as frozen pellets (–80°C). All the steps of this purification procedure should be done at 4°C.

1. Transfer the sepharose 4B beads coupled with the LA/C4 antibody in a suitable column. The procedure described here is for 15 mL of sedimented resin. Equilibrate the column with a least 3 volumes (i.e., 45 mL) of ATP-B.

2. Weigh a frozen pellet of insect cells expressing TTL and let it thaw at 4°C in EB containing 0.02 mM PMSF. We usually use 50 mL of EB for 10 g of insect cell pellet.
3. Centrifuge this solution for 45 min at 100,000 × *g*, 4°C.
4. Take the supernatant, volume of which is around 40 mL, and add PMSF (0.02 mM, final concentration).
5. Load the supernatant onto the column. This loading step is fractionated (volume of fractions: 5 mL). Wait for 5 min between each fraction in order to let TTL bind to the antibody (see Note 16).
6. Rinse the affinity column with ATP-B containing freshly added 0.02 mM PMSF. We usually rinse the column with 20 volumes of ATP-B.
7. Wash the column with about 30 volumes of WB.
8. Reequilibrate the column with 5–10 volumes of B.
9. Check the $OD_{280\ nm}$, which needs to be close to zero.
10. A stepwise elution of TTL is then performed by fractions of half a column volume of EB. The majority of TTL is eluted after 8–10 fractions. The beginning of the peak of elution is expected in the third fraction.
11. Check the $OD_{280\ nm}$ of the fractions and pool the five most concentrated ones.
12. Dialyze overnight against 1 L of DB.
13. The affinity column can be rinsed in freshly prepared 6 M urea MEM buffer (at least 3 volumes), reequilibrated in MEM or PBS, and stored for further use in the same buffer containing 0.02% NaN_3 (see Note 17).
14. The next day, change twice the dialysis buffer.

3.2.2. Concentration and Storage

1. Concentrate the dialyzed TTL to reach a (circa) 4 mg/mL concentration. We usually use spin concentrators. This concentration step should be as short as possible and should not last more than 2 h.
2. Distribute TTL in small aliquots.
3. Freeze the aliquots in liquid nitrogen.
4. Store frozen at −80°C.

3.2.3. Check of Purity

The yield of different purification steps and the degree of purity of isolated TTL can be easily checked by SDS–PAGE analysis on a 10% SDS–polyacrylamide gel. A single band at 40 kDa should be observed.

3.3. Assay of TTL Activity

To obtain large amount of affinity-purified tyrosinated tubulin, it is required to load maximally tyrosinated tubulin onto the immunoaffinity column. Thus, TTL concentration and activity have to be checked in order to determine optimal conditions for tubulin tyrosination.

Tyrosine incorporation into PC-tubulin can be monitored with [^{3}H] tyrosine (see Notes 18 and 19).

3.3.1. Pre-prep

The tyrosination reaction is performed in a 30°C bath. Because of their instability, however, all the reagents have to be thawed, prepared, and stocked on ice, with the exception of the stock tyrosine solution which is kept in DMSO at room temperature. Prepare the disk filters; separate them and dispose the needed quantity onto a Whatman paper sheet. We usually prepare 12 disk filters to assay TTL activity in duplicates at time points 0, 2, 5, 10, 20, and 30 min.

1. Prepare a concentrated reagent mixture (CRM) by mixing an equal amount of 100 mM DTT, 500 mM ATP, 2.5 M $MgCl_2$, and 20 mM tyrosine. A typical experiment, with six time points, requires circa 30 μL of CRM.
2. To a V Volume of CRM, add the same V volume of KCl 2.5 M and the same V volume of [^{3}H] tyrosine (specific activity circa 50 Ci/mmole). The final mixture is called CRM*.
3. Prewarm a bath incubator at 30°C.
4. Thaw PC-tubulin (see Note 20). Transfer PC-tubulin in MME buffer and eliminate GTP by gel filtration on a spin column. Adjust the concentration to 5.4 mg/mL by adding MME buffer.
5. Thaw an aliquot of TTL.

3.3.2. Kinetics of Tyrosination Reaction

1. On ice, in a tube suitable for bath incubation, mix PC-tubulin, CRM*, and TTL. For one time point, (50 μL final volume), mix 42 μL of PC-tubulin + 6 μL CRM* + 2 μL TTL. The final concentrations are thus: 4.5 mg/mL PC-tubulin, 200 μM tyrosine, 0.04 μCi/μL of [^{3}H] tyrosine, 1 mM DTT, 5 mM ATP, 25 mM $MgCl_2$, 100 mM KCl (see Note 21), and 2 μL of TTL, which should correspond to 0.01 U (see Note 22). A typical kinetic, with six time points including zero, thus requires 6 × 50 μL of the above described mixture.
2. Immediately after adding TTL (see Note 23), collect 2 × 20 μL aliquots of the solution, place them on two disk filters, and transfer the filters into a 5–10 mL 12% TCA bath (see Note 24) for each filter. This will give the T_0 point of the kinetics.
3. Put the tube containing the reaction mixture into the bath incubator and proceed as described above for each kinetic point. The plateau should be reached, at least, after 15–20 min of incubation.
4. Slowly agitate the vials containing the filters in 12% TCA for 5 min.

5. Replace four times the TCA baths with fresh ones, with a 5-min agitation between each as described above.
6. Transfer the filters in 5 mL of ethanol.
7. Remove ethanol and let dry the filters.
8. Transfer the filters into scintillation vials.
9. Add scintillating liquid appropriate for tritium and count on a suitable instrument.

3.4. PC-Tubulin Recycling

1. PC-tubulin is usually stored at a concentration higher than 5 mg/mL in PME buffer containing 1 mM GTP and 5–10 mM $MgCl_2$. We usually recycle 250–300 mg of PC-tubulin, but the procedure could be adapted to smaller quantities. Thaw PC-tubulin and add glycerol (25% final concentration) and enough GTP to be 1 mM (final concentration). Mix by gentle vortexing.
2. Incubate at 37°C in a water bath for 30 min.
3. Layer on a warm cushion of 60% glycerol in PME. For a V volume of assembled PC-tubulin, use the same V Volume of glycerol cushion.
4. Spin for 45 min to 1 h at 100,000 × *g* at 37°C.
5. After centrifugation, a whitish pellet of polymerized tubulin can clearly be seen at the bottom of the tube. Aspirate the supernatant and rinse the supernatant cushion interface with warm PME.
6. Aspirate completely and rinse the tubulin pellet twice with warm PME (half a tube volume) to remove any residual glycerol. Do this while the tube containing the pellet is held in a 37°C water bath.
7. Resuspend pellet in a small volume of ice-cold PME. Begin with a volume corresponding to 2 pellet volumes. It is possible to add a little bit more PME later if the pellet is too hard to depolymerize.
8. Incubate the tube containing the chunky pellet on ice for half an hour. Gently resuspend the solution (avoid the formation of foam) using a pipette to break up the chunks. During this cold depolymerization, keep resuspending every 2–3 min. An ice-cold dounce can also be used.
9. Spin at 100,000 × *g* at 4°C for 15 min to clarify the depolymerization mix.
10. Collect supernatant on ice and measure $OD_{280\ nm}$ of a 1/100 dilution in PME. Calculate the concentration of tubulin using an extinction coefficient at 280 nm of 115,000 M^{-1}/cm.
11. Adjust tubulin concentration to 10 mg/mL with PME buffer.

3.5. Purification of Tyrosinated, Detyrosinated, and Δ2-Tubulin

As stated above, this purification uses affinity chromatographic procedure to isolate tyrosinated tubulin from nontyrosinatable (Δ2-tubulin). Maximally tyrosinated tubulin is loaded onto a column of immobilized YL½. Δ2-tubulin is not retained, whereas tyrosinated tubulin is eluted in a salt-containing buffer (Fig. 1). The relative proportion of each species is around 50%. Tyrosinated tubulin is rapidly desalted using spin columns. After concentration, the preparation of tyrosinated tubulin is split into two parts. One part is treated with CPA to obtaine detyrosinated tubulin. Tyrosinated and detyrosinated tubulins are then transferred into D_2O-based PME buffer, aliquoted, frozen in liquid nitrogen, and stored at −80°C.

The key factors to obtain good yield of assembly-competent pure tubulin isoforms are:

- PC-tubulin of good quality, assembly competent. If any doubt, do not hesitate to recycle it (see above protocol).
- Tubulin is a labile molecule that converts to a nonpolymerizing state within hours at 4°C. The faster the overall procedure is done, the better the yield will be. If the manipulator is not trained or if Δ2-tubulin is not needed, we recommend to discard the tubulin not retained onto the column and to focus on the purification of tyrosinated tubulin.

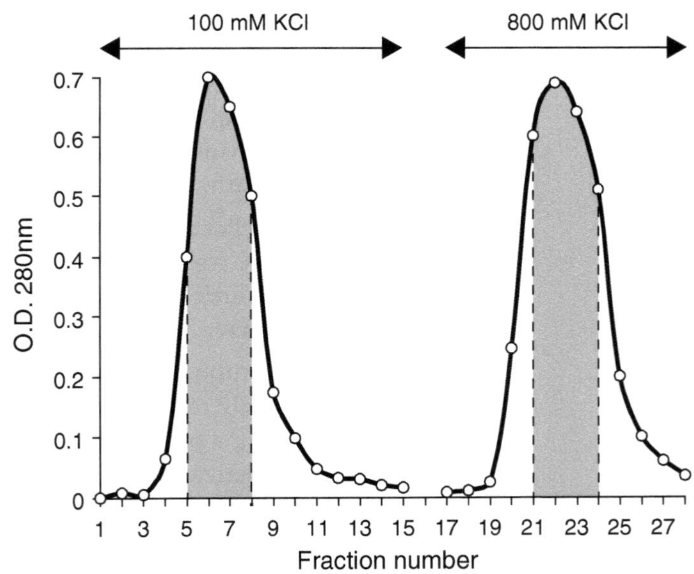

Fig. 1. Affinity chromatography of tyrosinated and nontyrosinated tubulin species. Pure tubulin was maximally tyrosinated, as described, exchanged in PME buffer and loaded onto a YL½ immunoafffinity column. The column was further eluted with at least 5 volumes of PME buffer containing 0.1 M KCl, as indicated. A stepwise elution of tyrosinated tubulin was then performed with PME buffer containing 0.8 M KCl. Fractions were eluted directly onto spin columns equilibrated in PME buffer for immediate desalting. The protein elution is followed by OD $_{280nm}$ measurement. The gray areas indicate the fractions that are usually pooled for further concentration.

– Highly saline buffer, such as the buffer used to elute tyrosinated tubulin, rapidly denatures tubulin. We, thus, recommend to directly collect the eluted fractions onto the spin columns.

3.5.1. Maximal Tyrosination

The procedure is described for a purification using a 20-mL affinity column.

1. Mix 5 mL of PC-tubulin (10 mg/mL) with 228 μL of CRM (see Subheading 3.3.1, step 1, Note 25), 228 μL of purified TTL (this volume has been determined after checking TTL activity), and 228 μL of 2.5 M KCl.
2. Incubate for 30 min in water bath at 30°C.
3. After this incubation step, tubulin tyrosination should be maximal. During this step, however, tubulin has been warmed with 100 mM KCl, which often leads to a partial denaturation of tubulin. If it is the case, the sample has a whitish aspect.
4. Spin at 100,000 × *g* at 4°C for 15 min to clarify the sample.
5. Eliminate the tyrosination reagents (especially DTT) by transferring the sample into PME buffer using disposable spin columns.
6. Measure $OD_{280\ nm}$ of a 1/100 dilution in PME in order to check the tubulin concentration.

3.5.2. Purification of Δ2-Tubulin

1. Gather and equilibrate in PME buffer enough spin columns for rapid buffer exchange or for rapid desalting of eluted samples.
2. All the procedure has to be done in a cold room (4°C).
3. Load the sample at a concentration of 2 mg tubulin per mL of gel onto the 20-mL column. This loading step is fractionated (volume of fractions: ≈ 2–3 mL). Wait for 5 min between each fraction in order to let tyrosinated tubulin bind to the antibody.
4. Elute stepwise nontyrosinated tubulin – which should be composed mostly of Δ2-tubulin if the previous tyrosination step was maximal (see Note 26) – by adding fractions of 5 mL of 0.1 M KCl–PME buffer (Fig. 1). Follow the elution by reading the $OD_{280\ nm}$ of each fraction. The protein elution peak should begin circa in the fourth fraction, corresponding to the second-column volume.
5. Pool the fractions which are the most concentrated (see Note 27) and transfer them in PME buffer using spin columns. Read the $OD_{280\ nm}$ of pooled fractions (see Note 28).
6. Concentrate the pool to circa 5 mg/mL using disposable devices, such as Centriprep 30 (see Note 29).
7. Transfer into D_2O-based PME buffer using spin columns.
8. Add 1 mM GTP and 10 mM $MgCl_2$ (final concentrations).
9. Aliquot, freeze in liquid nitrogen, and store at –80°C.

3.5.3. Affinity Purification of Tyrosinated Tubulin

1. All the procedure continues in a cold room (4°C).
2. After nontyrosinated tubulin elution (see above), wash extensively the column (5 volumes) with 0.1 M KCl–PME buffer, until the $OD_{280\ nm}$ is close to zero.
3. Add 2.5 mL fractions of 0.8 M KCl–PME buffer (Fig. 1) and immediately read the $OD_{280\ nm}$ of eluted fractions.
4. As soon as the $OD_{280\ nm}$ rises (usually at the beginning of the second-column volume), collect the eluted fractions directly onto suitable desalting spin column. We usually collect 16 × 2.5 mL fractions to cover the majority of elution peak. The columns are spun after the collection of all 16 fractions.
5. Read the $OD_{280\ nm}$ of spin columns' flow through. Pool the fractions which are the most concentrated (see Note 27) and read the $OD_{280\ nm}$ of pooled fractions.
6. Concentrate the pool to circa 5 mg/mL using disposable devices, such as Centriprep 30.
7. Transfer into D_2O-based PME buffer (see Note 30) using spin columns.
8. Add 1 mM GTP and 10 mM $MgCl_2$ (final concentrations).
9. Aliquot, freeze in liquid nitrogen, and store at −80°C.

3.5.4. Preparation of Detyrosinated Tubulin

Detyrosinated tubulin can be obtained from tyrosinated tubulin by incubation with CPA. CPA cleaves easily aromatic amino acids, but very slowly acidic amino acids (see Note 31). This detyrosination step is usually done after the concentration of tyrosinated tubulin or after its transfer into D_2O-based PME buffer.

1. Dilute fresh CPA to 200 μg/mL (see Note 32).
2. Add 1% of diluted CPA (2 μg/mL final concentration) to tyrosinated tubulin.
3. Incubate for 10 min at 30°C in water bath.
4. Stop the reaction by adding 2% of 1 M DTT (20 mM final concentration, see Note 33).
5. Transfer into D_2O-based PME buffer (see Note 30) using spin columns.
6. Add 1 mM GTP and 10 mM $MgCl_2$ (final concentrations).
7. Aliquot, freeze in liquid nitrogen, and store at −80°C.

3.5.5. Check of Isoforms' Purity

Purity of tubulin isoforms can easily be analyzed in western blots with specific antibodies commercially available (see Note 34).

4. Notes

1. Prior to pH adjustment, PIPES forms a precipitate which is progressively solubilized at rising pHs.
2. The components of this buffer take a long time to dissolve. It is, thus, advised to prepare this buffer at least the day before its use. Moreover, the final pH of the buffer is a little bit more acidic than that of the BB. We did not find crucial to rise up the pH to 6.8.
3. Any other monoclonal antibody specific of tyrosinated tubulin can be used.
4. It is important to use PMSF-treated CPA as untreated CPA can have additional endoprotease activity. We use CPA from Sigma (C9268).
5. The use of freshly activated gels was found to be a prerequisite to obtain reproducible results. To avoid hazardous manipulation of CNBr, it is possible to use commercially available activated resins, but results are somewhat erratic.
6. Adjust proportionally the quantities of reagents for activation of greater amounts of sepharose 4B. The molarity of the NaOH solution depends on the amounts of sepharose 4B and CNBr added; it should vary between 2 M (for 5–10 mL of sepharose 4B and 1–3 g of CNBr) and 8 M (for 100–200 mL of sepharose 4B and 20–30 g of CNBr).
7. This can be done either using a Buchner on vacuum flask or a series of low-velocity centrifugation to decant the resin/aspiration of the supernatant/addition of water and suspension steps. In all cases, care has to be taken not to let the resin dry.
8. If you do not have a balance under a ventilated hood, it is possible to weigh CNBr by using a preweighted closed vial. In this case, take the CNBr out of its original container under the hood, put it into the vial, close the vial, and weigh. Repeat this procedure until the desirable weight is obtained.
9. The pH of the suspension must not decrease below 9; otherwise, toxic cyanidric acid would be liberated.
10. This procedure has to be done at room temperature. We have observed bad yields of coupling when the activation was done in cold water.
11. Coupling yield is also bad when the pH is kept higher than 11.
12. The volume of wash should be at least 10–15 times that of the packed sepharose.

13. At this step, there is no longer any hazard regarding to CNBr and the resin can be manipulated outside of the hood. Packed sepharose is obtained by a rapid transfer of activated sepharose from the Buchner to a centrifuge tube, a low-velocity centrifugation, and supernatant discard.
14. The entire procedure of washing, adding the IgG solution, and mixing should be as rapid as possible (less than 5 min). It is important that these procedures be performed rapidly and the temperature be lowered, since the "activated" sepharose is unstable.
15. Pure active TTL can also be obtained after expression in bacteria as a GST fusion protein and purification on a sepharose 4B Glutathione column. Other procedures of bacterial expression would most probably be adequate.
16. It is unproductive to extend the duration of this step. For example, we have observed that an overnight incubation, with the aim to raise the level of bound TTL, led to inactive purified TTL.
17. We observed that the capacity of the column decreases with its use. We recommend preparing a fresh column after 3–4 uses.
18. Instead of using GF/C filters, it is also possible to assay the incorporation of ^{3}H-Tyrosine into tubulin following Sandoval and Cuatrecasas' procedure (24).
19. An alternative to the use of radioactivity is to follow tyrosine incorporation by quantitative western blot with specific antibodies.
20. If it is not wished to later purify Δ2-tubulin and if very extensive tubulin purity is not a major concern, microtubule protein (MTP), i.e., microtubule-associated protein (MAP) (the so-called 1×, 2×, or 3×), can be used instead of PC-tubulin.
21. Although such a high concentration of KCl can partially denature tubulin during the subsequent 30°C incubation, it avoids tubulin assembly. Tubulin assembly is proscribed for tyrosination reaction, as TTL is able to tyrosinate only tubulin dimers (3, 25) and does not work on microtubules.
22. 1 U of enzyme is defined as the amount that incorporates 1 nmol of tyrosine into tubulin in 1 min under the conditions used for reaction (21). In practice, we assay different concentrations (volumes) of purified TTL in order to find the concentration which allows reaching the plateau of maximal tyrosination after 15 min of 30°C incubation. Of course, to be able to vary the volumes of TTL, it is necessary to consequently adapt the concentration of stock solutions of other reagents in order to keep the same final reaction volume.

23. This step has to be done rapidly, as we have noticed that a good TTL preparation yields to a TTL that is able to tyrosinate tubulin even at 4°C.
24. 6-well microplates, disposable vials, or vials for scintillation counting can be used.
25. This CRM is made with a tyrosine stock solution of 50 mM. The final tyrosine concentration is, thus, 500 μM.
26. TTL is not retained on the affinity column and is eluted together with nontyrosinated tubulin. It is also the case for MAPs when MTP is used instead of PC-tubulin (see Note 20).
27. It is recommended not to try to increase the yield by pooling less-concentrated fractions, as it will lengthen the concentration time and may thus lead to denatured tubulin. Gray areas in Fig. 1 indicate the fractions that we usually pool.
28. When pure Δ2-tubulin is not needed, it is advised to go directly on the next step, i.e., "purification of tyrosinated tubulin." If one wants to save Δ2-tubulin and also purify tyrosinated tubulin, it is advised to continue the procedure while concentrating on Δ2-tubulin.
29. To obtain completely pure Δ2-tubulin (regarding the different isoforms), Δ2-tubulin from the first peak can be subjected to a second tyrosination cycle and rechromatographed on the affinity column (18).
30. We have found that D_2O is a powerful stabilizing agent for affinity-purified tyrosinated tubulin. This form of tubulin loses polymerization capacity upon freezing in an H_2O-based PME buffer while it remains fully functional in a D_2O-based PME buffer. D_2O can be provided by Eurisotop (reference D216L).
31. We have tried to generate Δ2-tubulin from detyrosinated tubulin incubated for a long time (even overnight) at 30°C with CPA, re-adding fresh CPA from time to time, with no success.
32. CPA is not a solution, but a suspension. It is, thus, needed to vortex the stock and the intermediate solutions before aspirating samples of CPA.
33. The action of DTT on CPA is irreversible. Thus, when detyrosinated tubulin is used in different experiments, there is no possible effect of CPA on other proteins.
34. YL½, tyrosinated-tubulin monoclonal antibody, is supplied by Abcam. Detyrosinated- and Δ2-tubulin polyclonal antibodies are supplied by AbCys/Eurobio.

References

1. Barra, H. S., Arce, C. A., and Argarana, C. E. (1988) Posttranslational tyrosination/detyrosination of tubulin. *Mol Neurobiol 2*, 133–53.
2. Westermann, S., and Weber, K. (2003) Post-translational modifications regulate microtubule function. *Nat Rev Mol Cell Biol 4*, 938–47.
3. MacRae, T. H. (1997) Tubulin post-translational modifications--enzymes and their mechanisms of action. *Eur J Biochem 244*, 265–78.
4. Lafanechere, L., Courtay-Cahen, C., Kawakami, T., Jacrot, M., Rudiger, M., Wehland, J., Job, D., and Margolis, R. L. (1998) Suppression of tubulin tyrosine ligase during tumor growth. *J Cell Sci 111 (Pt 2)*, 171–81.
5. Mialhe, A., Lafanechere, L., Treilleux, I., Peloux, N., Dumontet, C., Bremond, A., Panh, M. H., Payan, R., Wehland, J., Margolis, R. L., and Job, D. (2001) Tubulin detyrosination is a frequent occurrence in breast cancers of poor prognosis. *Cancer Res 61*, 5024–7.
6. Erck, C., Peris, L., Andrieux, A., Meissirel, C., Gruber, A. D., Vernet, M., Schweitzer, A., Saoudi, Y., Pointu, H., Bosc, C., Salin, P. A., Job, D., and Wehland, J. (2005) A vital role of tubulin-tyrosine-ligase for neuronal organization. *Proc Natl Acad Sci USA 102*, 7853–8.
7. Peris, L., Thery, M., Faure, J., Saoudi, Y., Lafanechere, L., Chilton, J. K., Gordon-Weeks, P., Galjart, N., Bornens, M., Wordeman, L., Wehland, J., Andrieux, A., and Job, D. (2006) Tubulin tyrosination is a major factor affecting the recruitment of CAP-Gly proteins at microtubule plus ends. *J Cell Biol 174*, 839–49.
8. Badin-Larcon, A. C., Boscheron, C., Soleilhac, J. M., Piel, M., Mann, C., Denarier, E., Fourest-Lieuvin, A., Lafanechere, L., Bornens, M., and Job, D. (2004) Suppression of nuclear oscillations in Saccharomyces cerevisiae expressing Glu tubulin. *Proc Natl Acad Sci USA 101*, 5577–82.
9. Steinmetz, M. O., and Akhmanova, A. (2008) Capturing protein tails by CAP-Gly domains. *Trends Biochem Sci 33*, 535–45.
10. Weisbrich, A., Honnappa, S., Jaussi, R., Okhrimenko, O., Frey, D., Jelesarov, I., Akhmanova, A., and Steinmetz, M. O. (2007) Structure-function relationship of CAP-Gly domains. *Nat Struct Mol Biol 14*, 959–67.
11. Peris, L., Wagenbach, M., Lafanechere, L., Brocard, J., Moore, A. T., Kozielski, F., Job, D., Wordeman, L., and Andrieux, A. (2009) Motor-dependent microtubule disassembly driven by tubulin tyrosination. *J Cell Biol 185*, 1159–66.
12. Konishi, Y., and Setou, M. (2009) Tubulin tyrosination navigates the kinesin-1 motor domain to axons. *Nat Neurosci 12*, 559–67.
13. Cai, D., McEwen, D. P., Martens, J. R., Meyhofer, E., and Verhey, K. J. (2009) Single molecule imaging reveals differences in microtubule track selection between Kinesin motors. *PLoS Biol 7*, e1000216.
14. Roll-Mecak, A., and Vale, R. D. (2008) Structural basis of microtubule severing by the hereditary spastic paraplegia protein spastin. *Nature 451*, 363–7.
15. Paturle-Lafanechere, L., Edde, B., Denoulet, P., Van Dorsselaer, A., Mazarguil, H., Le Caer, J. P., Wehland, J., and Job, D. (1991) Characterization of a major brain tubulin variant which cannot be tyrosinated. *Biochemistry 30*, 10523–8.
16. Paturle-Lafanechere, L., Manier, M., Trigault, N., Pirollet, F., Mazarguil, H., and Job, D. (1994) Accumulation of delta 2-tubulin, a major tubulin variant that cannot be tyrosinated, in neuronal tissues and in stable microtubule assemblies. *J Cell Sci 107 (Pt 6)*, 1529–43.
17. Lafanechere, L., and Job, D. (2000) The third tubulin pool. *Neurochem Res 25*, 11–8.
18. Paturle, L., Wehland, J., Margolis, R. L., and Job, D. (1989) Complete separation of tyrosinated, detyrosinated, and nontyrosinatable brain tubulin subpopulations using affinity chromatography. *Biochemistry 28*, 2698–704.
19. Banerjee, A. (1999) A monoclonal antibody to alpha-tubulin: purification of functionally active alpha-tubulin isoforms. *Biochemistry 38*, 5438–46.
20. Wehland, J., Schroder, H. C., and Weber, K. (1986) Isolation and purification of tubulin tyrosine ligase. *Methods Enzymol 134*, 170–9.
21. Schroder, H. C., Wehland, J., and Weber, K. (1985) Purification of brain tubulin-tyrosine ligase by biochemical and immunological methods. *J Cell Biol 100*, 276–81.
22. Weingarten, M. D., Lockwood, A. H., Hwo, S. Y., and Kirschner, M. W. (1975) A protein factor essential for microtubule assembly. *Proc Natl Acad Sci USA 72*, 1858–62.
23. Cuatrecasas, P. (1970) Protein purification by affinity chromatography. Derivatizations of agarose and polyacrylamide beads. *J Biol Chem 245*, 3059–65.
24. Sandoval, I. V., and Cuatrecasas, P. (1976) Protein kinase associated with tubulin: affinity chromatography and properties. *Biochemistry 15*, 3424–32.
25. Wehland, J., and Weber, K. (1987) Turnover of the carboxy-terminal tyrosine of alpha-tubulin and means of reaching elevated levels of detyrosination in living cells. *J Cell Sci 88 (Pt 2)*, 185–203.

Chapter 6

αβ-Tubulin and Microtubule-Binding Assays

Jaime N. Campbell and Kevin C. Slep

Abstract

Dynamic instability is a hallmark of the microtubule cytoskeleton. In order to regulate microtubule dynamics in vivo, a varied host of microtubule-associated proteins are mobilized to promote microtubule nucleation, growth, stabilization, catastrophe, depolymerization, rescue, and severing. To confer these various functions, cytoskeletal regulators have highly tuned affinities for tubulin, recognizing the unpolymerized αβ-tubulin heterodimer, dynamic microtubule lattice, stabilized microtubule lattice, or a combination therein. The protocols presented here probe αβ-tubulin and microtubule binding using gel filtration and cosedimentation, respectively.

Key words: Tubulin, Tubulin-binding protein, Microtubule-associated protein, Gel filtration, Microtubule cosedimentation assay, Dynamic microtubules, Taxol

1. Introduction

The microtubule is a polarized structure characterized by a minus end, the site of nucleation, and a plus end, where the majority of dynamic growth and depolymerization occurs (1). GTP-bound αβ-tubulin is incorporated into the growing microtubule lattice while GDP-bound αβ-tubulin is released as microtubules depolymerize (2). Regulators of microtubule dynamics recognize features of free αβ-tubulin and/or the microtubule lattice (3–6). To assay binding between a protein of interest and the various forms of αβ-tubulin (free versus lattice-bound), two methods are discussed: (1) gel filtration to detect αβ-tubulin binding and (2) cosedimentation to detect binding between a protein of interest and static or dynamic microtubules.

Gel filtration assays use depolymerized, heterodimeric αβ-tubulin and are conducted at 4°C to prevent polymerization. In this assay, the elution profiles of tubulin alone, tubulin-binding

Anne Straube (ed.), *Microtubule Dynamics: Methods and Protocols*, Methods in Molecular Biology, vol. 777,
DOI 10.1007/978-1-61779-252-6_6, © Springer Science+Business Media, LLC 2011

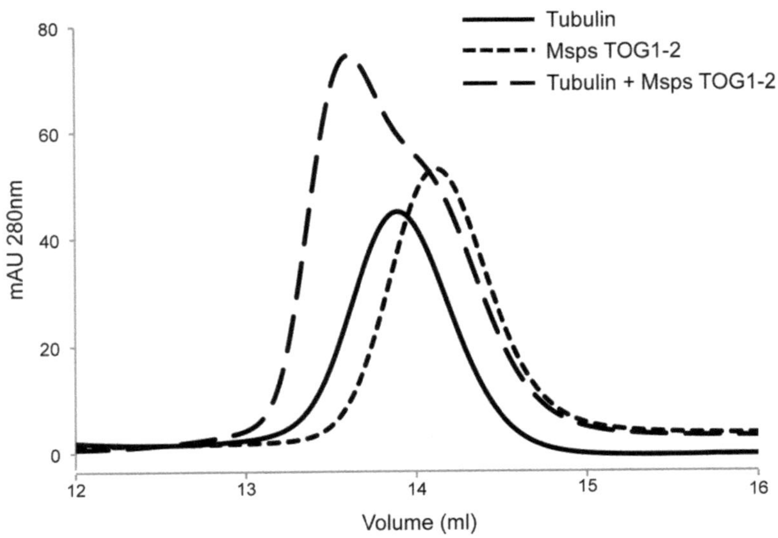

Fig. 1. Gel filtration-binding assay showing the interaction between the first two TOG domains from *Drosophila* Minispindles (Msps) and αβ-tubulin. αβ-tubulin, with a molecular weight of 110 kDa, elutes at 13.9 ml when run alone (*solid line*, injected amount = 50 μl of 20 μM tubulin). Msps TOG1-2, with a molecular weight of 57 kDa, elutes at 14.2 ml when run alone (*short dashed line*, injected amount = 50 μl of 40 μM Msps TOG1-2). When 20 μM tubulin is incubated with 40 μM Msps and run over gel filtration, a peak elutes at 13.6 ml, earlier than either individual component elutes, indicative of complex formation (*long dashed line*, injected amount = 50 μl) (11). Excess Msps creates a right shoulder to the complex peak. A Superdex 200 10/300 GL column was used.

protein alone, and a mixture of tubulin and tubulin-binding protein are compared (Fig. 1). The size-exclusion chromatographic step separates proteins and protein complexes based on their relative size and shape. Larger proteins and protein complexes elute earlier than smaller proteins or protein complexes. A shift in the elution profile of the components to a higher molecular weight as compared to the summated elution profiles of the individual components is indicative of complex formation. The relative stoichiometric shift can be monitored through fractionation and SDS-polyacrylamide gel electrophoresis (SDS-PAGE) analysis.

Microtubule cosedimentation assays are used to determine if a protein has microtubule-binding properties. A centrifugation step pellets taxol-stabilized microtubules or dynamic microtubules through a 40% glycerol cushion. Proteins that bind the microtubule are carried through the glycerol cushion and cosediment with the microtubules. Proteins that fail to bind the microtubule remain layered above the glycerol cushion. SDS-PAGE analysis of the supernatant and pellet fractions is used to assay microtubule-binding activity (Fig. 2). When examining dynamic microtubules, the relative portion of polymerized tubulin in the pellet versus nonpolymerized tubulin in the supernatant can be assayed to determine the effect a microtubule regulator has on polymerization.

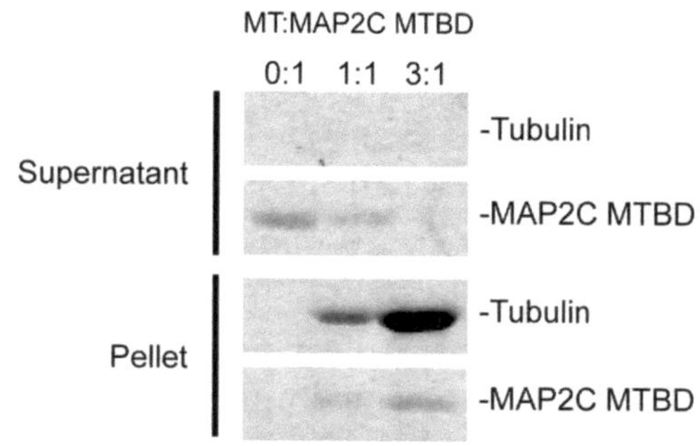

Fig. 2. The microtubule-binding domain of MAP2C (MAP2C MTBD) cosediments with taxol-stabilized microtubules (MTs). 6 μM MAP2C MTBD was incubated alone and in the presence of 6 and 18 μM taxol-stabilized microtubules representing 0:1, 1:1, and 3:1 MT:MAP2C MTBD ratios, respectively. Incubated samples were centrifuged over a 40% glycerol cushion and the supernatant and pellet fractions analyzed by SDS-PAGE and Coomassie staining. All tubulins sedimented to the pellet fraction. MAP2C is present in the supernatant in the absence of taxol-stabilized microtubules, but cosediments as the concentration of taxol-stabilized microtubules is increased.

2. Materials

2.1. Gel Filtration Tubulin-Binding Assay

1. Purified αβ-tubulin (bovine or porcine) at a concentration of 20 μM or higher (Cytoskeleton, Inc., Denver, CO) (see Note 1).
2. Purified tubulin-binding protein at a concentration of 20 μM or higher.
3. Amicon Ultracentrifugal filters (Millipore, Billerica, MA), 0.5, 5, or 15 ml capacity.
4. Protein denaturation solution: 6.67 M Guanidinium hydrochloride, 22.2 mM phosphate buffer pH 6.5 (for 100 ml solution, add 0.19 g Monosodium phosphate, monohydrate, and 0.16 g disodium phosphate, heptahydrate). Filter using a 0.22-μm filter.
5. Running buffer: 20 mM Pipes, 200 mM KCl, 1 mM $MgCl_2$, 50 μM GTP (Sigma, St. Louis, MO), 0.1% β-Mercaptoethanol (v/v), adjust pH to 6.5 using KOH. Filter the running buffer using a 0.22-μm filter (see Note 2).
6. Superdex 200 10/300 GL gel filtration column (GE Healthcare, Piscataway, NJ) (see Note 3).
7. GTP stock solution: 100 mM.

2.2. Taxol-Stabilized Microtubule Cosedimentation Assay

1. Items 1–4 from Subheading 2.1.
2. 2× BRB80 buffer: 160 mM K-Pipes, 2 mM $MgCl_2$, 2 mM EGTA, adjust pH to 6.8 using KOH.
3. GTP stock: 100 mM GTP (Sigma).

4. Dithiothreitol (DTT) stock: 1 M DTT (Sigma).
5. Taxol stock solutions: 20 μM, 200 μM, and 2 mM Paclitaxel (Sigma) in DMSO (Sigma). A master stock solution at 10 mM can be prepared in DMSO and stored at −20°C.
6. 7×20-mm polycarbonate ultracentrifuge tubes (Beckman Coulter, Inc., Brea, CA, item 343775).
7. Taxol–glycerol cushion solution: 1× BRB80 buffer supplemented with 40% glycerol and 20 μM taxol (see Note 4).

2.3. Dynamic Microtubule Cosedimentation Assay

1. Items 1–6 from Subheading 2.2.
2. DMSO (Sigma).
3. GTP–glycerol cushion solution: 1× BRB80 buffer supplemented with 40% glycerol and 1 mM GTP (see Note 4).

2.4. SDS-Polyacrylamide Gel Electrophoresis

1. 5× SDS gel loading buffer: 250 mM Tris–HCl pH 6.8, 10% SDS (w/v), 50% glycerol, 0.02% bromophenol blue (w/v), and 10% beta-mercaptoethanol (v/v). Solid SDS is a respiratory irritant, and a mask should be worn when handling it.
2. Molecular weight markers: PageRuler Plus Prestained Protein Ladder (Fermentas Life Sciences, Burlington, ON).
3. SDS-Polyacrylamide Tris–HCl Gel: 15% resolving gel or a 10–20% linear gradient gel (BioRad, Hercules, CA) (see Note 5).
4. Tris–glycine SDS running buffer: 25 mM Tris, 192 mM Glycine, 0.1% SDS, and pH 8.3.
5. Coomassie stain: 0.1% Coomassie blue (w/v), 10% acetic acid (v/v), and 40% methanol (v/v).
6. Destain solution: 10% acetic acid (v/v) and 20% methanol (v/v).
7. Kimwipes (Fisher Scientific).

3. Methods

3.1. αβ-Tubulin Gel Filtration-Binding Assays

1. In order to maintain αβ-tubulin in a heterodimeric, nonpolymerized, nonaggregated state, it is important to carry out all steps in this protocol at 4°C.
2. Independently exchange stocks of tubulin and tubulin-binding protein into running buffer at 4°C. This may be achieved through a tenfold dilution of the protein stock into running buffer followed by a concentration step using an Amicon Ultracentrifugal filter (Millipore), concentrating to the original volume by centrifugation at 3,000 × *g*, and then repeating the dilution and concentration step a second time. Use a centrifugal filter with a nominal molecular weight limit that is less than half the protein's molecular weight.

3. Centrifuge the exchanged protein stock solutions at 50,000 × *g* for 20 min at 4°C to remove aggregates. Remove supernatant and transfer protein stocks to a clean tube.
4. Determine the concentration of tubulin and tubulin-binding protein using the protein's extinction coefficient at 280 nm. The extinction coefficient (ε_{280}) of GTP-bound αβ-tubulin is 115,000 M^{-1} cm^{-1} (7). The extinction coefficient of the tubulin-binding protein can be calculated using the ExPASy ProtParam tool (http://ca.expasy.org/tools/protparam.html) or approximated using the equation $\varepsilon_{280} = (\#\ Trp)(5{,}690) + (\#\ Tyr)(1{,}280) + (\#\ Cystine)(120)\ M^{-1}\ cm^{-1}$ (8). The ε_{280} of any cofactors should be added to the extinction coefficient in accord with binding stoichiometry. To each of the three 1.5-ml tubes, add 90 μl of protein denaturation solution and 8 μl of water. To the first tube, add 2 μl of running buffer. To the second tube, add 2 μl of concentrated tubulin. To the third tube, add 2 μl of concentrated tubulin-binding protein. Mix the contents of each tube. Place the contents of tube one into a 100-μl, 1-cm path length quartz cuvette and blank a spectrophotometer at 280 nm against this solution. Once the spectrophotometer is blanked, measure the A_{280} of the denatured tubulin and tubulin-binding protein solutions, respectively. The A_{280} measurement should be in the range of 0.05–1.00 AU. The volume of protein diluted into denaturation solution can be adjusted to obtain an A_{280} value in this range. Calculate the amount of protein in the stock solution using the Beer–Lambert Law for a 50-fold dilution: protein concentration = $(A_{280})(50)/(\varepsilon_{280}\ M^{-1}\ cm^{-1})(1\ cm)$.
5. Insert the chromatography system's pump A lead into the running buffer and exchange pump A into running buffer.
6. Preclean a 50-μl injection loop by flushing with 10 ml of running buffer. Attach the loop to the injection valve.
7. Without a column attached, flush the system with 10 ml of running buffer by setting a flow rate on pump A of 1 ml/min. After 10 ml of running buffer have flowed through the system, reduce the flow to 0.1 ml/min and set the system pressure limit at 1 MPa. Attach the Superdex 200 10/300 GL gel filtration column to the chromatography system between the injection valve and the UV monitor. An active flow attachment process reduces the introduction of air bubbles onto the column. It is best to have a spring-loaded syringe filled with column storage solution attached to the column outlet line to create a reverse flow during the attachment procedure. Once the column's inlet line is connected, immediately remove the spring-loaded syringe to prevent pressure buildup. Attach the column outlet line to the UV monitor and stop the flow from pump A.

8. Equilibrate the Superdex 200 10/30 column with two-column volumes of running buffer (one-column volume = 24 ml) at a flow rate of 0.5 ml/min with a pressure limit cutoff set at 1 MPa. Monitor the UV trace to ensure that the buffer has exchanged and a baseline at 280 nm has been established.
9. Make protein samples for gel filtration runs. Using 1.5-ml tubes, make 120 μl protein solutions: Tube 1: 20 μM tubulin; Tube 2: 20 μM tubulin-binding protein; and Tube 3: 20 μM tubulin + 20 μM tubulin-binding protein. This 1:1 tubulin:tubulin-binding protein stoichiometry can be adjusted in attempts to saturate binding or to meet non-1:1 binding stoichiometries. Use running buffer to adjust the final volume of protein samples to 120 μl. If tubulin binding is GTP-dependent, the concentration of GTP in the 120 μl solution can be increased from 50 μM to 1 mM by adding 1.2 μl of a 100 mM GTP solution. Prepare protein samples 30 min prior to injection, allowing them to incubate at 4°C (see Note 6).
10. Fit a 5-ml syringe with an injection needle and load 5 ml of running buffer. Set the chromatography injection valve to the load position. Flush the 50-μl loop with 4 ml of running buffer to clean and equilibrate the loop.
11. Fit a 1-ml syringe with an injection needle and load the 120 μl of protein sample (see Note 7). Do not draw air into the syringe. Set the chromatography injection valve to the load position. Load the 120-μl protein solution into the 50-μl loop and leave the syringe mounted in the injection port (see Note 8).
12. Run a Superdex 200 10/300 GL column method using a 0.5 ml/min flow rate on pump A and a pressure limit of 1 MPa. Inject the contents of the 50-μl loop using 0.5 ml of running buffer. Monitor absorbance at 280 nm (see Note 9). Begin to collect eluate in 0.2 ml fractions 20 min post injection (once 10.5 ml have flowed over the column). Stop collecting fractions 50 min post injection (25.5 ml point). The total volume of the run should be 1.4-column volumes (33.6 ml).
13. Repeat steps 10 through 12 for all samples. The A_{280} chromatographic trace for each sample can be exported to a graphing program, plotted and overlayed for comparison.
14. Comparative analysis of sample fractions using SDS-PAGE: To collected fractions, add 50 μl of 5× SDS gel loading buffer. Fractions can be stored at −20°C or used immediately for SDS-PAGE analysis. Choose a window of fractions across all runs that cover the range of peak elution. Heat the samples at 95°C for 5 min. Follow the procedure in Subheading 3.4 for running and analyzing SDS-polyacrylamide gels.

3.2. Taxol-Stabilized Microtubule Cosedimentation Assay

1. Preparation of taxol-stabilized microtubules: Dilute tubulin to 2.2 mg/ml (20 μM) in BRB80 buffer. Add GTP and DTT to a final concentration of 1 mM each. Incubate at 4°C for 5 min and then warm to 37°C. Introduce taxol stepwise as follows: add 1/100 volume of 20 μM taxol in DMSO. Mix gently and incubate at 37°C for 5 min. Add 1/100 volume of 200 μM taxol in DMSO. Mix gently and incubate at 37°C for 5 min. Add 1/100 volume of 2 mM taxol, mix, and incubate at 37°C for 15 min. Taxol-stabilized microtubules are now formed and can be stored at room temperature (see Note 10).
2. Exchange the microtubule-binding protein under investigation into BRB80 buffer using an Amicon Ultracentrifugal filter as outlined in step 2, Subheading 3.1. Centrifuge the protein at 100,000 × *g* at 25°C to clarify. Take the supernatant and determine the protein concentration following the method outlined in step 4, Subheading 3.1.
3. Warm a Beckman TLA100 fixed angle rotor and a Beckman Optima TL ultracentrifuge to 25°C. An equivalent rotor (fixed angle or swinging bucket) or centrifuge may be substituted.
4. Using 1.5-ml tubes, create 120 μl samples containing 10 μM taxol-stabilized microtubules (60 μl of product from step 1, Subheading 3.2) and 10 μM microtubule-binding protein. Other taxol-stabilized microtubule:microtubule-binding protein ratios can be used to promote microtubule saturation and/or determine an apparent affinity(9). Add taxol to maintain a final concentration of 20 μM and add 2× BRB80 and water to reach a final volume of 120 μl in 1× BRB80. Create control samples of taxol-stabilized microtubules alone and microtubule-binding protein alone. Incubate samples at 25°C for 20 min.
5. For each sample, fill a 7 × 20-mm polycarbonate ultracentrifuge tube with 150 μl of the taxol–glycerol cushion solution and equilibrate at 25°C (see Note 11).
6. Once the microtubule–microtubule-binding protein incubation is complete, layer 100 μl of the protein sample above the 150 μl glycerol cushion. Save the remaining 20 μl of protein sample for SDS-PAGE analysis, designated as "load."
7. Insert the centrifuge tubes into the TLA100 rotor (see Note 12). Centrifuge the samples at 100,000 × *g* for 30 min at 25°C.
8. Remove 50 μl off the top of the supernatant and save this for SDS-PAGE analysis, designated as "supernatant." Remove the remaining 50 μl of solution atop the glycerol cushion and discard. Wash the interface three times with 100 μl of 1× BRB80 at 25°C.

9. Remove the 150 μl glycerol cushion, keeping the pipet tip on the opposite face of the tube from the pellet.
10. Resuspend the pellet in 100 μl of 1× BRB80 supplemented with 20 μM taxol and save this sample, designated as "pellet."
11. To each sample (load, supernatant, and pellet), add 0.25 volumes of 5× SDS gel loading buffer. Heat the samples to 95°C for 5 min. The samples can now be analyzed by SDS-PAGE following the protocol outlined in Subheading 3.4.

3.3. Dynamic Microtubule Cosedimentation Assay

1. Exchange the microtubule-binding protein under investigation into BRB80 buffer using an Amicon Ultracentrifugal filter as outlined in step 2, Subheading 3.1. Centrifuge the protein at 100,000×*g* at 37°C to clarify. Take the supernatant and determine the protein concentration following the method outlined in step 4, Subheading 3.1.
2. Warm a Beckman TLA100 fixed angle rotor and a Beckman Optima TL ultracentrifuge to 37°C. An equivalent rotor (fixed angle or swinging bucket) or centrifuge may be substituted.
3. Using 1.5-ml tubes, create 120 μl protein samples in 1× BRB80 supplemented with 5% DMSO and 1 mM GTP. Protein samples include Tube 1: 20 μM tubulin; Tube 2: 20 μM microtubule-binding protein; and Tube 3: 20 μM tubulin and 20 μM microtubule-binding protein. The concentration of microtubule-binding protein can be varied in this assay. Incubate samples at 37°C for 20 min to initiate microtubule polymerization.
4. For each sample, fill a 7×20-mm polycarbonate ultracentrifuge tubes with 150 μl of the GTP–glycerol cushion solution (see Note 11). Insert the centrifuge tubes into the 37°C TLA100 rotor (see Note 12). Warm the rotor and tubes to 37°C.
5. Once the microtubule–microtubule-binding protein incubation is complete, layer 100 μl of the protein sample above the 150 μl glycerol cushion. Save the remaining 20 μl of protein sample for SDS-PAGE analysis, designated as "load."
6. Insert the centrifuge tubes into the TLA100 rotor (see Note 12). Centrifuge the samples at 100,000×*g* for 30 min at 37°C.
7. Remove 50 μl off the top of the supernatant and save this for SDS-PAGE analysis, designated as "supernatant." Remove the remaining 50 μl of solution atop the glycerol cushion and discard. Wash the interface three times with 100 μl of 1× BRB80 at 37°C.
8. Remove the 150 μl glycerol cushion, keeping the pipet tip on the opposite face of the tube from the pellet.

9. Resuspend the pellet in 100 μl of BRB80 supplemented with 5% DMSO and 1 mM GTP and save this sample, designated as "pellet."
10. To each sample (load, supernatant, and pellet), add 0.25 volumes of 5× SDS gel loading buffer. Heat the samples to 95°C for 5 min. The samples can now be analyzed by SDS-PAGE following the protocol outlined in Subheading 3.4.
11. The amount of tubulin in the pellet corresponds with the fraction of polymerized tubulin while the amount in the supernatant represents nonpolymerized tubulin, representative of the critical concentration. The level of tubulin in the supernatant and pellet can be quantitated and analyzed to determine if a microtubule or tubulin-binding protein shifts tubulin toward a polymerized or depolymerized state.

3.4. SDS-Polyacrylamide Gel Electrophoresis

1. Load an SDS-polyacrylamide Tris–HCl gel (15% resolving gel or 10–20% linear gradient gel) onto a gel rig apparatus and fill reservoirs with Tris–glycine SDS running buffer (see Notes 5 and 13).
2. Remove the comb from the wells and wash the wells with running buffer by pipeting up and down repeatedly in each well. Load equal volumes of protein samples into respective wells, including one well for molecular weight markers. Run the gel at 150 V until the dye front reaches the bottom of the resolving gel. For optimal separation of tubulin (55 kDa) and tubulin-binding protein, use the Prestained Molecular Weight Markers as a guide to stop the electophoresis.
3. Analyze the gel by Coomassie staining, silver staining, or Western blot. For Coomassie staining, remove the gel from its case, transfer to a clean container, and incubate in Coomassie stain for 1 h on a rotating platform. Remove the Coomassie stain and add destain solution. Add a Kimwipe to promote the destaining process.
4. Once destained sufficiently, the gel can be scanned and analyzed for tubulin binding. Quantitative measurements can be made using the Image J program (NIH) to measure band intensity.

4. Notes

1. Tubulin can be purified from bovine or porcine brain (10). If investigations are anticipated to require gram quantities of tubulin over the course of a year, purifying tubulin can be a cost-effective measure.

2. The gel filtration running buffer can be modified to suit the solubility requirements of the tubulin-binding protein by altering the buffer, pH, or salt composition and concentration.
3. A Superose 6 gel filtration column (GE Healthcare) or equivalent may substitute. Ensure that the individual components and complex are within the separation range of the column. Molecular weight standards should be run over the column to determine the column's resolution and to calibrate the column. The chromatography system can also be used in-line with a static light-scattering unit to determine the mass of a complex (11).
4. 40% glycerol can be substituted with 40% sucrose.
5. SDS-PAGE gels can be readily created in the laboratory following standard procedures.
6. 120 μl of protein solution is used to ensure that the 50-μl injection loop is completely filled. This amount can be reduced if reagents are limiting.
7. A disposable 1-ml syringe may be used and is listed here due to common availability. A smaller, Hamilton syringe is an alternative option.
8. If the syringe is removed from the injection port, sample can drain from the injection loop before being injected onto the column.
9. To specifically monitor a tubulin shift, tubulin can be doped 1:100 with rhodamine-labeled tubulin (Cytoskeleton, Inc.). A multiwavelength monitor can be used to detect absorbance at 280 nm (total protein with 280 nm absorbance character) and 500 nm (rhodamine-labeled tubulin) simultaneously. If excess tubulin-binding protein masks the location of tubulin, this method enables distinct determination of the tubulin elution profile. Fluorophores can also be covalently attached to the tubulin-binding protein or genetically engineered, such as an EGFP fusion, monitored at 460 nm.
10. If the microtubule-binding protein is expected to recognize a feature at the end of the microtubule, the number of microtubule ends can be increased by shearing the taxol-stabilized microtubules by passing them through a 26- or 30-gauge needle (9).
11. An additional tube may need to be used to balance the rotor.
12. Mark the outermost segment of the centrifuge tube's rim with a lab marker to designate the side of the tube that contains the pellet.
13. If multiple samples are analyzed, it may be advantageous to skip wells to avoid cross contamination during gel loading or to group the supernatant samples and pellet samples independently with an empty well between them.

Acknowledgments

The authors would like to thank Dr. Stephen Rogers for advice. This work was supported by a Basil O'Connor Research Starter Grant from the March of Dimes.

References

1. Desai A and Mitchison TJ (1997) Microtubule polymerization dynamics. Annu Rev Cell Dev Biol. **13**: 83–117.
2. Margolis RL (1981) Role of GTP hydrolysis in microtubule treadmilling and assembly. Proc Natl Acad Sci USA **78**: 1586–1590.
3. Brouhard GJ, Stear JH, Noetzel TL et al (2008) XMAP215 is a processive microtubule polymerase. Cell **132**: 79–88.
4. Vitre B, Coquelle FM, Heichette C et al (2008) EB1 regulates microtubule dynamics and tubulin sheet closure in vitro. Nat Cell Biol. **10**: 415–421.
5. Roll-Mecak A and Vale RD (2008) Structural basis of microtubule severing by the hereditary spastic paraplegia protein spastin. Nature, **451**: 363–367.
6. Moores CA, Hekmat-Nejad M, Sakowics R and Milligan RA (2002) A mechanism for microtubule depolymerization by KinI kinesins. Mol Cell **9**: 903–909.
7. Budde PP, Desai A and Heald R (2006) Analysis of microtubule polymerization in vitro and during the cell cycle in Xenopus egg extracts. Methods **38**: 29–34.
8. Gill SC and von Hippel PH (1989) Calculation of protein extinction coefficients from amino acid sequence data. Anal Biochem. **182**: 319–326.
9. Pasqualone D and Huffaker TC (1994) STU1, a suppressor of a beta-tubulin mutation, encodes a novel and essential component of the yeast mitotic spindle. J Cell Biol. **127**: 1973–1984.
10. Borisy GG, Marcum JM, Olmsted JB et al (1975) Purification of tubulin and associated high molecular weight proteins from porcine brain and characterization of microtubule assembly in vitro. Ann N Y Acad Sci. **253**: 107–132.
11. Slep KC and Vale RD (2007) Structural basis of microtubule plus end tracking by XMAP215, CLIP-170, and EB1. Mol Cell **27**: 976–991.

Chapter 7

Analysis of Microtubule Assembly Kinetics Using Turbidimetry

Felicia Gaskin

Abstract

Turbidity measurements are rapid and reliable methods to follow the effects of drugs, proteins, nucleotides, metals, and other factors on the assembly kinetics of tubulin into microtubules in vitro and have been in use since 1974. Improvements in spectrophotometers and software have resulted in the use of less protein, and more curves can be analyzed continuously and almost simultaneously.

Key words: Tubulin, Assembly, Turbidimetry, Microtubules, Kinetics

1. Introduction

Turbidity measurements are frequently used to follow the assembly and disassembly of tubulin in vitro. Microtubules assembled in vitro are a good approximation to the long rod limit when light (320–600 nm) is used to measure turbidity. Berne has shown that for very long rod-like particles with a small diameter compared to the wavelength (λ) of the incident light, the turbidity is proportional to λ^{-3} (1). Three tests were described in a 1974 paper by Gaskin et al. that show that the turbidity of in vitro-assembled microtubules is proportional to the weight of assembled tubulin in microtubules (2). They included the following.

1. A plot of turbidity (absorbance) as a function of λ showed A was proportional to $\lambda^{-3.3}$.
2. Sonication to 20% of the initial length resulted in 80% of the initial A_{350}.

Anne Straube (ed.), *Microtubule Dynamics: Methods and Protocols*, Methods in Molecular Biology, vol. 777,
DOI 10.1007/978-1-61779-252-6_7, © Springer Science+Business Media, LLC 2011

3. After polymerization, a curve of A_{350} versus initial concentration (C_o) coincided with a curve of mass of pelleted microtubules versus C_o (see Note 1).

It has been well-documented that high concentrations of microtubules and nonmicrotubule structures, such as zinc-induced sheets of tubulin, do not scatter light in proportion to the weight of tubulin in polymers. For example, see Gaskin and Kress (3). However, assembly kinetics can be informative in studies of promoters and inhibitors of assembly of tubulin into sheets, microtubules, bundles, etc. In order to use turbidity to analyze tubulin assembly, it is important to do negative-staining electron microscopy to verify the polymers formed (4) (see Note 2).

2. Materials

2.1. Tubulin or Microtubule Protein Preparations

Tubulin preparations with or without microtubule-associated proteins (MAPs), which are competent to assemble (see Note 3). Purified tubulin (1–5 mg/ml) and/or tubulin plus MAPs (0.5–2 mg/ml), which we call microtubule protein (MTP), can be used and assembly is promoted:

(a) By changing the temperature, i.e., from 4 to 37°C and/or

(b) By adding a necessary requirement for assembly, such as GTP, to 1 mM.

(c) For purified tubulin, buffer containing 30% glycerol and a higher concentration of $MgCl_2$ or 20 μM taxol.

2.2. Assembly Buffers

1. Mes A buffer: 0.1 M 2-(*N*-morpholino)-ethane sulfonic acid, pH 6.6, 1 mM EGTA, 0.5 mM $MgCl_2$.
2. Mes A buffer with 60% glycerol: 0.1 M 2-(*N*-morpholino)-ethane sulfonic acid, pH 6.6, 1 mM EGTA, 0.5 mM $MgCl_2$, 60% glycerol.
3. Mes A buffer with 30% glycerol: An equal volume of 60% glycerol in Mes A buffer is added to Mes A buffer.
4. 100 mM ATP stock solution.
5. 100 mM $MgCl_2$ stock solution.
6. 100 mM GTP stock solution.

2.3. Spectrophotometer

Assembly is monitored with a recording spectrophotometer preferably with an automatic thermostatable sample changer which can monitor 4–5 samples simultaneously or with a 96-well plate spectrophotometer with appropriate filters and a heating element. Larger quantities of tubulin are required for the conventional spectrophotometer, and the matched cuvettes may require

0.3–1 ml of tubulin. As little as 50 μl can be used with the 96-well recording spectrophotometer. Turbidimetric measurements are usually done at A_{350} to maximize experimental measurements, although other wavelengths can be used (see Note 4).

3. Methods

The time course of A_{350} assembly is measured after changing the temperature or adding a promoter of tubulin assembly. The time course of assembly will show a lag, an initial rate of assembly after the lag, and final $A_{350,}$ which should be constant for 20 min, suggesting that steady state is reached. The lag decreases and the rate of assembly increases as the temperature is increased. The $t_{1/2}$ (time to reach 50% polymerization) can also be used to compare assembly curves. Thus, the assembly can be compared with and without a protein, drug, nucleotide, or other compounds of interest. It is important that additions are added with the same buffer or solvent as the compound of interest. A_{350} is determined after subtracting the turbidity of the sample under nonassembly conditions from the assembled value.

Although 37°C is frequently used to follow assembly, tubulin is unstable at higher temperatures and forms irreversible aggregates. Thus, it may be better to follow assembly at 35°C or even 30°C. Final levels of polymerization should be similar in this temperature range.

Disassembly of microtubules can be followed using A_{350} as a function of time after adding an inhibitor or by decreasing the temperature.

3.1. Analysis of Tubulin Assembly

A turbidity experiment to analyze microtubule assembly kinetics is illustrated in Fig. 1. This experiment looks at the effect of 1 mM ATP on the assembly kinetics at four different concentrations of purified tubulin.

1. Cofactors were removed from the Mono Q fast-performance liquid chromatography (FPLC)-purified tubulin preparation using 0.5 ml (~10 mg/ml) of tubulin in a Crowe-Englander-type thin-film dialyzer and two changes of 500 ml of Mes A buffer at 4°C over a 2 h period (5).
2. The sample was centrifuged at 260,600 × *g* for 6 min at 4°C to remove aggregates.
3. The supernatant was used.
4. A small part of the supernatant was diluted 1:40 with Mes A buffer and A_{280} determined. The extinction coefficient for tubulin is 115. Thus, the tubulin concentration in mg/ml is A_{280} tubulin (less A_{280} buffer) × 0.87 × dilution factor for a 1-cm path length cuvette.

5. An HPLC analysis of the supernatant showed 1.02 mol of GTP (presumably, 1.00 mol at the nonexchangeable site and 0.02 mol at the exchangeable site).
6. Tubulin was diluted with Mes A buffer to make 7 mg/ml tubulin A.
7. 1 volume of Mes A buffer with 60% glycerol was added to make 3.5 mg/ml tubulin A in Mes A with 30% glycerol. Tubulin A will not assemble at 37°C, even with 3 mM $MgCl_2$. However, it will assemble if 1 mM GTP is added.
8. For the Assembly Experiment, the following samples were prepared:
   (a) Sample 1–300 μl tubulin A.
   (b) Sample 2–257 μl tubulin A plus 43 μl Mes A with 30% glycerol.
   (c) Sample 3–214 μl tubulin A plus 86 μl Mes A with 30% glycerol.
   (d) Sample 4–129 μl tubulin A plus 171 μl Mes A with 30% glycerol.
9. To each sample, 3 μl 100 mM ATP and 8 μl 100 mM $MgCl_2$ were added.
10. The samples were vortexed well and kept at 4°C for 25 min (see Note 5).
11. Assembly was followed by putting samples into cuvettes that were thermostated to 37°C.
12. Assembly was followed for 45 min (see Fig. 1).

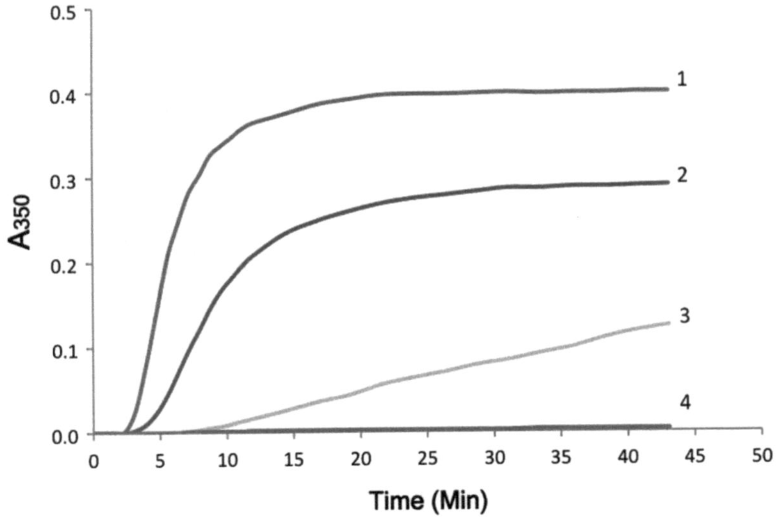

Fig. 1. A turbidity experiment to analyze the effect of ATP on tubulin assembly kinetics at 37°C with four different concentrations of tubulin. Protein concentrations were 3.5, 3.0, 2.5, and 1.5 mg/ml for curves 1, 2, 3, and 4, respectively. Assembly was done in Mes A buffer with 30% glycerol, 1 mM ATP, 3 mM $MgCl_2$, 1 mM EGTA, and pH 6.6.

Curves 1 and 2 show typical assembly kinetics for tubulin assembly into microtubules. Tubulin at 3.5 mg/ml has a shorter lag (2.4 vs. 3.5 min) and $t_{1/2}$ (5.4 vs. 8.7 min) than for 3.0 mg/ml. A_{350} for assembly at equilibrium was higher for 3.5 mg/ml than at 3.0 mg/ml (0.40 vs. 0.29). Curve 3 (2.5 mg/ml) shows slow assembly, and A_{350} does not approach a constant value during this time frame. Curve 4 (1.5 mg/ml) shows no assembly. Microtubules were the predominant structures found after 45 min. (Curves 1, 2, and 3). After 10 min at 4°C, all three returned to the starting A_{350} showing that the microtubules disassemble (not shown). However, there was no reassembly at 37°C. This study shows that under these conditions with 1 mM ATP, tubulin can assemble only once (see Note 6). If 1 mM GTP is then added, tubulin will assemble, so tubulin is still active.

3.2. Determination of Critical Concentration

Tubulin can be in equilibrium with microtubules. Analysis of assembly at several C_o of tubulin can be used to determine the critical concentration (C_c) for assembly. A plot of final A_{350} versus C_o should be a straight line and not go through the origin. The value of C_c is the intercept on the x axis and is a measure of unpolymerized tubulin. Although it is tempting to try to calculate a C_c for the experiment in Fig. 1, it is not valid since the tubulin is not in equilibrium with microtubules as tubulin will not reassemble. Also, curve 3 has not reached an apparent equilibrium in 45 min. However, if the experiment was done with 1 mM GTP, there is a tubulin–microtubule equilibrium. A C_c of 0.3 mg/ml has been reported (6).

4. Notes

1. A more recent paper by Hall and Minton describes a detailed analysis of using turbidity as a probe of tubulin polymerization kinetics using a theoretical and experimental reexamination (7). For those interested in distributions of microtubules that are not long relative to the wavelength of light, the simulation work should permit a valid interpretation.
2. Ten microliter of sample is put on a Formvar film on a copper grid for 1 min and stained by rinsing with six drops of 1% uranyl acetate and blotting dry with filter paper. Polymers are analyzed by electron microscopy.
3. Initial experiments can be done with an MTP which is prepared from homogenates by assembling microtubules at 37°C, centrifugation to pellet microtubules, and using 4°C to disassemble the microtubules in the resuspended pellets, followed by centrifuging in the cold to remove aggregates that do not disassemble (8). This MTP preparation contains

tubulin and MAPs and is more stable than purified tubulin. However, it also contains trace enzymatic activities, such as NDPK, adenylate kinase, and ATPases, which may interfere with the analysis. Tubulin can be purified from MTP using phosphocellulose chromatography (9) or FPLC (6). Purity should be checked by SDS-PAGE and enzymatic measurements. Both tubulin and MTP can be frozen in liquid nitrogen and stored for years. The protein can be subjected to another cycle of assembly–disassembly before use. Glycerol cushions used in centrifugation or thin-film dialysis (Crowe-Englander-type microdialyzer) for 2–3 h can be used to remove cofactors. Purified tubulin without MAPs requires a higher concentration for assembly and is less stable in buffers. Assembly can be followed using 30% glycerol in buffer (6) or taxol (10), which promotes tubulin assembly and stabilizes microtubules.

This book contains a chapter by Gell et al. on how they purify tubulin from porcine brain. Similar methodology is used, although the buffers are different. If assembly-competent MTP or tubulin is purchased, conditions for assembly should be followed as described by the company.

4. If using a new system with unknown cofactors, it is a good idea to follow absorption as a function of λ from 600 to 320 nm before and after polymerization. There should be no suggestion of absorption under nonassembly conditions, and the log A versus log λ curve after polymerization should be a straight line and approach λ^{-3}. A is determined by subtracting the initial A from the final A.
5. During this time period, the A_{350} for nonassembled tubulin can be determined at 4°C. Or if a lag time is expected before assembly, the A_{350} of the lag time can be subtracted from A_{350} during polymerization, which is what was done in this experiment.
6. It is unlikely that trace NDPK activity is responsible for these findings (NDPK could transfer a P to GDP at the exchangeable site) and GTP at the exchangeable site will promote assembly. NDPK is a stable enzyme and should remain active during assembly, disassembly, and reassembly. But tubulin assembles only once with ATP.

References

1. Berne BJ (1974) Interpretation of the light scattering from long rods. J Mol Bio 89:755–758
2. Gaskin F, Cantor CR, Shelanski ML (1974) Turbidimetric studies of the in vitro assembly and disassembly of porcine neurotubules. J Mol Biol 89:737–755
3. Gaskin F, Kress Y (1977) Zinc-ion induced assembly of tubulin. J Biol Chem 252: 6918–6924

4. Gaskin F (1982) Techniques for the study of microtubule assembly in vitro. Methods Enzymol 85:433–439
5. Xu S, Gaskin F (1998) Probing the ATP binding site of tubulin with thiotriphosphate analogues of ATP. Biochima et Biophysica Acta 1383:111–122
6. Roychowdhury S, Gaskin F (1986) Separation of assembly-competent tubulin from brain microtubule protein preparations using a fast-performance liquid chromatography procedure. J Neurochem 46:1399–1405
7. Hall D, Minton AP (2005) Turbidity as a probe of tubulin polymerization kinetics: A theoretical and experimental re-examination. Anal Biochem 345:198–213
8. Shelanski ML, Gaskin F, Cantor CR (1973) Microtubule assembly in the absence of added nucleotides. Proc Natl Acad Sci USA 70:765–768
9. Weingarten MD, Lockwood AH, Hwo S-Y, Kirschner MW (1975) A protein factor essential for microtubule assembly. Proc Natl Acad Sci USA 72:1858–1862
10. Schiff PB, Fant J, Horwitz SB (1979) Promotion of microtubule assembly in vitro by taxol. Nature 277:665–667

Chapter 8

Isolation of Centrosomes from Cultured Cells

Simone Reber

Abstract

The centrosome is a unique organelle that nucleates and organizes the interphase microtubule array and facilitates bipolar spindle assembly during mitosis. Isolated centrosomes are ideal starting materials for biochemical and structural studies as well as for functional assays to study microtubule-dependent processes. In fact, they are hitherto the only system to study radial microtubule arrays in vitro. This chapter provides a practical method and the rationale of isolating centrosomes from adherent cultured cells by density-gradient centrifugation. It further describes how to evaluate the activity and function of centrosomes by microtubule nucleation and immunofluorescence assays and how to measure microtubule dynamics nucleated from isolated centrosomes in *Xenopus* egg extracts.

Key words: Centrosome, γ-tubulin, Microtubules, Microtubule dynamics, *Xenopus* egg extracts

1. Introduction

Throughout the cell cycle, the centrosome is the dominant microtubule-organizing center and thus involved in many fundamental microtubule-dependent processes, including mitosis, cytokinesis, intracellular transport, and regulation of cell shape and motility. Structurally, the centrosome consists of two centrioles, the mother and the daughter centrioles, both embedded into a pericentriolar matrix (PCM). It is duplicated once per cell cycle as bipolar spindle assembly requires two functional centrosomes.

The importance of the centrosome as a central cell-organizing center was already appreciated more than one century ago (1). Still, there is an ongoing interest in understanding the molecular nature and function of centrosomes, especially since centrosomal abnormalities are linked to aneuploidy, chromosomal instability, and thus cancer development (2).

Anne Straube (ed.), *Microtubule Dynamics: Methods and Protocols*, Methods in Molecular Biology, vol. 777,
DOI 10.1007/978-1-61779-252-6_8,

The successful isolation of centrosomes (or their functional equivalent) has been described from a number of species, including yeast (3), surf clam oocytes (4), mitotic sea urchin eggs (5), the early *Drosophila* embryo (6), cultured *Xenopus* cells (7), and mammalian tissue culture cells (8–11). The purification of centrosomes from Chinese hamster ovary (CHO) cells, as described in this chapter, was pioneered by Blackburn and colleagues (12). However, the procedure described herein is largely based on the protocols by Mitchison and Kirschner (8) as well as Moudjou and Bornens (10). Below, I deconstruct the rationale of the protocol, facilitating potential adaptations for your cell line of choice.

The challenge when applying this method to a previously characterized or novel cell line is to obtain sufficiently pure centrosomes that still faithfully reconstitute all centrosome-dependent activities. The treatment of cells with Nocodazole and Cytochalasin B depolymerizes microtubules and actin and consequently weakens the association of the centrosome with the nuclear envelope and helps to release the centrosome. Moreover, this treatment reduces the contamination with tubulin, microtubule-associated proteins (MAPs), and actin. Before the cells are lysed in a low ionic strength buffer in the presence of detergent, it is important (1) to remove all salts and metal ions (by washing with sucrose) and (2) to be fast (i.e., washes should happen in less than 1 min per plate) as the cells swell and eventually burst. The first centrifugation serves as an initial cleanup of the cell lysate, as cell debris and chromatin are spun down. The second centrifugation concentrates the centrosomes onto a Ficoll cushion while the final density-gradient centrifugation further concentrates the partially purified centrosomes into a sharp fraction according to the centrosomes' sedimentation velocity. In some protocols, a continuous sucrose gradient (20–62.5%) is used, whereas other protocols prefer a discontinuous gradient (40, 50, 70%). Here, I describe the use of a discontinuous gradient, as in my hands, it proved to be faster and more robust. The high sucrose concentration in peak fractions stabilizes centrosomal proteins and serves as a cryoprotectant when freezing. Finally, the isolated centrosomes can be analyzed by Western blot, functional (e.g., microtubule nucleation) as well as structural (localization of structural centrosome components by immunofluorescence) assays.

2. Materials

2.1. Cell Culture

1. CHO cells.
2. Cytochalasin B (Sigma C6762); stock solution 10 mg/ml in DMSO.

3. MEM α medium, supplemented with 10% fetal bovine serum (FBS) (dialyzed), 0.2 U/ml penicillin, 200 ng/ml streptomycin, 2 mM L-glutamine.
4. Nocodazole (Sigma M1404); stock solution 10 mg/ml in DMSO.
5. Incubator preset to 37°C and 5% CO_2.
6. Tissue culture dishes (245 × 245 × 25 mm) (Nunc #166508).

2.2. Isolation of Centrosomes

The following solutions can be prepared one day before the actual isolation of centrosomes. Store all solutions at 4°C!

1. 1 mM Tris–HCI, pH 8.0 (before use, add 8 mM 2-mercaptoethanol).
2. 1× PBS: 137 mM NaCl, 8.1 mM Na_2HPO_4, 2.7 mM KCl, 1.5 mM KH_2PO_4.
3. 8% Sucrose in 0.1× PBS (w/w) (see Note 1).
4. 8% Sucrose in ddH_2O (w/w).
5. 50 × PE: 500 mM PIPES, pH to 7.2 with KOH; 50 mM EDTA.
6. Ficoll cushion: 20% Ficoll (Sigma F8016) (w/w) in 1× PE, 0.1% NP-40 (Sigma NP40S); stir at RT for several hours to dissolve; before use, add 8 mM 2-mercaptoethanol.
7. Lysis buffer: 1 mM Tris, pH to 8.0 with HCl; 0.5% NP-40; before use, add 8 mM 2-mercaptoethanol and protease inhibitors (SigmaFAST™ Protease Inhibitor Cocktail Tablet; Sigma S8830).
8. Standard solution to calibrate the refractometer: 10% Ficoll in 1× PE, 0.1% NP-40.

Sucrose density-gradient solutions.

9. 40% Sucrose in 1× PE, 0.1% NP-40 (w/w).
10. 50% Sucrose in 1× PE, 0.1% NP-40 (w/w).
11. 70% Sucrose in 1× PE, 0.1% NP-40 (w/w) (see Note 2).
12. BECKMAN Preparative Ultracentrifuge Optima LE-80K.
13. JS-13.1, Swinging Bucket Rotor (Beckman Coulter).
14. SW 28 Ti, Swinging Bucket Rotor (Beckman Coulter).
15. Heraeus Megafuge 1.0R.
16. COREX® II Centrifuge Tubes (30 ml).
17. Tubes, Thinwall, Ultra-Clear™, 38.5 ml (Beckman #344058).
18. Pasteur pipette.
19. Refractometer.
20. Rotating platform.
21. Vacuum trap.

2.3. Centrosome Analysis

2.3.1. Western Blot Analysis

1. Mouse anti-γ-tubulin monoclonal antibody (SIGMA T6557).
2. Standard materials, solutions, and equipment for SDS-polyacrylamide gel electrophoresis (PAGE) and Western blotting.

2.3.2. Structural and Functional Analysis

1. Labeled tubulin (see Note 3).
2. Mounting medium.
3. Mouse anti-γ-tubulin monoclonal antibody (Sigma T6557) or any other antibody of interest.
4. $NaBH_4$ pellets, Cobalt (Doped) (Sigma-Aldrich #247677).
5. Secondary antibody labeled with a fluorescent dye, e.g., Alexa Fluor® 594 donkey anti-mouse IgG (Sigma A21203).
6. Tubulin (see Note 3).
7. 1× PE: 100 mM PIPES, pH to 7.2 with KOH; 10 mM EDTA.
8. 5× BRB80: 400 mM PIPES, pH to 6.8 with KOH; 5 mM $MgCl_2$, 5 mM EGTA.
9. Cushion: 1× BRB80, 25% glycerol, 1 mM GTP.
10. Fix solution: 1× BRB80, 10% glycerol, 0.1% TX-100, 0.25% glutaraldehyde (25% in H_2O, EM grade; Sigma G5882), 1 mM GTP.
11. PBSTS: 1× PBS, 0.1% TX-100, 10% FBS.
12. *Xenopus* egg extract (see Note 4).
13. Coverslips (∅ 12 mm) (see Note 5).
14. Flat-bottom PP Tubes (Greiner bio-one GmbH #863504).
15. Heraeus Megafuge 1.0R.
16. Microscope slides (see Note 5).
17. Needle (18 G×1.5 In.).

2.4. Measuring Microtubule Dynamics

1. Labeled tubulin (see Note 3).
2. Antifading mix: 5 μl of catalase solution (10 mg/ml in 1× BRB80, 50% glycerol; store at −20°C), 5 μl of glucose oxidase solution (10 mg/ml in 1× BRB80, 50% glycerol; store at −20°C), 5 μl of 1 M glucose (1 M in ddH_2O; store at −20°C).
3. Saturated hemoglobin solution: 30–40 mg/ml in ddH_2O; spin and pipette off the supernatant. Prepare freshly.
4. *Xenopus* egg extract (see Note 6).
5. Coverslips (22×22 mm) (see Note 5).
6. Image Analysis Software.

7. Inverted fluorescence microscope (e.g., Axiovert 200; Carl Zeiss, Inc.) equipped with a
   (a) 100× Plan Apochromat NA 1.4 oil immersion objective lens.
   (b) CCD camera (e.g., CoolSNAP; Roper Scientific).
   (c) Long-pass rhodamine filter.
8. Microscope slides (see Note 5).

3. Methods

3.1. Cell Culture

Expand the CHO cells to 15 24.5 × 24.5-cm plates, which should be near confluency on the day of use. On the day of centrosome isolation, exchange CHO growth medium with medium containing Nocodazole and Cytochalasin B (final concentrations of 10 and 5 μg/ml, respectively). Return the plates to the 37°C incubator for 90 min.

3.2. Isolation of Centrosomes

All solutions and equipment should be at 4°C and all work should be done at 4°C.

1. Before starting, remember to add 2-mercaptoethanol to the 1 mM Tris washing solution and 2-mercaptoethanol and protease inhibitors to the lysis buffer. Just before pouring the gradient, add 28 μl 2-mercaptoethanol per 50 g sucrose solution.
2. Prepare the discontinuous sucrose gradient by progressively layering less-dense sucrose solutions upon one another. Therefore, pipette 5 ml 70%, 3 ml 50%, and 3 ml of 40% sucrose solution in a Thinwall, Ultra-Clear Beckman tube (max volume 38.5 ml).
3. Bring one cell culture dish at a time to the cold room, aspirate the medium, and successively wash the cells as fast as possible with approximately 60 ml per plate of
   (a) 1× PBS
   (b) 8% Sucrose in 0.1× PBS
   (c) 8% Sucrose in ddH_2O
   (d) 1 mM Tris, pH 8.0, 8 mM ß-ME
4. Finally, add 30 ml lysis buffer and shake each plate on a rotating platform for at least 20 min.
5. Combine the cell lysates from all plates, add 1/50 volume of 50× PE, and spin in 30 ml COREX tubes in a JS-13.1 rotor at 3,000 rpm (1,500 × *g*) for 3 min at 4°C.
6. Prepare fresh COREX tubes with a 2 ml Ficoll cushion. Collect the supernatants after the first spin and carefully apply

onto the Ficoll cushions. Alternatively, the supernatants can be underlaid with the Ficoll cushion using a syringe.

7. Spin at 12,700 rpm in a JS-13.1 rotor (25,000 × *g*) for 15 min at 4°C, no brake.
8. Aspirate the supernatant until you reach approximately 2 ml above the Ficoll cushion. Carefully collect the final 2 ml above the Ficoll cushion with a Pasteur pipette. The centrosomes should be concentrated in this layer. Pool all interfaces.
9. To get a reasonable density of centrosomes, it is important to layer all centrosomes onto a single sucrose gradient; therefore, make sure to not exceed the total volume of the Ultra-Clear Beckman tube (max volume 38.5 ml, cushion 11 ml).
10. Before applying the centrosomes onto the sucrose density gradient, make sure that the Ficoll concentration is less than 10% by using a refractometer. If the Ficoll concentration is >10%, dilute with lysis buffer supplemented with 1× PE. This is essential to ensure that the centrosomes will layer onto the sucrose gradient.
11. Spin in an SW 28 Ti rotor at 27,000 rpm (130,000 × *g*) for 90 min at 4°C, no brake.
12. Fractionate the gradient manually from either the bottom (by puncturing the bottom of the tube) or the top or by using an automated fraction collector. Collect 0.5 ml fractions and determine the sucrose density by refractometry. Check the fractions between 40 and 60% sucrose for centrosomes by Western blot analysis (as described in Subheading 3.3.1).
13. Pool the centrosome-containing fractions (see Note 7) and snap-freeze 20-μl aliquots in liquid nitrogen. Store for long term at −80°C.

3.3. Centrosome Analysis

To determine in which fraction the centrosomes peak, take a 10 μl sample of each fraction in the 40–60% sucrose range and process it for SDS-PAGE and subsequent Western blot analysis.

3.3.1. Western Blot Analysis

In a NUPAGE 4–12% Bis-Tris gel (Invitrogen), γ-tubulin travels at 51 kDa (HiMark™ Pre-stained Protein Standard, Invitrogen). If you are interested in the enrichment throughout the isolation process, samples at different steps of the purification procedure can be taken and probed against γ-tubulin (see Fig. 1).

3.3.2. Structural and Functional Analysis

The centrosomes can be assayed functionally by an in vitro polymerization reaction or by incubating them in *Xenopus* egg extracts. In either case, the samples can be further processed for immunofluorescence to visualize a protein of interest.

1. Set up one of the following reactions:

   In vitro polymerization reaction

   (a) Tubulin (3 mg/ml).

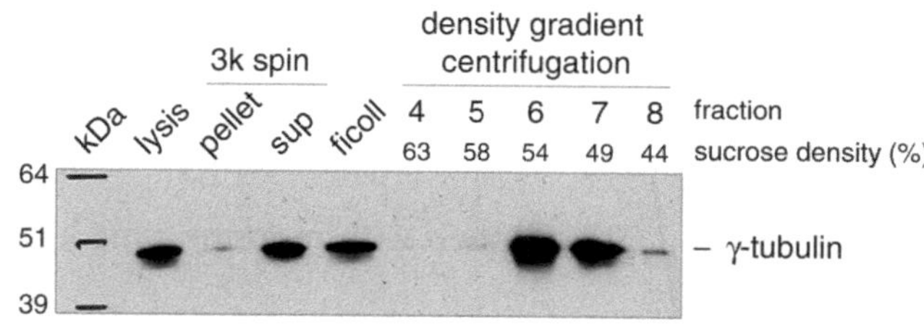

Fig. 1. Centrosomal isolation profile. Throughout the centrosome isolation procedure, samples were taken at the indicated steps and processed for Western blot analysis; 5 μl of each sample were separated on a 4–12% Bis-Tris gel and probed against γ-tubulin.

(b) Cy3-labeled tubulin (0.25 mg/ml at a 1.5 labeling ratio; see Note 3).

(c) GTP (1 mM).

(d) Centrosomes (1 μl of the peak fraction).

(e) 1× BRB80 (ad 20 μl).

(f) Incubate at 37°C for 20 min.

Xenopus egg extract reaction

(a) Cy3-labeled tubulin (0.25 mg/ml at a 1.5 labeling ratio).

(b) Centrosomes (1 μl of the peak fraction).

(c) *Xenopus* egg extract (ad 20 μl).

(d) Incubate at 18°C for 20 min.

2. Place a coverslip into a flat-bottom tube and overlay with 3 ml cushion (1× BRB80, 25% glycerol, 1 mM GTP).
3. Stop the in vitro/*Xenopus* egg extract reaction by adding 1 ml fix solution/20 μl reaction.
4. Spin 1 ml over the glycerol cushion onto a coverslip at 3,000 rpm (1,700 × *g*) for 15 min at 20°C in a clinical centrifuge.
5. Aspirate the solution and most of the cushion, and leave about 0.5 ml.
6. Pick up the coverslip by pushing with a needle from the bottom of the plastic tube.
7. Fix in cold methanol at −20°C for 5 min.
8. Quench 2× with 0.1% $NaBH_4$ in 1× PBS for 10 min.
9. Wash 3× with 1× PBS.

   Process for immunofluorescence:
10. Block in PBSTS for 60 min at RT or overnight at 4°C.
11. Incubate with first antibody (e.g., anti-γ-tubulin diluted 1:1,000 in PBSTS) for 4 h at RT or overnight at 4°C.
12. Wash 3× with 1× PBS.

13. Incubate with second antibody (e.g., Alexa Fluor® 594 donkey anti-mouse IgG diluted 1:500 in PBSTS) for 60 min at RT in the dark.
14. Wash 3× with 1× PBS.
15. Embed with mounting medium onto a microscope slide.

3.4. Measuring Microtubule Dynamics

Ideally, the microscope room or, if available, an incubation chamber at the microscope is set to 20°C.

3.4.1. Time-Lapse Microscopy

1. Set up the following reactions:
   (a) 10 μl CSF extract
   (b) 1 μl purified centrosomes
   (c) 0.25 mg/ml Cy3-labeled tubulin (1.5 labeling ratio)
   (d) 0.5 μl of saturated hemoglobin solution
   (e) 0.33 μl of antifading mix
2. Gently mix the reaction and keep at 20°C.
3. Squash 2.7 μl of the mixture under a 22 × 22-mm coverslip.
4. Locate a single centrosome (as shown in Fig. 2)
5. Determine the minimum exposure time.
6. Start recording time lapses, ideally in 2 s intervals for at least 1 min (it might be necessary to refocus while imaging) and compile the images into stacks.

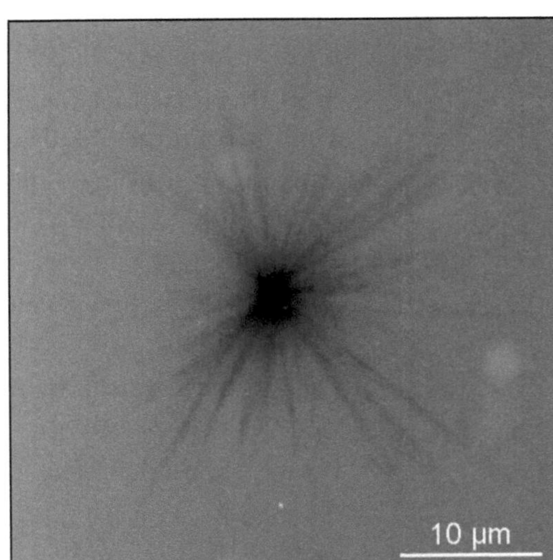

Fig. 2. Dynamic measurements of centrosomally nucleated microtubules in metaphase. CSF-arrested egg extracts were supplemented with isolated centrosomes and Cy3-labeled tubulin. Still image of fluorescence recording from time-lapse video; note that the fluorescent image is shown as an inverted image making the (+) end tracking of microtubules more convenient. *Scale bar*: 10 μm.

3.4.2. Image Analysis

Use an appropriate imaging analysis software, e.g., Image J or MetaMorph, to track the microtubules. Track the position of the centrosome in the first frame of the stack and subsequently the (+) end of a given microtubule at least over 20 frames (i.e., 40 s). The centrosome is set as the origin of the microtubule while in each following frame, the (+) end's pixel location is noted. The distance between the two points measured, i.e., the centrosome and the microtubule (+) end, thus simply represents the microtubule length. Pixel location over time consequently represents microtubule growth, shrinkage, or pausing. Finally, for each microtubule tracked, you obtain a file with two columns containing (1) time in seconds and (2) length of microtubules in pixels (ideally converted into μm). These coordinates can be read by a Matlab macro (see Note 8), which will calculate microtubule dynamics that are characterized by four parameters: (1) rate of growth (μm/min); (2) rate of shrinkage (μm/min); (3) catastrophes defined as the frequency of transition from growth to shrinkage (min^{-1}); and (4) rescues defined as the frequency of transition from shrinkage to growth (min^{-1}). In addition, the pausing time is given. In order to get reasonable statistics, I recommend analyzing at least 60 microtubules for a minimum of 30 s each.

4. Notes

1. Some of the solutions in this protocol are w/w. This means for a 20% sucrose solution, you weigh out 20 g sucrose and then add buffer until the weight is 100 g.
2. 70% sucrose does not go into solution completely until it is heated.
3. Porcine tubulin can be ordered (Sigma T6954) or prepared as described in (13). Labeled tubulin can be ordered (Cytoskeleton, Inc. TL331M) or prepared as described in (14). A labeling ratio of 1.5 means ~1.5 fluorophores per tubulin dimer.
4. CSF-arrested *Xenopus* egg extract is prepared as described in (15). Extract prepared by this protocol is arrested in M-phase. If you are interested in interphase processes, an interphase egg extract can be prepared as described in (16).
5. Coverslips and microscope slides should be washed in ethanol, rinsed in ddH_2O, dried on a Whatman paper, and kept dust-free in a tissue culture dish.
6. Microtubule assembly from centrosomes can also be observed in BRB80 buffer in the presence of tubulin as described in (17).
7. To determine the centrosome concentration, spin the centrosomes onto coverslips (as described in Subheading 3.3.2.), assume a quantitative sedimentation, and count the number

of centrosomes in a chosen area. A good yield is 1×10^{16} centrosomes/ml.

8. The Matlab macro was developed by Francois Nedelec (EMBL, Heidelberg) and is available online (http://www.cytosim.org/).

Acknowledgments

I thank Alex Bird, Tim Noetzel, and Zoltan Maliga for technical advice and helpful comments on the manuscript and Andrea Zinke for technical assistance. I am grateful to Tony Hyman for giving me the opportunity to work in his lab and for excellent scientific guidance.

References

1. Boveri, T (1901) Über die Natur der Centrosomen. Jena. Z. Med. Naturw. 28, 1–220.
2. Nigg EA (2006) Origins and consequences of centrosome aberrations in human cancers. Int J Cancer. 119:2717–2.
3. Wigge PA, Jensen ON, Holmes S, Soues S, Mann M and Kilmartin JV (1998) Analysis of the Saccharomyces spindle pole by matrix assisted laser desorption/ionization (MALDI) mass spectrometry. J. Cell Biol. 141:967–977.
4. Palazzo RE, Vogel JM (1999) Isolation of centrosomes from Spisula solidissima oocytes. Methods Cell Biol. 61:35–56.
5. Thompson-Coffe C, Coffe G, Schatten H, Mazia D, Schatten G. (1996) Cold-treated centrosome: Isolation of centrosomes from mitotic sea urchin eggs, production of an anti-centrosomal antibody, and novel ultrastructural imaging. Cell Motility and the Cytoskeleton 33:197–207.
6. Moritz M and Alberts BM (1999) Isolation of centrosomes from Drosophila embryos. Methods Cell Biol. 61:1–12.
7. Wiese C, Mayers JR, and Albee AJ (2009) Analysis of Centrosome Function and Microtubule Dynamics by Time-Lapse Microscopy in *Xenopus* Egg Extracts. In: Gavin RH (ed) Cytoskeleton Methods and Protocols, Methods in Molecular Biology, vol. 586.
8. Mitchison TJ, Kirschner MW (1986) Isolation of mammalian centrosomes Methods Enzymol. 134:261–8.
9. Bornens M, Paintrand M, Berges J, Marty MC, Karsenti E (1987) Structural and chemical characterization of isolated centrosomes. Cell Motil Cytoskeleton. 8(3):238–49.
10. Moudjou, M. and Bornens, M. (1998) Method of centrosome isolation from cultured animal cells. In: Cells, JE (ed), Cell Biology, a Laboratory Handbook, Second Edition, Academic Press Inc., San Diego, Vol. 2, pp. 111–119.
11. Bornens M, Moudjou M. (1999) Studying the composition and function of centrosomes in vertebrates. Methods Cell Biol. 61:13–34.
12. Blackburn GR, Barrau MD and dewey WC (1978) Partial purification of centrosomes from Chinese hamster ovary cells. Experimental Cell Research 113:183–187
13. Castoldi M, Popov AV (2003) Purification of brain tubulin through two cycles of polymerization-depolymerization in a high-molarity buffer. Protein Expr Purif. 32(1):83–8.
14. Hyman A, Drechsel D, Kellogg D, Salser S, Sawin K, Steffen P, Wordeman L, Mitchison T (1991) Preparation of modified tubulins. Methods Enzymol. 196:478–85.
15. Murray AW (1991) Cell cycle extracts. In: Kay BK and Peng HB (eds) *Xenopus laevis*: Practical uses in cell and molecular biology. Academic Press Inc., San Diego, Methods Cell Biol. Vol. 36. pp. 581–605.
16. Powers M, Evans EK, Yang J, Kornbluth S. (2001) Preparation and use of interphase *Xenopus* egg extracts. Curr Protoc Cell Biol; Chapter 11:Unit 11.10.
17. Kinoshita K, Arnal I, Desai A, Drechsel DN, Hyman AA. (2001) Reconstitution of physiological microtubule dynamics using purified components. Science 294(5545):1340–3.

Chapter 9

Preparation of Dual-Color Polarity-Marked Fluorescent Microtubule Seeds

Miho Katsuki, Etsuko Muto, and Robert A. Cross

Abstract

Assaying microtubule dynamics in vitro requires stabilized nucleation centers, a method to immobilize individual microtubules onto a surface, and a specialized microscope to image the microtubule. Microtubules are polar structures with different dynamic properties at the plus and minus ends. However, the dynamics of the two ends can be modified by the addition of other proteins, such as microtubule plus-end-tracking proteins (+TIPs), so that it becomes impossible to distinguish the microtubule polarity by measuring the differences in the dynamic properties of the ends alone.

In this chapter, we describe a method for labeling tubulin protein with *N*-hydroxysuccinimide ester fluorescent dyes, enabling the formation of dual-color polarity-marked stable microtubule seeds that can be immobilized onto a microscopic cover glass for imaging by fluorescence microscopy. These seeds create functional nucleation centers for the growth of dynamic microtubules.

Key words: *N*-hydroxysuccinimide ester, Succinimidyl ester, Tubulin, Microtubule, In vitro microtubule dynamics assay, Polarity-marked microtubules, Guanylyl-(alpha, beta)-methylene-diphosphonate

1. Introduction

In earlier works, fragments of axonemes were used as immobilized nucleation centers for microtubule formation, together with VE-DIC microscopy to image dynamic microtubules formed from unlabeled tubulin (1). However, for many experiments, the use of immobilized seeds composed of pure tubulin is preferable. Meanwhile, the development of total internal reflection fluorescence (TIRF) microscopy has enabled the visualization of fluorescent dynamic microtubules against a background of fluorescent tubulin in solution (2).

Anne Straube (ed.), *Microtubule Dynamics: Methods and Protocols*, Methods in Molecular Biology, vol. 777,
DOI 10.1007/978-1-61779-252-6_9,

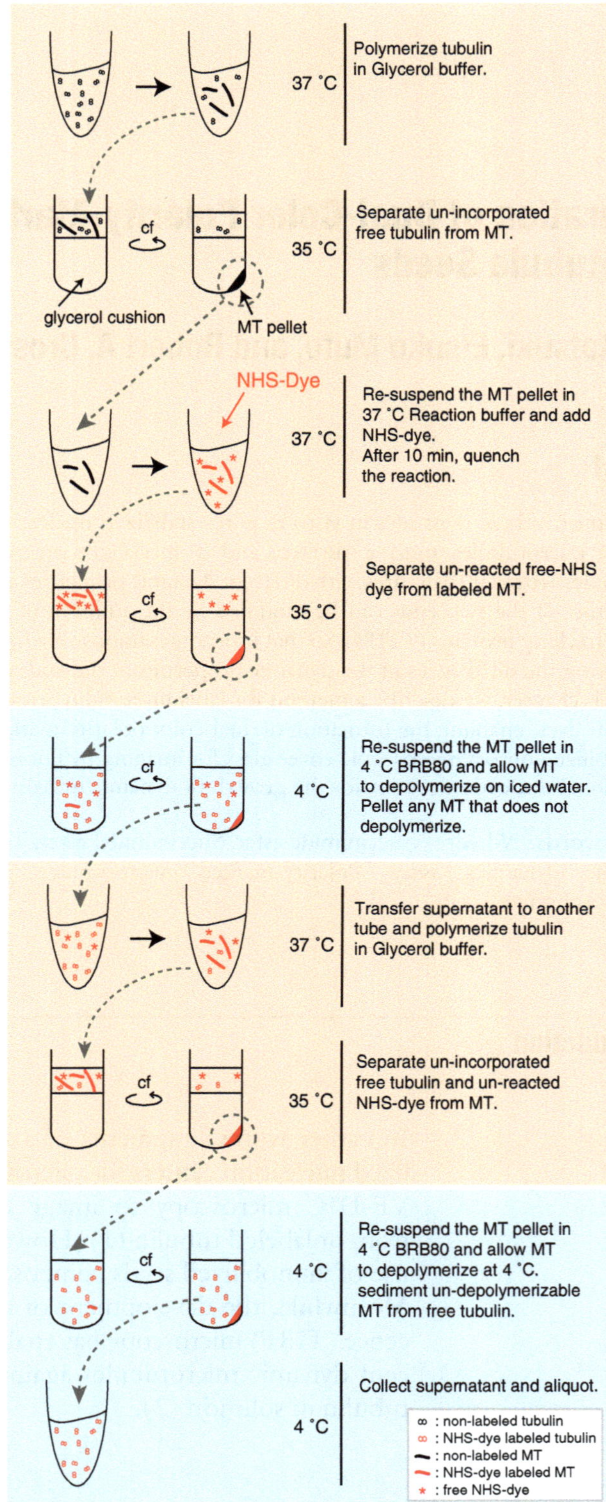

Fig. 1. Flowchart of the labeling procedure.

N-hydroxysuccinimide (NHS) ester reacts with accessible primary amine groups on proteins and links to them covalently. Tubulin can be chemically modified without any obvious loss of function using various NHS-conjugated chemicals, such as NHS fluorescent dyes or NHS biotin, which are commercially available. Tubulins modified by NHS-conjugated chemicals are also commercially available, although the variety is limited. In addition to published articles, many excellent protocols for the modification of tubulins are also available on the Web, such as the "Mitchison Lab protocols" (http://mitchison.med.harvard.edu/protocols.html).

This protocol (see Fig. 1 for overview) is based on the method of Hyman et al. (3). Using their method, we usually obtain tubulin with 50–100% of the tubulin dimers labeled. Before the labeling reaction, tubulin is polymerized in glycerol-containing MES or PIPES buffer at pH 6.8 to form microtubules (see Note 1); otherwise, the tubulin would lose the ability to polymerize as the dye may link to sites on the tubulin molecule that are critical for polymerization. Then, the microtubules are centrifuged, resuspended in pH 8.6 buffer, and labeled by mixing with NHS-conjugated dye. After the labeling reaction, two cycles of polymerization and depolymerization are conducted to select active tubulin with a recovery of about 40% of the initial tubulin. The labeled tubulin can be stored at −80°C (for up to 12 months) or in liquid nitrogen (indefinitely).

Our protocol for dual-colored polarity-marked stable microtubule seeds is based on the method of Howard and Hyman (4). A short core seed labeled with the first fluorescent dye is assembled with guanylyl-(alpha, beta)-methylene-diphosphonate (GMPCPP) replacing GTP. GMPCPP is only slowly hydrolyzed in buffers containing potassium ion, so it stabilizes the microtubule (5). Tubulin containing a second fluorescent dye is assembled on this core, again using GMPCPP to stabilize the elongated microtubule seed. Since the microtubule plus end assembles more rapidly than the minus end, seeds are formed with a longer extension of the second color of tubulin from the core seed at the plus end. Polarity can be tested using a microtubule motility assay and a motor protein, such as kinesin, of known directionality on microtubules. Using this protocol, 83% of microtubules have the longer extension of the second color at the microtubule plus end (6). The polarity-marked dual-color microtubule seeds are used on the same day they are made. Upon storage, the microtubules appear to anneal, causing loss of polarity marking.

2. Materials

2.1. Stock Solutions

1. 1 M MES buffer, pH 6.8 (Sigma M2933) or 1 M PIPES buffer, pH 6.9 (Melford B2004) (see Note 1), pH adjusted with KOH at 25°C, filtered, and then stored at 4°C. The water used in this protocol is ultrapure with a resistance of >18 MΩ. All solutions are filtered through a 0.22-μm pore size membrane (Millipore Express PLUS).
2. 0.3 M HEPES buffer, pH 8.6 (Sigma H2933), pH adjusted with KOH at 25°C, filtered, and then stored at 4°C.
3. 1 M $MgCl_2$ solution (Sigma M1028), store at 4°C.
4. 0.1 M EGTA solution, pH 6.8 (Sigma E4378), pH adjusted with KOH at 25°C, filtered, and then stored at 4°C.
5. 1 M K-glutamate (Sigma G1501). pH of the solution is usually about pH 7.1 without adjustment at 25°C. Filter and store at 4°C.
6. 0.1 M GTP, pH 7.0 (Jena Bioscience NU-1012), pH adjusted with KOH at 4°C, and then stored at –80°C in small aliquots. Each aliquot is used on the day it is thawed.
7. Dry dimethylsulfoxide (DMSO) (Sigma 27,685-5). DMSO is taken from the bottle just before use using a syringe and a needle.
8. 0.2 M GMPCPP solution (Jena Bioscience NU-405). You can request the form of compound to be supplied. We usually order it as 0.5 ml of 0.2 M solution. Store at –80°C in small aliquots.

2.2. Working Solutions

1. Glycerol buffer: 0.16 M MES, 12 mM $MgCl_2$, 1 mM EGTA, 1.6 mM GTP, 83% (v/v) glycerol (MP Biomedicals, 800689), pH 6.8 (see Note 2).
2. pH 8.6 glycerol cushion: 0.1 M HEPES, 1 mM $MgCl_2$, 1 mM EGTA, 1 mM GTP, 60% (v/v) glycerol, pH 8.6.
3. Reaction buffer: 0.1 M HEPES, 1 mM $MgCl_2$, 1 mM EGTA, 1 mM GTP, 40% (v/v) glycerol, pH 8.6.
4. Quenching buffer: 0.16 M MES, 1 mM $MgCl_2$, 1 mM EGTA, 1 mM GTP, 0.1 M K-Glutamate, 40% (v/v) glycerol, pH 6.8.
5. pH 6.8 glycerol cushion: 80 mM MES, 1 mM $MgCl_2$, 1 mM EGTA, 1 mM GTP, 60% (v/v) glycerol, pH 6.8.
6. BRB80: 80 mM MES, 1 mM $MgCl_2$, 1 mM EGTA, pH 6.8, 1 mM GTP, filtered.
7. 10–80 mM dye stock solution in dry DMSO (see Note 3).

2.3. Apparatus and Typical Reactive Dyes

1. TLA-120.2 and TLA-100.3 rotors (Beckman Coulter).
2. Optima MAX-XP benchtop ultracentrifuge (Beckman Coulter).
3. Fixed-angle rotor F-45-30-11 (Eppendorf).
4. Microcentrifuge 5417 R (Eppendorf).
5. Thickwall Polycarbonate 3.5-ml Tube (Beckman Coulter, 349622).
6. Thickwall Polycarbonate 1.0-ml Tube (Beckman Coulter, 343778).
7. Cy5 mono-Reactive Dye Pack (GE Healthcare, PA25001).
8. Cy5 NHS ester, 1 mg (GE Healthcare, PA15101).
9. Alexa Fluor® 488 carboxylic acid, succinimidyl ester "mixed isomers," 1 mg (Invitrogen, A-20000).
10. Alexa Fluor® 680 carboxylic acid, succinimidyl ester, 1 mg (Invitrogen, A-20008).
11. 5-(and-6)-Carboxytetramethylrhodamine, succinimidyl ester (5(6)-TAMRA, SE) "mixed isomers" (Invitrogen, C-1171).
12. 5-(and-6)-Carboxy-X-rhodamine, succinimidyl ester (5(6)-ROX, SE) "mixed isomers" (Invitrogen, C-1309).

3. Methods

3.1. Labeling Procedure

1. Thaw 1 ml of 15–25 mg/ml phosphocellulose-purified pig brain tubulin (see Note 1) in BRB80, add 0.6 ml of glycerol buffer, and mix gently by pipetting up and down using a blue (1,000 μl) pipette tip with the end cut off (as used on an air-displacement Gilson pipette). The solution will become turbid just after mixing because of microtubule polymerization. Layer the solution onto 1.3 ml of prewarmed (37°C) pH 8.6 glycerol cushion in a 3.5-ml ultracentrifuge tube. Seal the tube top with Parafilm, and then incubate at 37°C for 30 min to polymerize the microtubules.
2. Centrifuge polymerized microtubules at 55,000 rpm (128,405 ×*g*) and 35°C for 30 min in a TLA-100.3 rotor (see Note 4).
3. Wash the pellet of polymerized microtubules with prewarmed reaction buffer in the following two steps to avoid any carry-over of glycerol buffer. (1) Aspirate the supernatant above the glycerol cushion first, and then wash the tube wall and the interface between the supernatant and cushion by adding and removing 500 μl of reaction buffer. (2) Aspirate the cushion

layer carefully, and then rinse the tube wall and pellet surface by adding and removing 200 μl of warm reaction buffer (37°C), taking care to avoid any perturbation of the pellet.

4. Resuspend the microtubules with a small volume of prewarmed reaction buffer (pH 8.6, 37°C) so as to make the final tubulin concentration ~100 mg/ml (see Note 5). Using a cutoff pipette tip, pipette the microtubule solution until the pellet is broken into small pieces, followed by gently vortexing until no particles of the pellet remain in the tube. Since the microtubules are not very stable at pH 8.6, the temperature of the solution has to be maintained at 37°C until step 6.
5. Rapidly add the dye stock solution (10–80 mM in dry DMSO) in a volume 1/20–1/10 of the volume of the microtubule solution to give a 1:1–4:1 molar ratio of dye to tubulin heterodimer (see Note 6). Immediately after mixing, vortex the solution vigorously. Incubate at 37°C for 10 min and vortex every 2 min during incubation.
6. To stop the reaction, add 2 volumes of warm Quenching buffer to the reaction solution, mix well by vortexing, and layer the mixed solution onto 0.4 ml of warm pH 6.8 glycerol cushion in a 1-ml ultracentrifuge tube.
7. Sediment the labeled microtubules at 70,000 rpm (173,970 × *g*) for 30 min at 35°C in a TLA-120.2 rotor.
8. Wash the pellet of labeled microtubules with prewarmed (37°C) BRB80 using the two-step method described in step 3 to avoid any carryover of free dye and tubulin dimer.
9. Add 200 μl of BRB80 (4°C), break up the microtubule pellet by pipetting up and down using a cutoff blue (1,000 μl) pipette tip, and leave on ice for 5 min. Resuspend the pellet completely by pipetting intermittently, with the tube kept in iced water for another 10 min.
10. Centrifuge at 70,000 rpm (173,970 × *g*) for 15 min at 4°C in a TLA-120.2 rotor.
11. Add 0.5 volumes of glycerol buffer to the supernatant, mix the solution well with a cutoff pipette tip by pipetting up and down, and incubate at 37°C for 30 min. Layer the mixture onto 0.4 ml of warm pH 6.8 glycerol cushion in a 1-ml ultracentrifuge tube and centrifuge at 70,000 rpm (173,970 × *g*) and 35°C for 30 min in a TLA-120.2 rotor.
12. Repeat steps 8–10.
13. Collect the supernatant, dispense aliquots of 3–10 μl into ice-cooled tubes, and freeze in liquid nitrogen. The tubulin concentration is typically 15–25 mg/ml. Using several aliquots, the tubulin and dye concentrations are measured and the stoichiometry between dye and tubulin is calculated (see Note 7).

3.2. Dual-Color Polarity-Marked GMPCPP Microtubule Seeds

1. Prepare 30 μl of the seed-core mix, containing 30 μM tubulin labeled by the first color in BRB80 (see Note 8) and 1 mM GMPCPP (instead of 1 mM GTP), and leave on ice for 10 min to exchange GTP/GDP on ß-tubulin to GMPCPP and then incubate the seed-core mix at 37°C for 1 h (Fig. 2a).
2. 15 min before the end of the 1 h incubation at 37°C in step 1, prepare 100 μl of the elongation mix, containing 11 μM of the second colored tubulin, 0.5 mM GMPCPP, and 1 mM DTT in BRB80 (omit GTP), and incubate on ice for 10 min.
3. Centrifuge the elongation mix at 14,000 rpm (20,800 × *g*) for 5 min at 4°C in a 1.5-ml microcentrifuge tube in a F45-30-11 rotor using an Eppendorf microcentrifuge 5417R.

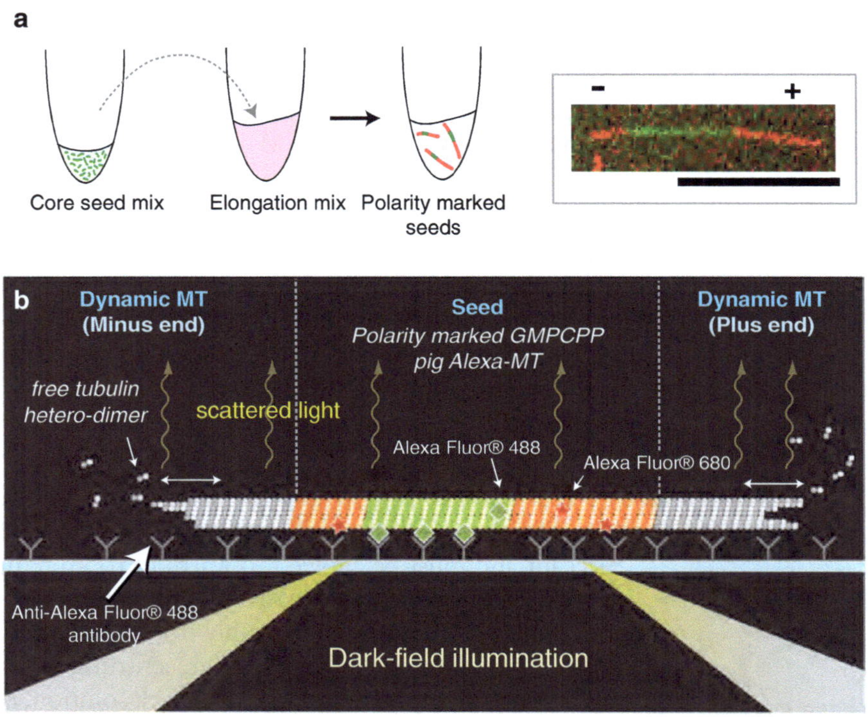

Fig. 2. (**a**) Schematic diagram of protocol for dual-color polarity-marked GMPCPP microtubule seeds. First-color microtubules are indicated in green and second-color microtubules are shown in red. Inset, fluorescence images of dual-color polarity-marked GMPCPP microtubule seed observed by fluorescence microscopy are shown. *Green*: Alexa 488; *Red*: Alexa 680. Polarity is indicated by "+" or "–". Plus end of the seed has a longer Alexa 680-labeled region. Scale bar, 10 μm. (**b**) Schematic diagram of an example application for dual-color polarity-marked GMPCPP microtubule seeds in a microtubule dynamics assay. A GMPCPP-stabilized polarity-marked microtubule seed is assembled from Alexa 488- and Alexa 680-labeled pig brain tubulin. Only the center of the seed is attached to the surface by anti-Alexa 488 antibody. Dynamic nonfluorescently labeled microtubules grown from the seeds are observed by dark-field illumination microscopy. This figure was originally published in Katsuki, M., Drummond, D. R., Osei, M., and Cross, R. A. "Mal3 masks catastrophe events in *Schizosaccharomyces pombe* microtubules by inhibiting shrinkage and promoting rescue." *J. Biol. Chem.* 2009; 284: 29246–29250. © the American Society for Biochemistry and Molecular Biology.

4. Immediately transfer the supernatant to another ice-cold Eppendorf microcentrifuge tube.
5. Incubate the 100 μl of elongation mix for 20 s at 37°C by immersing the bottom half of the microcentrifuge tube in a water bath.
6. Immediately add 2.5 μl of seed-core mix from step 1 to the 100 μl of elongation mix and gently mix by pipetting up and down about five times with a cutoff yellow (200 μl) tip.
7. Incubate the microtubule mix for at least 1 h at 37°C, and then keep at 25°C during the experiment.

4. Notes

1. In this protocol, we use MES buffer instead of PIPES buffer to obtain a high proportion of microtubules with 13 protofilaments (7). PIPES buffers can also be used. We use MAPs-free pig brain tubulin purified by either phosphocellulose chromatography (8) or 0.5 M PIPES at pH 6.8 with 10% DMSO (9). We do not find a significant difference in tubulin yield and labeling efficiency between the two types of preparation.
2. All working solutions are pH readjusted with KOH when they are made up from stock solutions. All working solutions are frozen and stored at −20°C. When they are thawed, they need to be thoroughly mixed. GTP, GMPCPP, and K-Glutamate are always freshly added to the buffers from stock solutions just before use.
3. The activated dye ester loses activity upon exposure to moisture. The dye powder can be stored at 4°C or −20°C with desiccant and must be allowed to warm to room temperature before opening the package to avoid absorbing moisture. Dye stock solution is always made freshly from powder on the day of the experiment, usually during the centrifugation stage in step 2. During preparation of the dye stock solution, the room atmosphere must be dry, so we try to label tubulin on a sunny dry day with the air conditioner on and fluorescent lights off. The powder is first suspended in 15 μl of dry DMSO, and the concentration of the dye is calculated from the optical density at its absorbance maxima wavelength and the dye extinction coefficient. The dye stock is then adjusted to the desired concentration by dilution with dry DMSO. Prepared dye stock is stored at room temperature in a dark place, such as in a drawer, until it is needed. The dye-containing solutions, such as those used during the reaction, polymerization, and depolymerization steps, are always kept in dark by covering tubes in foil.

4. Ultracentrifuge rotors must always be prewarmed or precooled to the temperature of the next run. We normally put the rotor in a polyethylene bag and submerge it in either a 37°C water bath to warm it up or in iced water for cooling to 4°C.
5. If 1 ml of the original solution contains 15 mg/ml of tubulin, the tubulin concentration would be 9.375 mg/ml after adding 0.6 ml of glycerol buffer. In these conditions, the steady state concentration of nonpolymerized tubulin during polymerization is usually about 1 mg/ml, so we estimate the polymerized tubulin to be 8.375 mg/ml, giving a quantity of 13.4 mg of polymerized tubulin in 1.6 ml. To prepare 100 mg/ml of microtubules, the microtubule pellet must be suspended in a total volume of 134 μl. Since the pellet has a volume of ~120 μl, only a small volume (~40 μl) of reaction buffer is added.
6. Labeling efficiency depends on the molar ratio of dye to tubulin dimmer in the reaction mixture; using a molar ratio of 1:1 achieves a labeling stoichiometry of ~50% (i.e., on average, 50% of tubulin heterodimers contain 1 dye molecule), and a ratio of 4:1 achieves a labeling stoichiometry of ~100%. Usually, making the dye stock solution from one vial of Cy5 mono-Reactive Dye Pack (five vials; GE Healthcare, Product Code: PA25001) achieves a 50% labeling efficiency, and either five vials of dye pack or an entire 1-mg container of Alexa Fluor dye (Invitrogen, A-20000, A-20008) produces 100% labeling. The dye concentration of the stock solution is adjusted to 10–80 mM to give a 10–80-fold molar excess over the microtubule solution (for tubulin heterodimer, 100 mg/ml ≈ 1 mM). When 10 mM of dye stock is diluted to 1/10 with the microtubule solution in step 5, this produces a molar ratio of 1:1 of dye to tubulin heterodimers.
7. The absorbance at 280 nm is used to measure the tubulin protein concentration. However, most dyes also absorb at 280 nm; therefore, the reading is adjusted to compensate for the dye component by the following formula: $OD_{280} = OD_{280.measured} - (OD_{dye} \times \text{dye correction factor})$. OD_{dye} is the absorbance of the dye at the wavelength used to calculate the dye concentration. The dye correction factor is supplied by the manufacturer and is the ratio of the dye OD_{280} to OD_{dye}. The corrected OD_{280} is used to calculate the tubulin protein contribution using the extinction coefficient 105,838 M^{-1}/cm for pig brain tubulin dimer. The dye concentration is determined from optical density at the wavelength near its absorbance maximum, OD_{dye}, using its extinction coefficient (provided by the dye manufacturer). The $OD_{280.measured}$ and OD_{dye} are conveniently measured using an absorbance scan of the labeled sample. The molar ratio of dye to tubulin heterodimer can then be calculated.

8. The microtubule-labeling stoichiometry, which is determined by the percentage of dye-conjugated tubulin dimers in total tubulin dimers in the polymerization mix, is adjusted according to the application. We usually use 10% Alexa 488-labeled tubulin for the core part of the seed and 94% Alexa 680-labeled tubulin for polarity-marking extensions of seeds. In this case, the higher percentage is required because of the poor sensitivity of microscope detection system for the long wavelength dye. Anti-Alexa Fluor 488, rabbit IgG fraction (Invitrogen A-11094), is used to attach the seed core to the glass surface, leaving the Alexa 680 extensions untethered (Fig. 2b). Dye color combinations and brightness can be modified depending on experimental conditions, optical filter sets of the microscope, and camera sensitivity. We normally use ~2% labeling stoichiometry for a TIRF microscope and ~10% for wide-field fluorescence microscopy.

Acknowledgment

We thank Dr Douglas Drummond for his comments on the manuscript. We gratefully acknowledge the Association for International Cancer Research (09-0221), Cancer Research UK (C19638/A6211), and Marie Curie Cancer Care for supporting this work.

References

1. Walker, R. A., O'Brien, E. T., Pryer, N. K., Soboeiro, M. F., Voter, W. A., Erickson, H. P., and Salmon, E. D. (1988) Dynamic instability of individual microtubules analyzed by video light microscopy: rate constants and transition frequencies, *J. Cell Biol. 107*, 1437–1448.
2. Bieling, P., Laan, L., Schek, H., Munteanu, E. L., Sandblad, L., Dogterom, M., Brunner, D., and Surrey, T. (2007) Reconstitution of a microtubule plus-end tracking system in vitro, *Nature 450*, 1100–1105.
3. Hyman, A., Drechsel, D., Kellogg, D., Salser, S., Sawin, K., Steffen, P., Wordeman, L., and Mitchison, T. (1991) Preparation of modified tubulins, *Methods Enzymol. 196*, 478–485.
4. Howard, J., and Hyman, A. A. (1993) Preparation of marked microtubules for the assay of the polarity of microtubule-based motors by fluorescence microscopy, *Methods Cell Biol. 39*, 105–113.
5. Hyman, A. A., Salser, S., Drechsel, D. N., Unwin, N., and Mitchison, T. J. (1992) Role of GTP hydrolysis in microtubule dynamics: information from a slowly hydrolyzable analogue, GMPCPP, *Mol. Biol. Cell 3*, 1155–1167.
6. Katsuki, M., Drummond, D. R., Osei, M., and Cross, R. A. (2009) Mal3 masks catastrophe events in Schizosaccharomyces pombe microtubules by inhibiting shrinkage and promoting rescue, *J. Biol. Chem. 284*, 29246–29250.
7. Ray, S., Meyhofer, E., Milligan, R. A., and Howard, J. (1993) Kinesin follows the microtubule's protofilament axis, *J. Cell Biol. 121*, 1083–1093.
8. Weingarten, M. D., Suter, M. M., Littman, D. R., and Kirschner, M. W. (1974) Properties of the depolymerization products of microtubules from mammalian brain, *Biochemistry 13*, 5529–5537.
9. Himes, R. H., Burton, P. R., and Gaito, J. M. (1977) Dimethyl sulfoxide-induced self-assembly of tubulin lacking associated proteins, *J. Biol. Chem. 252*, 6222–6228.

Chapter 10

Reconstitution and Quantification of Dynamic Microtubule End Tracking In Vitro Using TIRF Microscopy

Ivo A. Telley, Peter Bieling, and Thomas Surrey

Abstract

Several microtubule-associated proteins localize in living cells selectively to an extended region at the growing microtubule plus ends. Over the last years, these plus-end-tracking proteins, also called +TIPs, have attracted considerable interest because they are involved in a large variety of essential intracellular processes. GFP-labeled versions of EB proteins are also often used as markers for intracellular microtubule organization and dynamics. The mechanism of selective +TIP binding to the end region of growing microtubule was unkown. Recently, the phenomenon of end tracking was reconstituted in vitro from purified proteins, which allowed the identification of EB proteins as the minimal core of the plus-end-tracking system and the dissection of the molecular mechanism of end tracking by these proteins. This in vitro reconstitution has started to be widely used for several +TIPs and promises to provide mechanistic insight into the functioning of the dynamic +TIP network at growing microtubule ends. Here, we describe the purification of EB1 and CLIP-170, the total internal reflection fluorescence microscopy assay to observe dynamic end tracking in vitro, and the quantitative analysis of fluorescent +TIP comet shape and of single +TIP molecule turnover at growing microtubule ends.

Key words: Microtubule end tracking, EB1, CLIP-170, +TIPs, TIRF microscopy, Single-molecule imaging

1. Introduction

Cytoplasmic linker protein, with molecular weight 170 kDa (CLIP-170), and end-binding protein 1 (EB1) have been the first proteins to be shown to track selectively growing microtubule ends (1, 2). In the meantime, a large collection of the so-called plus-end-tracking proteins (+TIPs) has been discovered (3). Until recently, the phenomenon of growing microtubule end tracking could only be observed in living cells transfected with GFP fusions

Anne Straube (ed.), *Microtubule Dynamics: Methods and Protocols*, Methods in Molecular Biology, vol. 777,
DOI 10.1007/978-1-61779-252-6_10, © Springer Science+Business Media, LLC 2011

of +TIPs which then give the visual appearance of fluorescent comets moving through the cell when observed by fluorescence microscopy (1–3). After these observations, an interesting debate started about various possibilities for the molecular mechanism underlying microtubule plus-end tracking (4, 5). The notion emerged that a protein network consisting of several such +TIPs might be necessary to give rise to end tracking (6). It became evident that an in vitro system was lacking in which plus-end tracking could be reconstituted (5) in order to dissect the functioning of the +TIP network. In contrast to loss-of-function studies or overexpression studies in living cells, such in vitro systems allow to perform experiments addressing in a directed manner the molecular mechanism underlying dynamic protein behavior. Importantly, the minimal set of activities required for a certain behavior can be unambiguously identified using such in vitro approaches.

The first in vitro reconstitution of plus-end tracking was achieved with fission yeast +TIPs (7). Total internal reflection fluorescence (TIRF) microscopy (8) was used to visualize the dynamic behavior of microtubules under conditions of dynamic instability and in the presence of purified +TIPs. These experiments demonstrated that the EB1 ortholog Mal3 was an autonomous end-tracking protein. Quantitative image analysis allowed to determine that Mal3 recognized a specific tubulin feature existing only at the end region of growing microtubules. It was the measurement of fast-binding/unbinding turnover in this end region of the growing microtubule by TIRF microscopy that led to the conclusion that Mal3 recognizes specific binding sites existing over an extended stretch at the growing microtubule end (7). Later, in vitro reconstitutions demonstrated that other purified EB proteins from different organisms also track growing ends autonomously (9–13), suggesting that this is a conserved property among all members of this protein family. In contrast to EB1, members of the CLIP-170 family were shown not to be autonomous end-tracking proteins in vitro (7, 9, 10). Instead, CLIP-170 depended on the presence of EB1 and also showed fast turnover at growing microtubule ends (9, 10). This agrees with results from careful in vivo analysis (14). The mechanism of end tracking of CLIP-170 family members is, however, not fully conserved between members of this protein family from different species. Whereas EB1 is sufficient for CLIP-170 to end track in vitro (9, 10), the fission yeast CLIP-170 ortholog Tip1 was shown to depend on both Mal3 and the kinesin Tea2 (7), a motor protein which does not exist in higher eukaryotes. These in vitro experiments contributed to solve a long-standing debate about the role of molecular motors for end tracking (15) and, most importantly, established the conserved nature of the end-tracking mechanism of EB proteins which is now regarded as the central core of the +TIP network.

We describe here the purification of the two most prominent +TIPs, full-length EB1 and full-length CLIP-170, and explain the assay which allows the observation of growing microtubule end tracking in vitro using these proteins. We emphasize the importance of high-quality purified protein not only to prevent artifacts, but also as a prerequisite for meaningful quantitative analysis of the recorded TIRF microscopy images. It is such analysis that finally allows conclusions about the molecular mechanism underlying the observed dynamic behavior of proteins. We, therefore, also summarize the most important steps of the data analysis of +TIP turnover and of comet shape.

2. Materials

2.1. Purification of EB1 and CLIP-170

2.1.1. Materials

1. Ni-Proteino resin (Machery-Nagel, 745200.30).
2. Empty chromatography flex columns (Kimble Kontes, 0.7 × 4 cm: 420401-0704, 1.0 × 10 cm: 420401-1010).
3. Minipuls 3 peristaltic pump (Gilson, F117800) with tubing.
4. FPLC System (GE Healthcare, 28-4062-64).
5. Superose 6 10/300 G (GE Healthcare, 17-5172-01).
6. HiLoad 16/60 Superdex 200 pg (GE Healthcare, 17-1069-01).
7. EmulsiFlex-C5 homogenizer (Avestin).
8. Millipore Steritop filters, 0.22 μm pore size (Millipore, SCGPU01RE).
9. Vivaspin 6 concentrators (Sartorius-Stedim, VS0601).
10. *Escherichia coli* BL21(DE3) CodonPlus-RIL.
11. LB agar plates containing 30 μg/ml kanamycin and 100 μg/ml chloramphenicol.
12. Bac-to-bac system (Invitrogen).
13. Sf9 insect cells.

2.1.2. Buffers and Solutions

1. *LB medium A* containing 30 μg/ml kanamycin, 100 μg/ml chloramphenicol, and 1% glucose.
2. *LB medium B* containing 30 μg/ml kanamycin and 100 μg/ml chloramphenicol.
3. *PBS buffer*: 10 mM NaP_i pH 7.4, 137 mM NaCl, and 3 mM KCl.
4. *Buffer A*: 50 mM KP_i pH 7.2, 400 mM NaCl, 2 mM $MgCl_2$, and 10 mM β-mercaptoethanol.
5. *Lysis buffer A*: Buffer A containing complete, EDTA-free protease inhibitors (Roche, 04693132001).
6. *Wash buffer A*: Buffer A containing 2.5 mM imidazole.

7. *Elution buffer A*: Buffer A containing 300 mM imidazole (see Note 1).
8. *Buffer B*: 50 mM KP_i pH 7.5, 500 mM NaCl, 2 mM $MgCl_2$, 10 mM β-mercaptoethanol, and 0.25% (w/w) Brij-35.
9. *Lysis buffer B*: Buffer B containing complete, EDTA-free protease inhibitors (Roche, 04693132001).
10. *Wash buffer B*: Buffer B containing 3 mM imidazole, 50 mM arginine, and 50 mM glutamate.
11. *Elution buffer B*: Buffer B containing 300 mM imidazole, 50 mM arginine, and 50 mM glutamate.
12. *Gel filtration buffer B*: Buffer B containing 50 mM arginine and 50 mM glutamate (see Note 1).

2.2. Total Internal Reflection Fluorescence Microscopy

2.2.1. Microscope

1. Motorized inverted microscope with two-channel TIR condenser (e.g., IX-81 and CMR-TIR-2L, Olympus).
2. TIR objective (e.g., PLApo 100× O-TIRFM, Olympus).
3. 488- and 561-nm solid-state lasers (e.g., 85BCD020 and 85YCA025, Melles Griot).
4. Two light guides (e.g., kineFLEX, Point Source).
5. Laser light attenuator (e.g., AOTF.nC-vis from AA Optoelectronics or simple neutral density filters).
6. Appropriate single-band filters and dichroic mirrors for GFP (e.g., LF488-A, Semrock) and Alexa-568 (e.g., LF561-A, Semrock) fluorescence imaging.
7. Cooled EM-CCD camera (e.g., Cascade II, Photometrics).
8. Incubation box around the microscope stage with temperature control.
9. *Optional:* High-speed emission filter wheel (e.g., HF-108, Prior Scientific), and dual-band dichroic mirror for GFP and Alexa-568 (e.g., FF505/606-Di01-25 × 36, Semrock).
10. *Optional:* Dual-view optics for synchronous two-color emission (e.g., OptoSplit II, Cairns).
11. Controller Software: MetaMorph (Molecular Devices) or μManager (supported by 100× Imaging).

The list describes essential components for TIR green and near-red fluorescence imaging. The optional components can increase the temporal resolution of time-lapse recording. A detailed description of our microscope can be found elsewhere (16).

2.2.2. Materials for Flow Chambers

1. Biotin-polyethylene glycol-functionalized glass (biotin-PEG glass) made from No. 1 coverslips, cut into ~8 × 10-mm pieces made according to (17) or from MicroSurfaces, Inc.
2. Poly-l-lysine (PLL)-PEG-passivated glass (PLL-PEG glass) made from either standard microscopy slides (72 × 26 mm) or

No. 1 coverslips (24 × 60 mm) by adding a drop of 2 mg/ml PLL-PEG onto the glass and letting it dry completely (17).

3. Tesa® double-sided tape (15 mm wide).
4. Whatman filter paper (GE Healthcare 1001 240).

2.2.3. Buffers and Solutions

1. *BRB80*: 80 mM PIPES adjusted to pH 6.8 with KOH, 1 mM $MgCl_2$, and 1 mM EGTA.
2. *Tubulin seed mix*: 15 μM unlabeled tubulin, 7.5 μM biotin-tubulin, 7.5 μM Alexa-568-labeled tubulin, and 0.5 mM guanosine-methyleno-triphosphate (GMP-CPP) (Jena Bioscience NU-405) in BRB80. For tubulin preparations, see Subheading 3.1.1.
3. *Assay buffer*: 80 mM PIPES (pH 6.8, KOH), 85 mM KCl, 85 mM KAcetate, 0.25% (w/v) Brij-35 (Pierce 20150), 4 mM $MgCl_2$, 1 mM EGTA, 1 mM GTP (Sigma G8877), 2 mM ATP (Roche 13906420) (if a molecular motor is used), and 10 mM β-mercaptoethanol (see Note 2).
4. *Block mix*: 1% Pluronic F-127 (Sigma P2443) and 50 μg/ml κ-casein (Sigma C04506) in assay buffer (freshly ultracentrifuged).
5. *NeutrAvidin mix*: 50 μg/ml NeutrAvidin (Invitrogen A2666) and 50 μg/ml κ-casein (Sigma C04506) in assay buffer (freshly ultracentrifuged).
6. *OS stock*: 30 mg/ml glucose oxidase and 6 mg/ml catalase in assay buffer (freshly ultracentrifuged at 80 krpm for 10 min in a TLA-100.2 rotor) (see Note 3).
7. *MC stock*: 2% (w/w) methylcellulose in MilliQ water (dissolve at 80°C and let cool down to 4°C).
8. *CAS stock*: 5 mg/ml β-casein (Sigma C6905) in assay buffer.
9. *Ensemble assay mix*: Assay buffer containing 15–66 μM "dimly" labeled tubulin (of which 10% is Alexa-568-labeled tubulin, see Subheading 3.1.1), 50 μg/ml β-casein (from CAS stock), 20 mM glucose, 320 μg/ml glucose oxidase, 55 μg/ml catalase (from OS Stock), and 0.1–0.25% methylcellulose (from MC stock).
10. *SingMol assay mix*: Assay buffer containing 35 μM "dimly" labeled tubulin (of which 10% is Alexa-568-labeled tubulin, see Subheading 3.1.1.), 50 μg/ml β-casein (from CAS Stock), 20 mM glucose, 320 μg/ml glucose oxidase, 55 μg/ml catalase (from OS Stock), and 0.1–0.25% methylcellulose (from MC Stock) (see Note 4).

2.3. Software for Data Analysis

1. Kalaimoscope (TransInsight GmbH) for particle tracking.
2. Matlab (The MathWorks) for nonlinear fitting procedures.
3. ImageJ (NIH) for basic image processing.

3. Methods

3.1. Proteins

3.1.1. Tubulin

Tubulin is purified according to a standard procedure from pig brain (18). Part of the purified tubulin is labeled with 6((biotinoyl)amino)hexanoic acid, succinimidyl ester (Invitrogen, B-1582), or Alexa-568 carboxylic acid, succinimidyl ester (Invitrogen, A20003) as described (19). The labeling ratio should be 0.3–0.8 fluorophores per tubulin dimer. Tubulin is flash frozen in small aliquots and stored in liquid nitrogen tanks.

3.1.2. EB1

1. Transform competent *E. coli* BL21(DE3) CodonPlus-RIL with pHAT2 carrying the full-length coding sequence of EB1 or EB1 fused to GFP (EB1-GFP) (10) and plate out on LB agar plates containing 30 μg/ml kanamycin and 100 μg/ml chloramphenicol. Incubate the plates overnight at 37°C.
2. In the morning, inoculate four small liquid cultures (5 ml LB medium A) in glass culture tubes with single bacterial colonies from the agar plates and incubate the cultures in an incubator shaking with 160 rpm at 37°C.
3. In the evening, dilute the two best growing cultures (as decided per visual inspection of the optical density) into 500 ml LB medium A prewarmed to 37°C, each in 3-l Erlenmeyer flasks. Then, incubate the 500 ml cultures in an incubator shaking with 160 rpm at 25°C overnight.
4. In the morning, dilute the two 500 ml cultures into 2.5 l LB medium B each and split the 3 l into three 5-l Erlenmeyer flasks. Then, incubate the cultures at 25°C shaking in an incubator and monitor the OD_{600}.
5. When an OD_{600} of 0.8 is reached, induce the expression of EB1 with 0.5 mM IPTG. Then, incubate for 8 h at 25°C.
6. Pellet the cells by centrifugation in a JLA 8.100 rotor at 4,000 × *g* for 20 min.
7. Discard the supernatant and resuspend the cell pellets in 35 ml ice-cold PBS buffer for each JLA 8.100 rotor bucket.
8. Pool the resuspended cells in an ice-cold JLA 16.125 rotor bucket.
9. Pellet the cells by centrifugation in a JLA 16.125 rotor at 4,000 × *g* for 20 min.
10. Discard the supernatant.
11. (Optional Step) Freeze the semidry cell pellet in liquid nitrogen and store at −80°C until lysis.
12. Resuspend the harvested cells in 70 ml ice-cold lysis buffer A and lyse them by passing them through an ice-cold high-pressure homogenizer twice.

13. Ultracentrifuge the lysate at 50,000 rpm (184,000×*g*) for 30 min at 4°C in a precooled Ti 70 rotor (Beckman).
14. Incubate clarified lysate (approximately 50 ml) for 1.5 h at 4°C with 1.5 g Ni-Proteino resin (Machery-Nagel). After binding of the protein, transfer the resin into an empty column.
15. Wash with at least 250 ml wash buffer A.
16. Elute protein with elution buffer A and pool the peak fractions.
17. Immediately pass the protein pool over a HiLoad Superdex 200 gel filtration column equilibrated in Buffer A.
18. Pool peak fractions, supplement the pool with glycerol to a final concentration of 20% (v/v), flash freeze 5-μl aliquots, and store in liquid nitrogen.

3.1.3. CLIP-170

1. Infect Sf9 cells with baculovirus derived from pFastBacHTa (using the Bac-to-Bac system (Invitrogen) following the manufacturer's instructions) carrying the complete coding sequence of CLIP-170 or GFP–CLIP-170 (10).
2. Grow 500 ml baculovirus-infected insect cells in a 3-l Erlenmeyer flask shaking with 188 rpm in an incubator at 27°C for 72 h after infection with virus.
3. Pellet the cells by centrifugation in a JLA-8.100 rotor at 1,000×*g* for 20 min.
4. Resuspend harvested cells in 50 ml ice-cold lysis buffer B.
5. Lyse cells by passing through an ice-cold high-pressure homogenizer twice.
6. Ultracentrifuge the lysate at 50,000 rpm (184,000×*g*) for 30 min at 4°C in a precooled Ti-70 rotor.
7. Incubate clarified lysates for 1.5 h at 4°C with 0.75 g Ni-Proteino resin. After binding of the protein, transfer the resin into an empty column.
8. Wash the column with at least 150 ml wash buffer B.
9. Elute protein with elution buffer B and pool the peak fractions.
10. Carefully concentrate the protein in a Vivaspin 6 concentrator in centrifugation steps of maximally 10 min at 4,000×*g* at 4°C until a concentration of 4 mg/ml is reached.
11. Ultracentrifuge the concentrated protein in a TLA-100.2 rotor at 80,000 rpm (228,000×*g*) for 15 min at 4°C.
12. Immediately pass the pool over a Superose 6 10/300 G gel filtration column equilibrated in gel filtration buffer B.
13. Pool peak fractions and keep on ice for up to 2 days or immediately freeze in liquid ethane.

3.2. TIRF End-Tracking Assay

3.2.1. Assembly of the Flow Chamber

Using a PLL-PEG-passivated glass slide, two strips of double-sticky tape, and a piece of biotin-PEG-functionalized glass, assemble a flow chamber as shown in Fig. 1a. A detailed description is found in (17).

3.2.2. Preparation of Biotinylated, GMP-CPP-Stabilized Microtubule Seeds

1. Prepare 40 μl of a Tubulin Seed Mix. Incubate at 37°C for 30 min.
2. Add 400 μl of prewarmed BRB80.
3. Spin in a tabletop centrifuge at 14,000 rpm for 8 min at room temperature, and thoroughly resuspend the pellet in 50 μl of prewarmed BRB80.

3.2.3. Filling the Flow Chamber

1. Flow in 50 μl Block Mix by using a pipette on one side of the flow chamber and absorbing the solution with filter paper on the other side. Incubate for 5 min at room temperature.
2. Wash the flow chamber with 2 × 50 μl assay buffer.
3. Flow in 50 μl NeutrAvidin Mix and incubate for 5 min at room temperature.
4. Wash with 2 × 50 μl assay buffer.

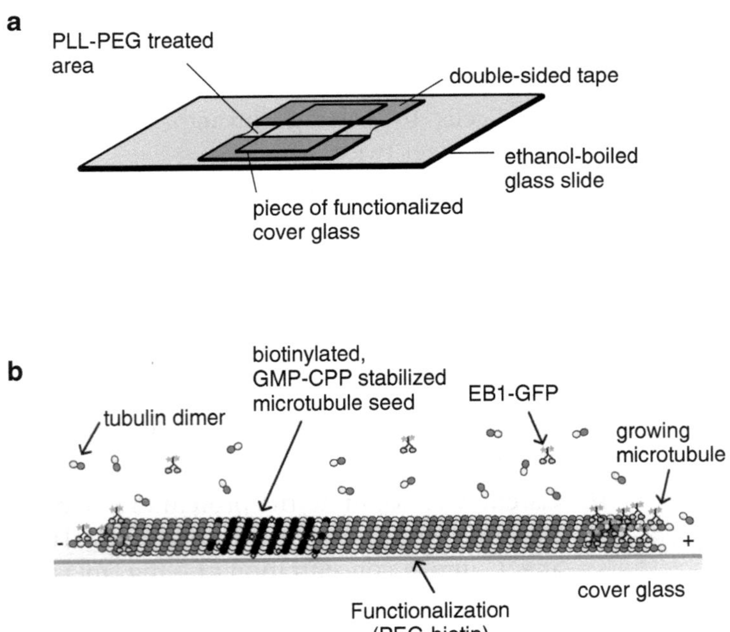

Fig. 1. Scheme of the flow cell and the setup. (**a**) Scheme of the flow chamber. Two strips of double-sided tape are placed at a distance of 5 mm between one ethanol vapor-cleaned and PLL-PEG-treated glass and one biotin-PEG-functionalized glass coverslip. (**b**) Cartoon of the microtubule end-tracking assay. A biotinylated and GMP-CPP-stabilized microtubule seed is immobilized via NeutrAvidin on functionalized glass and allowed to grow dynamically in the presence of free non-biotinylated tubulin and GTP in the solution. Imaging is performed with dual-color, time-lapse TIRF microscopy.

5. Dilute biotinylated microtubule seeds approximately 1:1,000 in assay buffer, wash the flow chamber with the microtubule dilution, and incubate for 5 min at room temperature (see Note 5).
6. Wash with 2 × 50 μl assay buffer.

7a. For ensemble imaging: Prepare 60 μl final sample mix on ice by diluting either EB1-GFP to a final concentration of 70–150 nM or unlabeled EB1 and GFP–CLIP-170 to final concentrations of 70–150 and 35–70 nM, respectively, in ensemble assay mix.

7b. For single-molecule imaging: Prepare 60 μl final sample mix on ice by diluting either unlabeled EB1 and EB1-GFP to a final concentration of 149 and 0.75 nM, respectively, or unlabeled EB1, unlabeled CLIP-170, and GFP–CLIP-170 to final concentrations of 150, 74, and 1 nM, respectively, in SingMol Assay Mix.

8. Flow in 50 μl of the final sample mix. Then, quickly seal the flow chamber with nail polish and start imaging.

3.2.4. TIRF Microscopy Imaging

Proceed to an imaging system consisting of an objective-based TIRF microscope equipped with a 60× or 100× 1.45 NA objective and a low-noise EM-CCD camera. Objective magnification and camera pixel size should be chosen such that the physical pixel size corresponds to 150–200 nm in object space. This allows spreading of the comet-shaped signal of +TIPs at growing microtubule ends over several pixels because the comet size is typically in the range of few micrometers. This guarantees sufficient spatial resolution for the digitally sampled comet signal which is important for robust quantitative analysis (see below). For our imaging system, an elongated comet-shaped +TIP accumulation can robustly be observed at microtubule growth velocities >2 μm/min. Lower microtubule growth velocities lead to shorter comets that cannot be detected reliably.

The following imaging options should be considered:

1. For ensemble imaging of +TIPs on microtubules growing with a velocity of 2–8 μm/min in vitro, acquire time-lapse movies at 3 s time intervals and 100 ms exposure time per frame for 5–6 min by switching the filter turret for separate green and near-red fluorescence. A time sequence and the corresponding kymograph (time-space plot) of GFP–CLIP-170 tracking selectively a growing microtubule end is shown in Fig. 2a. Good-quality protein is required for this experiment. Aggregating GFP–CLIP-170 causes artifacts, such as “backtracking”, as shown in Fig. 1b.
2. A high-speed emission filter wheel in combination with a dual-band dichroic mirror allows higher temporal resolution up to 2–3 frames per second depending on the controller.

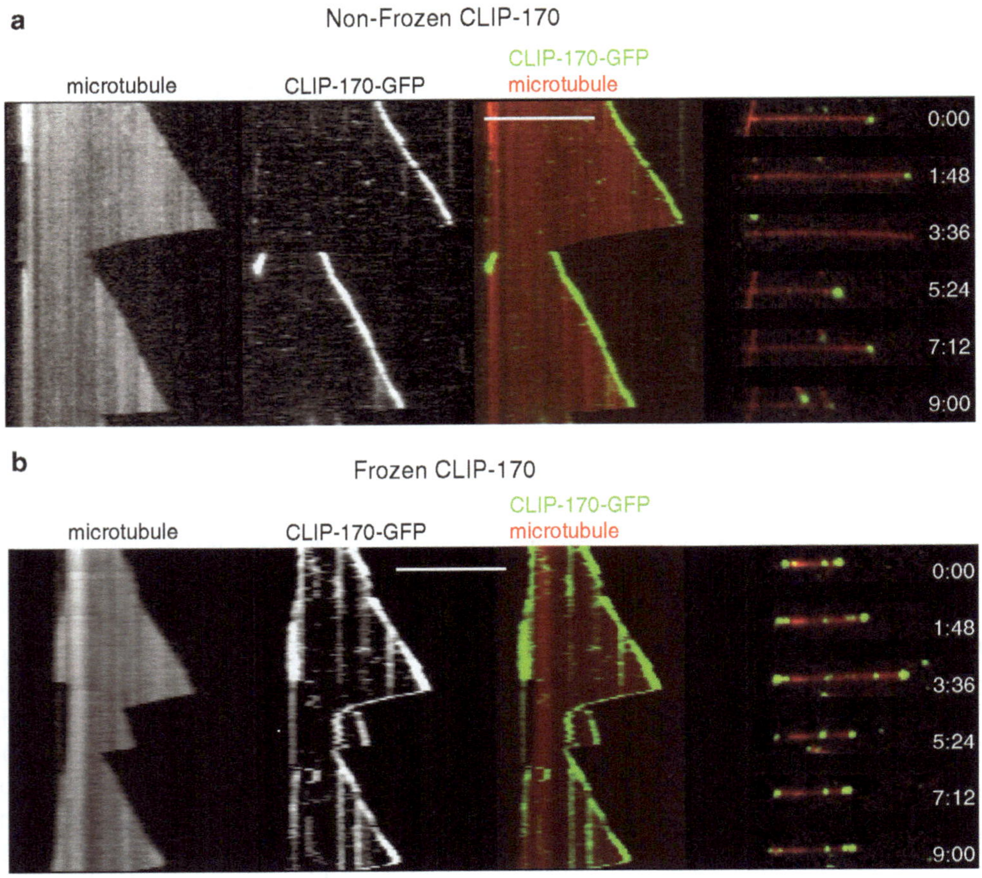

Fig. 2. Examples for good and bad CLIP-170 end tracking. Overlaid and single-channel TIRF microscopy images and corresponding time sequences of either nonfrozen (**a**) or flash-frozen (**b**) GFP-CLIP-170 (35 nM, *green*) in the presence of unlabeled EB1 (70 nM) interacting with dynamic Alexa 568-labeled microtubules (17.5 nM tubulin, *red*). The kymographs display a period of 9 min. Scale bars are 10 μm.

3. For one-color single molecule imaging start by taking an image of microtubules, then switching the filter turret to the GFP channel, and acquiring fast time-lapse images by continuously streaming for 1 min and finish by taking another image of the microtubules (Fig. 5a).
4. For synchronous two-color imaging, replace the high-speed emission filter wheel with a dual-view system, which splits and projects the two emission wavelengths onto one camera, in combination with the dual-band dichroic mirror.

3.2.5. Control Experiment

To test for protein solubility and potential aggregate formation in the assay buffer (see Note 2), fluorescently labeled +TIPs are imaged by TIRF microscopy between two fully passivated glass slides.

1. Assemble a flow chamber as described above, but using PLL-PEG-passivated glass for both the small cover glass and bigger slide.

2. Dilute the protein to a concentration of 0.2–2.0 nM in assay buffer and flow in 2 × 50 μl. Incubate for 3 min.
3. Wash the flow cell with assay buffer to remove unbound molecules.
4. Start imaging using the TIRF microscope as described above, but using the camera in continuous streaming mode (exposure time: 50–100 ms). Acquire movies at rates of 10–20 frames per second (see Note 6).

3.3. Data Analysis

3.3.1. Brightness Analysis

A key step for the in vitro reconstitution of microtubule end tracking is biochemical quality control of the protein in terms of oligomeric state and aggregation. Here, we describe a quantitative brightness analysis of fluorescent spots observed in the control experiment (see above and Fig. 3a), in which proteins in assay buffer are nonspecifically adhered to the passivated glass surface. The acquired microscopic image sequences (movies) are processed as follows:

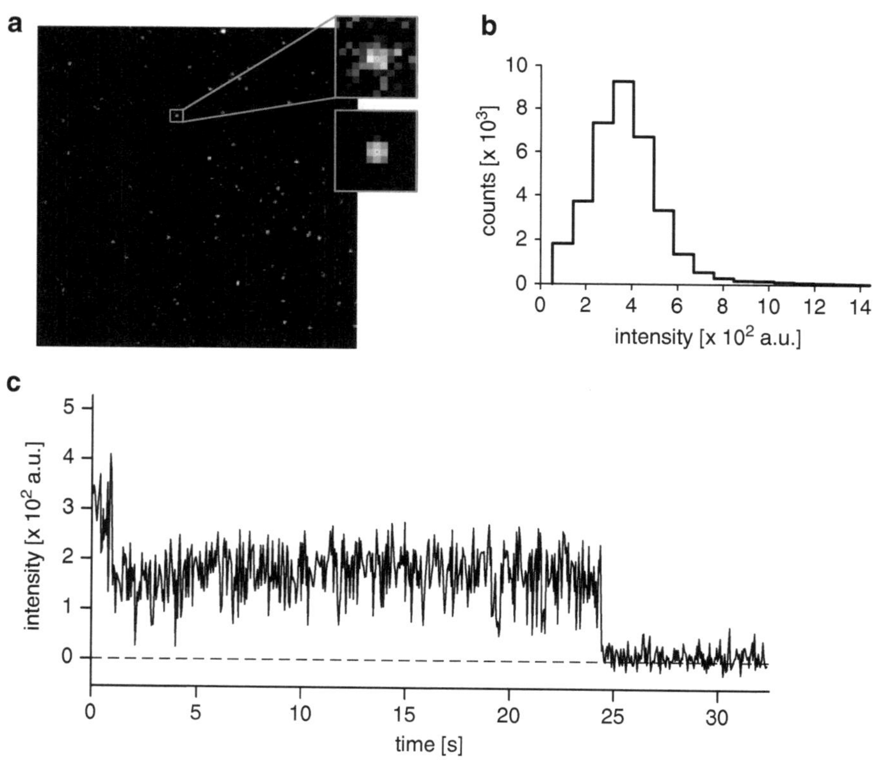

Fig. 3. Brightness analysis. (**a**) Overview image of a control experiment showing spot-like signals of GFP-CLIP-170-labeled protein at roughly even intensity, indicating absence of aggregates. Individual spots (*top inlet*) are fitted with a two-dimensional Gaussian or Lorentzian function (*bottom inlet*) to determine the background subtracted intensity. (**b**) Histogram of intensities with peak values at ~400 a.u., which corresponds to the fluorescence of two GFPs, and a small left tail most likely due to signal fusion (diffraction limit) and uneven illumination. (**c**) Two-step bleaching of individual spots with step amplitude of ~200, confirming the dimeric nature of the fluorescent protein.

1. For every frame, fit a 2-dimensional Gaussian or Lorentzian function to each spot-like signal (Fig. 3a) using an automated numerical procedure implemented in, e.g., "Kalaimoscope" (TransInsight GmbH) or the "object tracker" plug-in for ImageJ.

   This procedure generally involves the subtraction of the background signal and determination of the noise level using thresholding and/or statistical analysis (Poisson statistics, see also (20). The amplitude of fitted signals represents the brightness of spot signals.
2. Average the amplitude of each of the spots over 5–10 frames, and produce a histogram of these averaged amplitudes ("intensity distribution," Fig. 3b) (see Note 7).
3. A second criterion which can be used to test if fluorescently labeled +TIPs are single molecules or aggregates is "stepwise" bleaching of the GFP fluorophore. To this end, follow the brightness of individual spots over time and detect step-like intensity decreases with discrete amplitude (Fig. 3c). The step amplitude should be roughly half the value of peak intensity in the distribution (see Note 8).

3.3.2. Comet Shape Analysis

The fluorescence signal of +TIPs at growing microtubule ends is typically "comet"-shaped. From an intensity maximum at the microtubule end, the signal decays gradually along the microtubule (away from the growing end) (Fig 4a, b). It was observed that the length of the comet tail increases with the microtubule growth velocity (7, 10). Such a spatially decaying signal along the microtubule at a given time point is the consequence of a time-dependent decrease of the signal at a given location on the microtubule from its maximum value at the moment when this location is at the microtubule end (Fig. 4c). The older a specific location on the lattice is, the lower is the observed fluorescence signal (7). The first step toward the elucidation of the molecular mechanism underlying growing microtubule end tracking is the quantification of the comet shape, and thereby the quantification of the decrease of +TIP amount at a given microtubule position over time.

One-dimensional intensity profiles from line scans of individual comet tail signals along the microtubule show a roughly exponential decay in the presence of considerable noise which prevents robust fitting. To increase the signal-to-noise ratio, 50–100 such line scans can be averaged. To this end, individual comet signals have to be aligned with respect to the microtubule end which we assume to be represented by the location of the comet peak (Fig. 4c). Aligned line scans can then be used for averaging and/or exponential fitting (Fig. 4d). Finally, the spatial coordinate of the signal can be transformed into a time coordinate using the

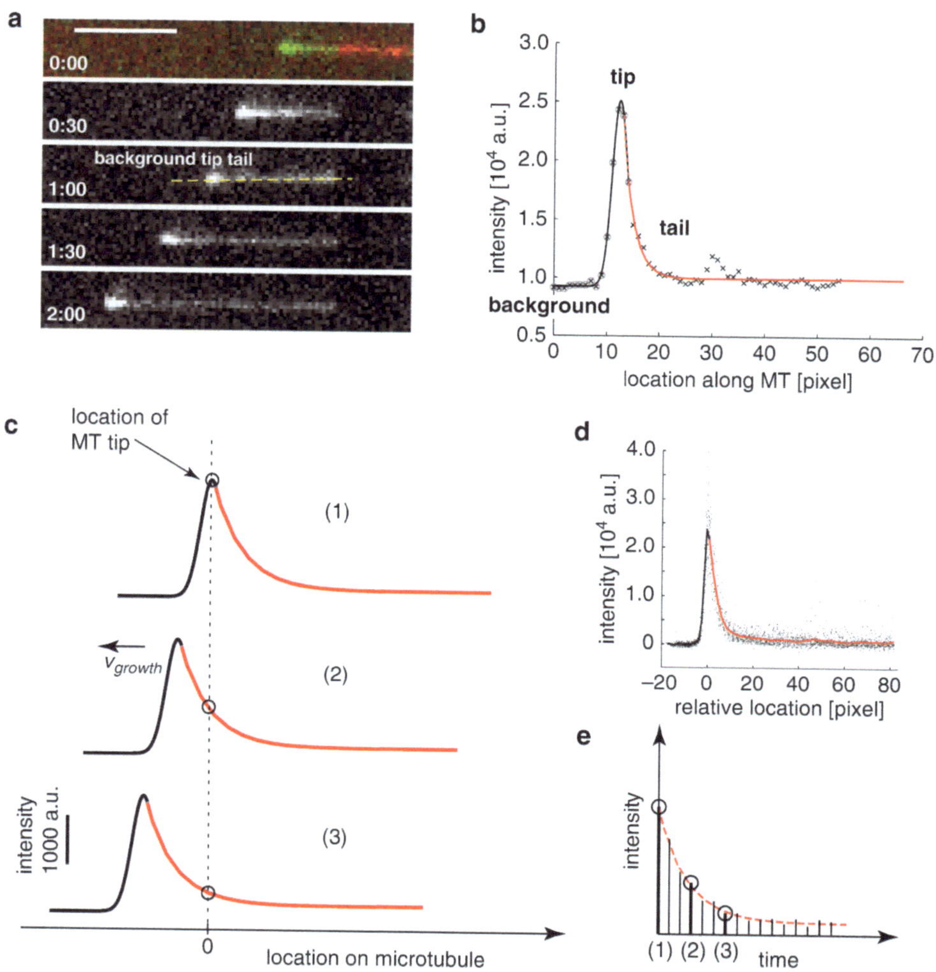

Fig. 4. Comet shape analysis. (**a**) Time sequence of TIRF microscopy images shown as overlay (*top*) of an Alexa-568-labeled microtubule (*red*) and the comet of end-tracking EB1-GFP (*green*), and the GFP signal alone in the lower gray-scale images. Bar is 5 μm, and time in min:sec. (**b**) Example of line scan providing a one-dimensional intensity profile along a microtubule as represented in A (*dashed line*) with overlaid fit, showing the basic properties of the comet. The data starts at background level (*left*) followed by a peak at the microtubule end (*circles, black curve*) and a long tail (*crosses, red curve*). (**c**) Scheme illustrating the basic idea for the interpretation of comet signals. While the microtubule is growing with v_{growth}, the intensity at the same location on the lattice that belonged to the tip initially decreases in an exponential manner to a base level slightly higher than background. Color code as in B. (**d**) Pooled data of several intensity profiles, and the overlaid curve representing the mean filtered data. (**e**) Mapping of intensity at the same site for the snapshots at three different time points as shown in C, illustrating the transformation of spatially localized data into temporal kinetic data.

growth velocity of the microtubule (Fig. 4e). This processed signal represents then the kinetic decay of the +TIP signal at a given site on the microtubule lattice.

Using the time-lapse movies acquired in the end-tracking assay, proceed as follows:

1. Generate line scans (profiles) of the GFP signal along microtubules (Fig. 4a, dashed line) with the spatial coordinate on

the x-axis and the intensity on the y-axis, either manually using ImageJ or by means of automated line detection (e.g., using Matlab) (see Note 9).

2. For each profile, find the location of the peak maximum value. Select the points that include the background, comet peak value, and two additional points from the comet tail (Fig. 4b, circles).

3. Fit a Gaussian function

$$g(x) = \mathrm{Bg} + A \cdot \exp\left(-\frac{(x - x_0)^2}{2\sigma^2}\right)$$

to this data selection (Fig. 4b, black) using nonlinear optimization on Matlab. Bg represents the background signal. Store the peak amplitude A for later calculations.

4. Shift the spatial coordinate by the intercept x_0, causing the x-axis origin to be located at the signal peak.

5. Overlay all shifted profiles to a point cloud. Note that the x-values are no longer integer pixel values. Calculate the mean intensity over all data points belonging to single-pixel-sized bins $[k, k+1]$ for $k = 0, 1, 2, \ldots$ on the x-axis ("mean filtering," see Fig. 4, outlined). Alternatively, "median filtering" can be applied in the same manner while being more robust against outliers.

6. Transform the spatial coordinate (x in integer units) into a temporal coordinate t by the formula

$$t = x \cdot \frac{S_{\mathrm{px}}}{v_{\mathrm{growth}}},$$

with S_{px} the pixel size in object space and v_{growth} the average microtubule growth speed.

7. Fit an exponential function

$$f(t) = I_{\mathrm{lattice}} + I_0 \cdot \exp\left(-\frac{t}{T}\right)$$

to the tail of the filtered data, starting at bin 1. The parameter T is the average decoration time which represents the time constant of the exponential signal decrease at a specific location on a growing microtubule (Fig. 4c, e).

8. The mean peak intensity can be calculated by averaging amplitudes A over all fitted profiles.

Different proposed molecular mechanisms for plus-end tracking make different predictions for the molecular process determining the decoration time (as extracted from the comet shapes and the microtubule growth velocity as described above). In the copolymerization mechanism, the time-dependent decrease of +TIP amount on the microtubule is a consequence of +TIP dissociation after +TIP loading together with tubulin to the very end of the growing microtubule. In the end recognition mechanism, this time-dependent decrease is a consequence of a transformation of high-affinity binding sites on tubulin, which has been freshly added to the microtubule end, to low-affinity binding sites on "aged" tubulin in the microtubule lattice distant from the growing end. In other words, according to the copolymerization mechanism, the decoration time is interpreted as the dissociation rate constant of +TIP from the microtubule, whereas it is interpreted as the lifetime of high-affinity binding sites in the microtubule end region according to the end recognition mechanism. Single-molecule imaging of +TIPs in the end region of the growing microtubule can test these predictions directly.

3.3.3. Analysis of Single +TIP Molecules at Growing Microtubule Ends

1. Select steadily growing microtubules based on visual inspection of the two microtubule images before and after the fast time-lapse +TIP imaging. Generate a kymograph of the GFP signal along the growing microtubule end.
2. Register the duration of individual binding events of single +TIP molecules (e.g., manually using ImageJ or any automatic tracking) within the tip region, i.e., from the distal end of the microtubule toward the seed with a distance determined by the comet shape profile (Fig. 5a, lines, ~1 μm). Binding events should be almost exclusively limited to this region, since +TIPs target the microtubule end region with high selectivity.
3. Plot a histogram of binding duration ("dwell time") and fit a single exponential function to determine the mean binding time (see Notes 10 and 11).

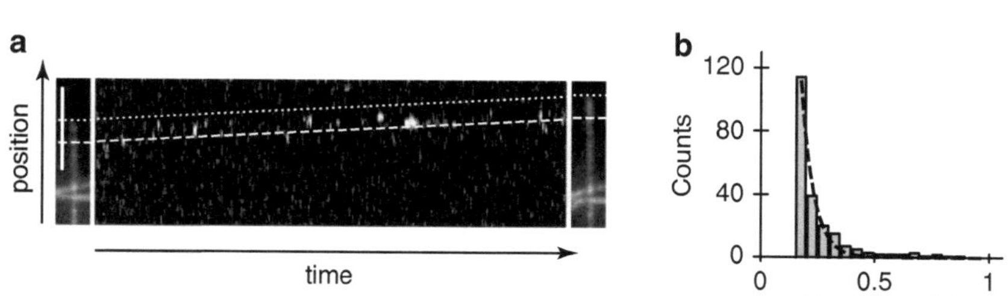

Fig. 5. Single-molecule analysis of +TIPs. (**a**) Kymograph of binding events of individual EB1-GFP molecules at the microtubule end region. Images of the growing microtubule before (*left*) and after (*right*) fast time-lapse imaging were used to identify where the moving end region is during fast time-lapse recording of the EB1-GFP signal (between *dotted* and *dashed lines*). (**b**) Histogram of dwell times of individual EB1-GFP molecules at the microtubule end with corresponding exponential fit (*dashed*). The mean dwell time from the fit is 0.055 s.

According to the copolymerization mechanism, the average dwell times of <0.25 s for both +TIPs discussed here suggest that the corresponding comet tail signal should have a length of <0.1 μm for a microtubule growth velocity of 3.3 μm/s (at 35 μM tubulin), which is not in agreement with the measurement of the typical length of comet tails under these conditions of ~0.6 μm. Instead, the measured fast single-molecule turnover at growing microtubule ends supports an end region recognition mechanism.

4. Conclusion

In vitro reconstitutions, such as the ones described above, are unique in being able to define the minimal composition of a system required to show a certain dynamic behavior. Comparison with in vivo data determines which aspects of the complex behavior observed in living cells have been reconstituted. Once the in vitro system is established, it is a powerful tool to dissect the functioning of the reconstituted system at the molecular level. Here, the biochemical quality of the purified proteins is critical. Protein aggregation can lead to artifacts, such as "backtracking" (Fig. 2b), i.e., tracking shrinking microtubule ends, because off rates can effectively be considerably reduced in aggregates. Furthermore, artificially increased dwell times hamper the analysis of molecular mechanisms, specifically for end-tracking proteins, the comparison between the ensemble decoration time and the single-molecule dwell time. Well-controlled in vitro experiments with microtubules exhibiting dynamic instability can, however, provide truly mechanistic insight into the different manners of how proteins interact with dynamic microtubule ends in order to perform their functions (7, 21).

5. Notes

1. Protein solubility should first be tested by ultracentrifugation of the protein in its own storage buffer. Protein aggregates of fluorescently labeled protein are also directly visible in the TIRF microscope, and the oligomeric state can be determined as described. If the protein forms aggregates in its own buffer, as indicated by the presence of a pellet after ultracentrifugation or the presence of heterogeneous, bright fluorescent spots in the TIRF microscope, the buffer needs to be optimized and the purification should be repeated using altered buffers. Once no aggregates are observed, one

should next test whether the protein aggregates upon flash freezing in liquid nitrogen or liquid ethane. Some proteins, like CLIP-170 for example, tend to form aggregates upon freezing in liquid nitrogen and sometimes to a lesser extent upon freezing in liquid ethane. Protein additives, such as mild detergents or arginine and/or glutamate, can prevent the formation of aggregates in some, but not all cases. For the case that the formation of aggregates upon freezing cannot be prevented, one needs to work with freshly gel-filtered, nonfrozen protein.

2. The most critical factor besides the quality of the proteins used (see above) is the choice of the assay buffer. The assay buffer has to fulfill two criteria: (1) it has to support microtubule growth and (2) all proteins (proteins of interest, tubulin, and oxygen scavengers) should be soluble over the full range of concentration used. It is important to iteratively explore a wide parameter space of buffer conditions and protein concentrations to gain a good overview of the range of conditions in which the protein is soluble and active. Otherwise, artifacts can be produced. For example, "backtracking" of aggregating CLIP-170 (Fig. 2b), somewhat reminiscent of "backtracking" of microspheres coated with multiple kinesins or microtubule-associated proteins (MAPs) (22, 23), is neither observed in vitro with nonaggregating CLIP-170 (Fig. 2a) nor in living cells (2).
3. The activity of the oxygen scavenger system is crucial. First, oxygen scavengers should be freshly prepared for best performance. In addition, enzymes from different suppliers can behave quite differently. The concentrations of glucose oxidase and catalase are optimized for the specific products used here and do not necessarily work for the same proteins from different suppliers. Furthermore, catalase at higher concentrations (>100 μg/ml) has a considerable capacity to block not only the surface, but also the interactions of motors and MAPs with microtubules.
4. The quality and purity of the water and containers need to be monitored. Taking care of the cleanliness of all water containers and final filtering of all buffers can improve the single-molecule imaging quality.
5. The microtubule density on the functionalized glass surface can be adjusted by varying the dilution of biotinylated microtubule seeds.
6. High-quality protein shows individual spot signals at similar brightness and uniform distribution over the surface (Fig. 3a). Large and very bright aggregates should not be observed. Ideally, each spot represents the signal of two GFPs of dimeric EB1 or CLIP-170.

7. In the brightness analysis, the intensity distribution should be monomodal and ideally a single Gaussian distribution. However, due to optical aberrations and nonuniform illumination settings of the microscope, a small (<5%) tail in the distribution may occur and is acceptable. Aggregates appear as larger and bright particles in the field of view and produce a broadened and often largely left-tailed or multimodal intensity distribution.
8. A reduction or lack of oxygen scavengers in the control experiment may significantly shorten the bleaching times and make it difficult to detect two steps in the step-bleaching analysis.
9. The line scans for the comet shape analysis should include some background signal (from solution) next to the peak signal of the comet (corresponding to the microtubule end) for sake of robustness and automation (Fig. 4a, b).
10. For the single-molecule analysis, it is important that the average bleaching time is much longer (>10×) than the average dwell time. The bleaching time can be determined as described (10). Otherwise, the mean dwell time needs to be corrected for bleaching.
11. When recording single-molecule behavior, a temporal detection limit defines the minimal duration of an event that can be observed. Nevertheless, a mean dwell time shorter than this detection limit can be determined by collecting a large number of events and fitting the resulting histogram which might contain data of only a part of the distribution. Refer to (16) for details.

Acknowledgments

We thank Stefanie Kandels-Lewis for generating the EB1 and CLIP-170 expression constructs and for protein expressions. We acknowledge financial support from the Deutsche Forschungsgemeinschaft (DFG), the STREP Active Biomics Network (the European Union), and the Swiss National Science Foundation (SNSF).

References

1. Mimori-Kiyosue, Y., Shiina, N., and Tsukita, S. (2000), The dynamic behavior of the APC-binding protein EB1 on the distal ends of microtubules. *Curr Biol* **10**: 865–8.
2. Perez, F., Diamantopoulos, G. S., *et al.* (1999), CLIP-170 highlights growing microtubule ends in vivo. *Cell* **96**: 517–27.
3. Akhmanova, A. and Steinmetz, M. O. (2008), Tracking the ends: a dynamic protein network controls the fate of microtubule tips. *Nat Rev Mol Cell Biol* **9**: 309–22.
4. Carvalho, P., Tirnauer, J. S., and Pellman, D. (2003), Surfing on microtubule ends. *Trends Cell Biol* **13**: 229–37.

5. Schuyler, S. C. and Pellman, D. (2001), Microtubule "plus-end-tracking proteins": The end is just the beginning. *Cell* **105**: 421–4.
6. Akhmanova, A. and Hoogenraad, C. C. (2005), Microtubule plus-end-tracking proteins: mechanisms and functions. *Curr Opin Cell Biol* **17**: 47–54.
7. Bieling, P., Laan, L., *et al.* (2007), Reconstitution of a microtubule plus-end tracking system in vitro. *Nature* **450**: 1100–5.
8. Axelrod, D. (2008), Chapter 7: Total internal reflection fluorescence microscopy. *Methods Cell Biol* **89**: 169–221.
9. Dixit, R., Barnett, B., *et al.* (2009), Microtubule plus-end tracking by CLIP-170 requires EB1. *Proc Natl Acad Sci USA* **106**: 492–7.
10. Bieling, P., Kandels-Lewis, S., *et al.* (2008), CLIP-170 tracks growing microtubule ends by dynamically recognizing composite EB1/tubulin-binding sites. *J Cell Biol* **183**: 1223–33.
11. Honnappa, S., Gouveia, S. M., *et al.* (2009), An EB1-binding motif acts as a microtubule tip localization signal. *Cell* **138**: 366–76.
12. Komarova, Y., De Groot, C. O., *et al.* (2009), Mammalian end binding proteins control persistent microtubule growth. *J Cell Biol* **184**: 691–706.
13. Zimniak, T., Stengl, K., *et al.* (2009), Phosphoregulation of the budding yeast EB1 homologue Bim1p by Aurora/Ipl1p. *J Cell Biol* **186**: 379–91.
14. Dragestein, K. A., van Cappellen, W. A., *et al.* (2008), Dynamic behavior of GFP-CLIP-170 reveals fast protein turnover on microtubule plus ends. *J Cell Biol.*
15. Maekawa, H. and Schiebel, E. (2004), CLIP-170 family members: a motor-driven ride to microtubule plus ends. *Dev Cell* **6**: 746–8.
16. Telley, I. A., Bieling, P., and Surrey, T. (2009), Obstacles on the microtubule reduce the processivity of Kinesin-1 in a minimal in vitro system and in cell extract. *Biophys J* **96**: 3341–53.
17. Bieling, P., Telley, I. A., *et al.* (2010), Fluorescence microscopy assays on chemically functionalized surfaces for quantitative imaging of microtubule, motor and +TIP dynamics. *Methods Cell Biol* **95**: 555–80.
18. Castoldi, M. and Popov, A. V. (2003), Purification of brain tubulin through two cycles of polymerization-depolymerization in a high-molarity buffer. *Protein Expr Purif* **32**: 83–8.
19. Hyman, A., Drechsel, D., *et al.* (1991), Preparation of modified tubulins. *Methods Enzymol* **196**: 478–85.
20. Helenius, J., Brouhard, G., *et al.* (2006), The depolymerizing kinesin MCAK uses lattice diffusion to rapidly target microtubule ends. *Nature* **441**: 115–9.
21. Brouhard, G. J., Stear, J. H., *et al.* (2008), XMAP215 is a processive microtubule polymerase. *Cell* **132**: 79–88.
22. Coue, M., Lombillo, V. A., and McIntosh, J. R. (1991), Microtubule depolymerization promotes particle and chromosome movement in vitro. *J Cell Biol* **112**: 1165–75.
23. Lombillo, V. A., Stewart, R. J., and McIntosh, J. R. (1995), Minus-end-directed motion of kinesin-coated microspheres driven by microtubule depolymerization. *Nature* **373**: 161–4.

Chapter 11

Force Generation by Dynamic Microtubules In Vitro

Svenja-Marei Kalisch, Liedewij Laan, and Marileen Dogterom

Abstract

Biopolymers are essential for cellular organization. They bridge the cell interior, forming a framework that is used as a reference for different cellular organelles. This framework, called the cytoskeleton, is not static but constantly reorganizes. The dynamics of the cytoskeleton allows the cell to rearrange its interior for various processes, such as cell division. This dynamic reorganization relies at least partly on forces that arise from the assembly and disassembly of cytoskeletal biopolymers. In many cases, these forces are generated when biopolymers interact with the cell boundary. This chapter focuses on force generation by and regulation of microtubules (MTs) that interact with growth-opposing barriers. We describe three in vitro assays that can be used to mimic MT interactions with the cell boundary. The essential components in each of our minimal systems are (functionalized) microfabricated barriers against which we grow MTs under different conditions. We describe in detail the different methods and assays necessary to realize these in vitro experiments.

Key words: Cytoskeleton, Microtubules, Microfabrication, Force, Glass barriers, Gold barriers, SU-8 barriers, Optical tweezers, Force feedback

1. Introduction

Microtubules (MTs) play an important role in cellular organization. Through interactions with the cellular cortex, pushing and/or pulling forces are generated that position various organelles in the cell (1–4) or physically segregate the chromosomes in dividing cells. In interphase *Schizosaccharomyces pombe* cells, motions of the nucleus are directly correlated with the dynamics of growing MTs interacting with the cell ends (5).

MTs are dynamic structures, possessing a property called "dynamic instability". (6, 7) It refers to the intrinsic ability of MTs to rapidly switch between a growing and a shrinking state,

Anne Straube (ed.), *Microtubule Dynamics: Methods and Protocols*, Methods in Molecular Biology, vol. 777,
DOI 10.1007/978-1-61779-252-6_11, © Springer Science+Business Media, LLC 2011

events that are termed catastrophes and rescues (8). Dynamic instability can be described by four intrinsic parameters: growth velocity, shrinkage velocity, catastrophe time (the average time an MT spends in the growing state until a catastrophe), and rescue time (the average time an MT spends in the shrinkage state until a rescue).

In vitro the parameters of dynamic instability can be regulated by temperature or tubulin concentration (9), but a cell cannot easily manipulate these parameters on a short timescale. Nevertheless, in cells, the dynamic instability parameters are quite variable and different from in vitro experiments (10). In recent years, it has become clear that cells exploit a large variety of proteins to regulate the parameters of dynamic instability both globally and locally to properly organize the MT cytoskeleton. Recently, many studies have been published which describe the regulation of MT dynamics by MT-associated proteins (MAPs) in vitro and in vivo (11–18). On the other hand, the cell can also physically regulate MT dynamics. Several in vitro studies using microfabricated barriers have directly shown that MT dynamics can be regulated while forces are generated (19–23). The next step would be to examine the combined effect of force and MAPs, globally and locally, on MT dynamics. First, we study the global effect of MAPs in an assay (Subheading 3.2.1), where MTs grow against glass barriers in the presence of MAPs freely floating in solution. Second, the local effect of proteins on MT dynamics is investigated by specifically binding MAPs to microfabricated barriers with which MTs interact (Subheading 3.2.2). Finally, to get a more complete picture of MT force generation and the consequent regulation of MT dynamics by force, including effects due to the presence of MAPs, we extended our assay with an optical trap (24, 25) as described in Subheading 3.3. However, our previous optical tweezers setup restricted us to the study of MT polymerization under increasing force conditions. To be able to exert constant forces, we extended it with a force-feedback algorithm (26–29). Besides, force-clamped optical tweezers allow us to decouple the influence of force on microtubule dynamics from the effect of MAPs.

2. Materials

Chemical reagents were obtained from Sigma and diluted in MRB80 buffer (see below) unless stated otherwise.

2.1. Microfabrication Materials

1. Chromosulfuric acid (Merck Eurolab).
2. S1813 (Microchem) is a positive near-UV photoresist. With a positive resist, the area of the photoresist that is exposed to

UV light becomes soluble to photoresist developer while the part of the photoresist that is unexposed remains insoluble to the developer. Store at room temperature.

3. SU-8 2005 (Microchem) is a negative-tone, epoxy-type, near-UV photoresist. With a negative resist, the area of the photoresist that is exposed to UV light becomes insoluble to photoresist developer while the part of the photoresist that is unexposed becomes soluble to the developer. Store at room temperature.
4. Gold, SiO, chromium (Kurt J. Lesker).
5. MF319 (Microchem) develops S1813 photoresist. Store at room temperature.
6. MRDEV600 (Shipley) develops SU-8 photo resist. Store at room temperature.
7. Gold etchant selectively dissolves gold and is compatible with both positive and negative photoresists.
8. Chromium etchant is a standard etchant for chromium and chromium oxide films. It is compatible with both positive and negative photoresists.
9. Hexamethyldisilazane (HMDS).
10. 200 mM 11-mercapto-1-undecanoic acid in ethanol: irritant to skin!
11. Pentafluorophenol (PFP) in ethanol.
12. 100 mM 1-ethyl-3-(3-dimethylaminopropyl)carbodiimide (EDAC) in ethanol (Molecular Probes): store at −20°C.
13. 0.1 mM biotin-PEO-LC-amine (Pierce Inc.): store at −20°C.
14. 0.1% Tween (Calbiochem) in MilliQ: store at room temperature.
15. 0.01% Triton X-100 (Plus One) in MilliQ: store at room temperature.
16. 0.2% agarose in MilliQ: store at room temperature.
17. 2 M NaCl in MilliQ: store at room temperature.
18. Ethanol, 2-propanol, acetone (Biosolve): store at room temperature.

2.2. Chemical Reagents for Assays

1. MRB80: 80 mM K-Pipes, 4 mM $MgCl_2$, 1 mM EGTA, pH = 6.8: store at 4°C.
2. 25 mg/ml (biotin-)BSA: store at −80°C.
3. 5 mg/ml κ-casein: store at −80°C.
4. 5 mg/ml streptavidin: store at −80°C.
5. 50 mM GTP: store at −80°C.

6. 50 mM ATP: store at −80°C.
7. 10 mM GMPCPP (Jena BioScience): store at −80°C.
8. 5 mg/ml rhodamine-/biotin- tubulin or 10 mg/ml unlabeled tubulin (Cytoskeleton, Denver): store at −80°C.
9. Oxygen scavenger system: 200 mM DTT (Roche), 10 mg/ml catalase, 20 mg/ml glucose oxidase, 2.5 M glucose: store at −80°C.
10. Tubulin mix: Variable tubulin/rhodamine tubulin concentration (12–20 μM), 1 mM GTP, 1 mM ATP, 0.8 mg/ml κ-casein, 0.1% methyl cellulose (4000cP), and 50× diluted oxygen scavenger system in MRB80.
11. 0.2 mg/ml PLL-PEG(-biotin) (SurfaceSolutions, Switzerland) in MilliQ: store at 4°C for a few weeks.
12. Methyl cellulose in MRB80: store at 4°C for a few weeks.

2.3. Biological Materials

1. MT seeds are polymerized from a tubulin mix in MRB80 containing biotin-labeled, rhodamine-labeled, and nonlabeled tubulin in a ratio of 2:1:10 (total of 50 μM tubulin) with 2 mM GMPCPP. The mix is incubated for 30 min at 36°C and can be kept at room temperature for a few days.
2. Centrosomes are purified from human lymphoblastic KE37 cell lines according to (30) with the generous help of Claude Celati. Store in MRB80 and 55% sucrose at −80°C.
3. Axonemes (a generous gift from Matt Footer) are purified from sea urchins according to (31) and stored following a method from (32), except that no glycerol is used. Store in aliquots at −80°C.
4. The dynein construct used in this study (a generous gift from Sam Reck-Peterson) is the biotinylated GST-dynein331 (VY268) construct (from now on, referred to as dynein). The purification methods of GST-dynein331 are described in (33). After the purification, the dynein is biotinylated with a halotag on the GST. It is tested in a gliding assay using a simple flow cell. The gliding speeds found are 130 ± 35(28) nm/s for dynein, similar to values previously described (33). Store at −80°C.

2.4. Laboratory Equipment

1. Evaporation chamber (home-built).
2. Spin-coater (Karl Suss Delta10).
3. Plasma-etcher (Oxford Plasma Plasmalab 80+ ICP).
4. Optical lithography system (Karl Suss MJB3) with Schott BG12 filter.
5. Optical tweezers setup: Due to its complexity, it is not listed here, see Subheading 3.3.

3. Methods

This chapter focuses on the fabrication procedure of three different structures that are developed to study the interaction of MTs with physical obstacles in living cells, such as organelles or the cortex. Every microfabricated structure has specific features designed for a specific experiment.

First, we use *glass barriers* to study the effect of force on MT dynamics, optionally combined with MAPs (34). MTs are grown from surface-attached stabilized MT seeds (Subheading 2.3) (19, 23) against the barrier (Fig. 1a and Subheading 3.1.1).

Second, *gold barriers* allow the attachment of proteins via thiol chemistry and the study of end-on interactions of dynamic MT ends with proteins, in our case, dynein. This type of experiments mimic MT–cortex or MT–kinetochore interactions (Fig. 1b and Subheadings 3.1.2–3.1.4).

Third, Fig. 1c (microfabrication described in Subheading 3.1.5) shows a cartoon of an ~7 μm-high *SU-8 barrier* that is used in optical trap experiments. In this experiment, MT dynamics, as well as MT-based pushing and pulling forces can be measured (24, 25, 36). MTs are grown from an axoneme attached to an optically trapped bead against an SU-8 photoresist barrier (37).

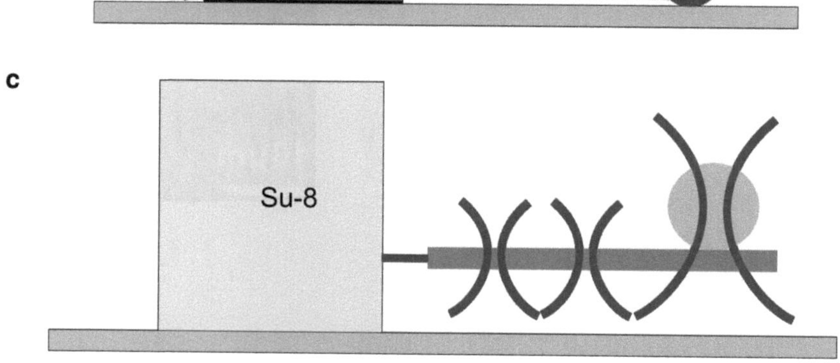

Fig. 1. Cartoons of assays exploiting microfabricated structures (side views). (**a**) MTs growing from stabilized MT seeds against glass (SiO) barriers. (**b**) MTs growing from centrosomes against gold barriers, consisting of a sandwich made of a thin chromium layer, a gold layer, and a thick chromium layer. (**c**) Axoneme–bead construct in an optical trap positioned against a barrier of SU-8 photoresist reproduced from (35), Copyright Elsevier 2010.

3.1. Fabrication Methods

Figure 2 presents glass barriers that are ~1.5 μm high and made of silicon monoxide (SiO).

3.1.1. Fabrication of Glass Barriers

1. Place coverslips in chromosulfuric acid for few hours. Rinse them 3× with MilliQ and blow-dry. Coat them with ~1.5 μm SiO under vacuum at evaporation rates of 2 Å/s (see Note 1). We determine the thickness of evaporated material with a sensor in the vacuum chamber.

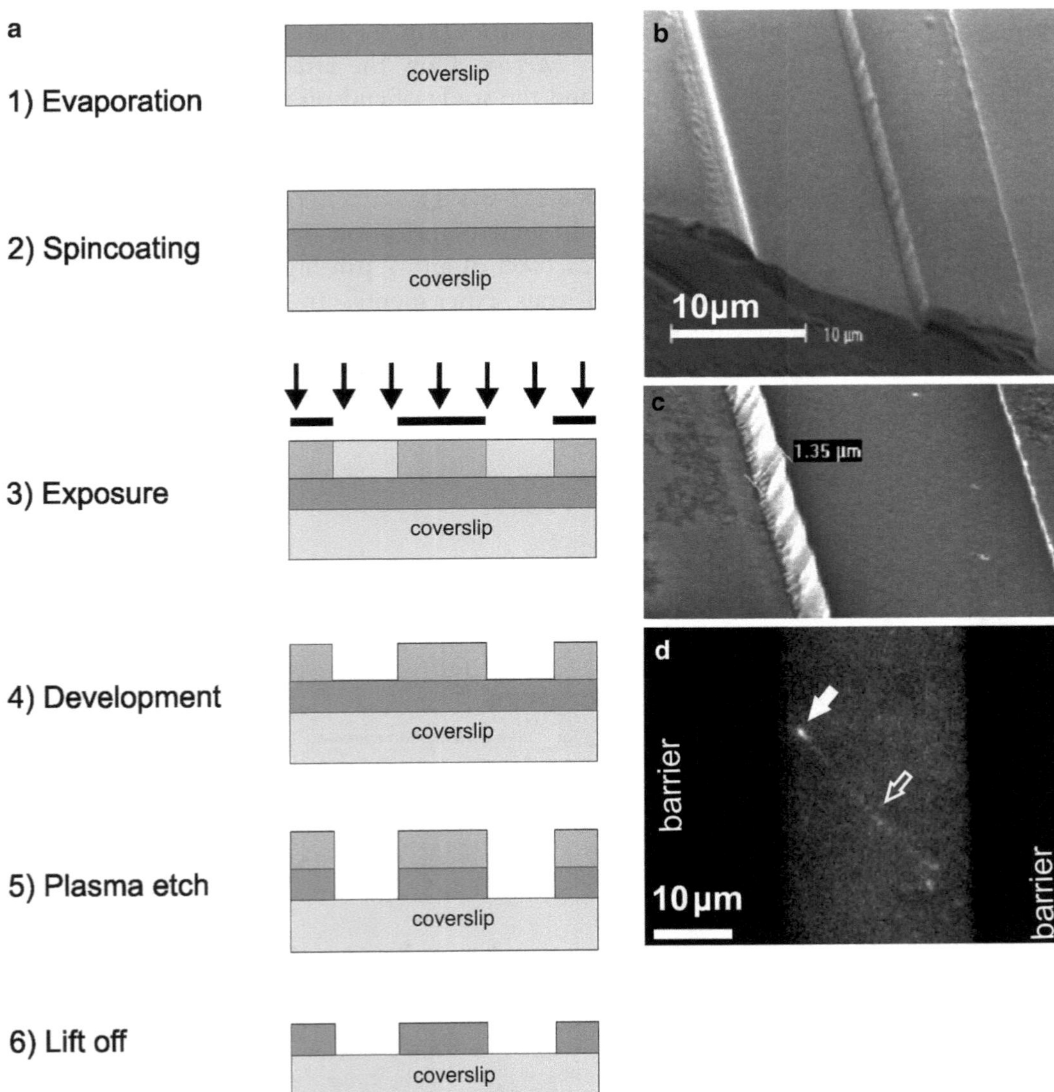

Fig. 2. Fabrication process of glass barriers. (**a**) Steps taken in the fabrication process as explained in the text. (**b**) A low-magnification scanning electron microscope (SEM) image of the glass barriers. (**c**) Higher magnification SEM image of the glass barriers. (**d**) Total Internal Reflection Fluorescence (TIRF) image of fluorescently labeled MAPs end tracking an MT +end (*white arrow*). The MT is grown from a stabilized seed (*open arrow*) reproduced from (35), Copyright Elsevier 2010.

2. Evaporate the samples with HMDS: Pour ~5 ml HDMS into a glass beaker. Place it together with the samples into a desiccator for ~10 min. The HDMS forms a thin layer on the SiO to improve the adhesion of the next layer of photoresist.
3. Spin-coat S1813 photoresist on top for approx. 40 s at 1,000 rpm. Soft-bake each slide at 115°C for 90 s on a hot plate.
4. Expose the slides to UV light through a chromium mask (see Note 2), exposure time 12 s. Post-exposure bake at 130°C for 30 min.
5. Develop the slides by swirling them in MF319 developer for 1 min. Rinse them afterward in MilliQ water and blow-dry with N_2.
6. Plasma-etch the coverslips in a mixture of a CHF_3-/SF_6-/Argon-plasma. We aim to etch through the evaporated layer of SiO (~1.5 μm high) which takes 1 h. To remove possible contamination of the plasma on the glass, end the process with 15 min of exposure to oxygen plasma.
7. Remove the remaining photoresist by sonication in acetone for 5 min.

3.1.2. Fabrication of Gold Barriers

1. Place coverslips in chromosulfuric acid for few hours. Rinse them 3× with MilliQ and blow-dry.
2. As shown in Fig. 3, coat coverslips with a sandwich of a thin chromium layer of 5 nm, a gold layer of 750 nm, and a thick chromium layer of 250 nm (see Note 1). We measure the thickness of evaporated material with a sensor in the vacuum chamber. The thin chromium layer ensures good adhesion of the gold while the thick chromium layer, which has an overhang, prevents MTs from growing over the barrier and enforces end-on contact between the MT and the barrier.
3. Evaporate the samples with HMDS: Pour ~5 ml HMDS into a glass beaker. Place it together with the samples into a desiccator for approx. 1 h. HMDS provides good adhesion for the photoresist to be spin-coated on top.
4. Spin-coat S1813 photoresist on top for approx. 40 s at 1,000 rpm. Soft-bake each slide at 115°C for 90 s on a hot plate.
5. Expose the slides to UV light through a chromium mask (see Note 2), exposure time 12 s. Post-exposure bake at 130°C for 30 min.
6. Develop the slides by swirling them in MF319 developer for 1 min. Rinse them afterward in MilliQ water and blow-dry with N_2.
7. Remove the remaining photoresist by sonication in acetone for 5 min.

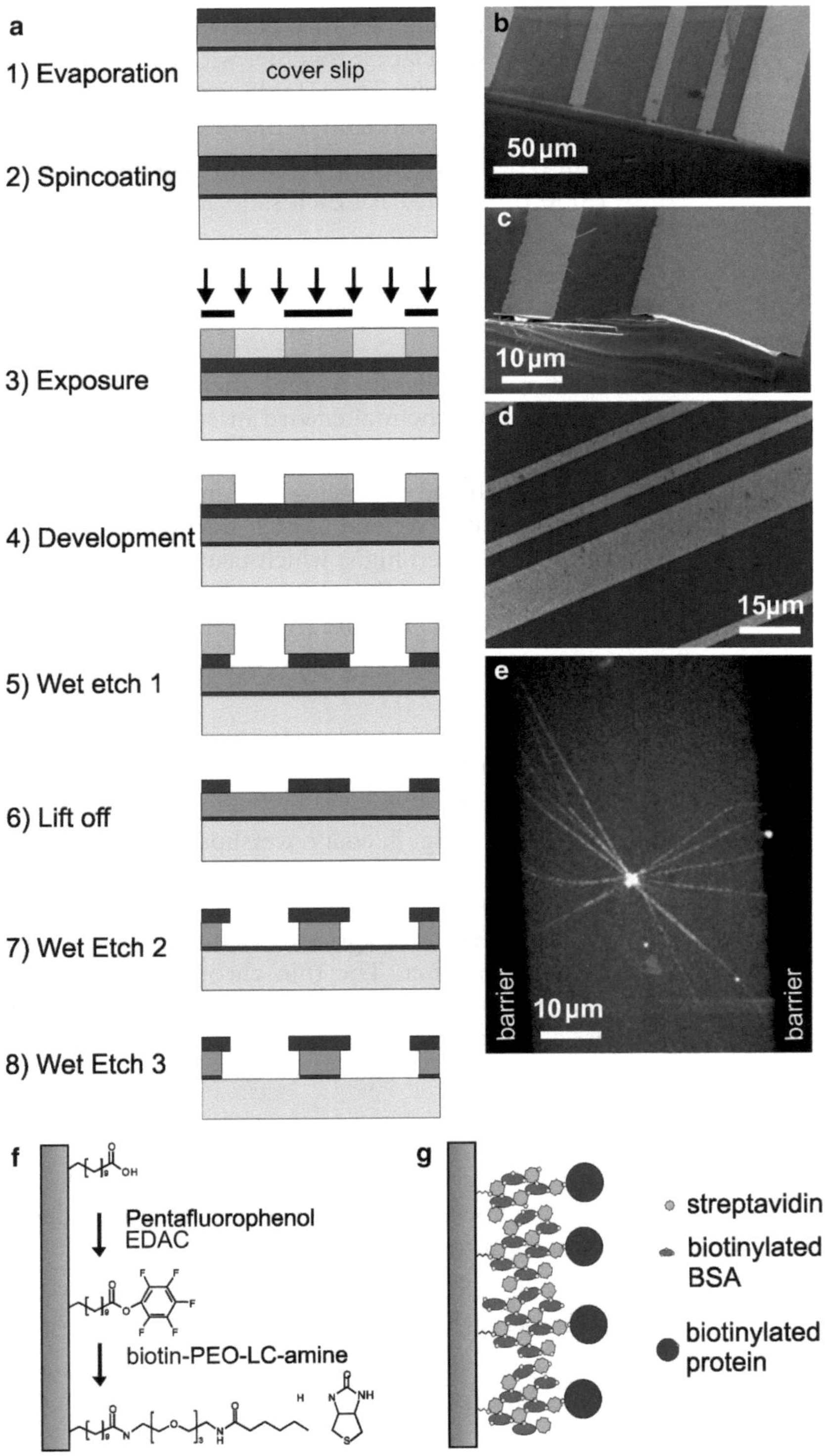

Fig. 3. Fabrication process of gold barriers. (**a**) Steps taken in the fabrication process as explained in the text. (**b**) A low-magnification scanning electron microscope (SEM) image of the gold barriers. (**c**) Higher magnification SEM image of the gold barriers with the chromium overhang on top. (**d**) Reflected light bright-field image of the barriers, image taken from above, showing the chromium layer(s). (**e**) Spinning-disk confocal images of MTs growing from a centrosome against functionalized gold barriers. (**f**) Cartoon of the chemical process involved in biotinylating the gold barrier. (**g**) Cartoon of a streptavidin–biotinylated-BSA layer assembled on a gold layer reproduced from (35), Copyright Elsevier 2010.

8. Excavate barriers by consecutive immersion in chromium etchant, acetone (to remove the photoresist), and gold etchant. Time the wet etch process by checking under the microscope that all gold or chromium between the structures is removed but the structures themselves are not lifted off. This leaves ~1 μm-high gold barriers (see Note 3).
9. Very briefly immerse samples in chromium etchant again to remove the thin layer of chromium at the bottom, but keeping the thick layer of chromium on top. Subsequently, rinse thoroughly with MilliQ (see Note 3).

3.1.3. Biotinylation of Gold Barriers to Allow for Specific Attachment of Proteins

We specifically label the gold barriers with biotin using thiol chemistry (38, 39) (Fig. 3f):

1. Clean a slide containing gold barriers by sonication in ethanol. Immerse it consecutively in the following liquids and rinse after each incubation three times with ethanol: 200 mM 11-mercapto-1-undecanoic acid in ethanol for 3 h, 100 mM EDAC and 200 mM PFP in ethanol for 20 min, and 0.1 mM biotin-PEO-LC-amine for 20 min.
2. To remove nonspecific interactions of the thiol groups with the glass surfaces, immerse each slide sequentially in 2 M NaCl for 7 min, 0.1% Tween in MilliQ for 15 min, 0.01% TritonX-100 for 7 min, and thoroughly rinse with MilliQ water.
3. Store the slides in ethanol; they can be used for several weeks.

3.1.4. Attachment of Biotinylated Proteins to Biotinylated Gold Barriers

Biotinylated dynein motor proteins are specifically attached to a gold surface via a biotin–streptavidin linkage and thorough blocking of other surfaces of the coverslip as shown in Fig. 3g. In between each of the following incubation steps, the sample is rinsed with three flow cell volumes of MRB80.

1. Block the glass surfaces with 0.2 mg/ml PLL-PEG for 5 min and subsequently 1 mg/ml κ-casein in MRB80.
2. Sequentially incubate the slides in the following solutions for 5 min each: A streptavidin mix (0.5 mg/ml streptavidin, 1 mg/ml κ-casein, 5 mg/ml BSA) and a biotinylated-BSA-mix (1.5 mg/ml biotinylated-BSA, 1 mg/ml κ-casein, 5 mg/ml BSA). Repeat again and finish with another incubation with the streptavidin mix. This results in a multilayer with three layers of streptavidin and two of biotinylated-BSA to enhance the number of binding sites (39).
3. Incubate dynein (20 nM dynein, 1 mg/ml κ-casein, 5 mg/ml BSA) for 5 min (in the case of control experiments without dynein, leave this step out).

3.1.5. Microfabricated Barriers for Optical Trap Experiments

To allow space for the trapped bead and minimize near-surface viscosity effects (40), the barriers need to be high enough. The bead–axoneme construct is positioned approximately 4 bead

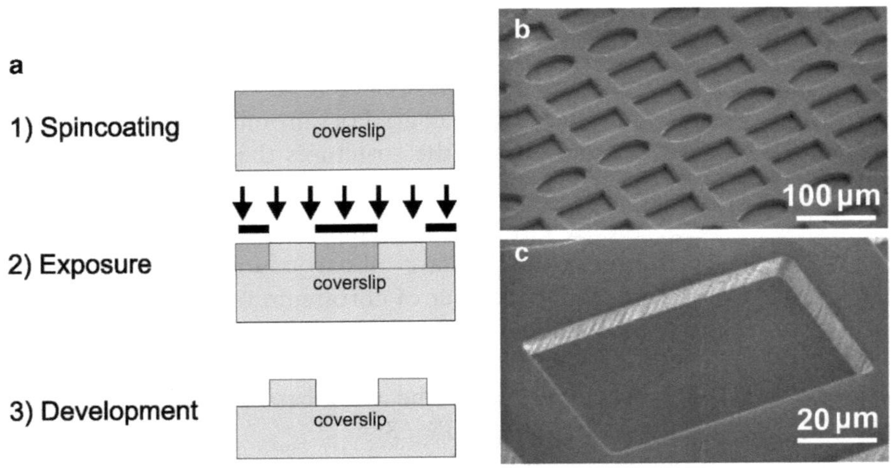

Fig. 4. Fabrication process of SU-8 chambers. (**a**) Steps taken in the fabrication process as explained in the text. (**b**) A low-magnification scanning electron microscope (SEM) image of the SU-8 chambers. (**c**) Higher magnification SEM image of the SU-8 chambers reproduced from (35), Copyright Elsevier 2010.

radii (~4 μm) above the bottom surface with an additional 3 μm barrier above to prevent the MTs from growing over. We produce ~7 μm-high chambers of 40 by 80 μm separated by 20 μm-wide interspacers (Fig. 4).

1. Spin-coat clean coverslips with SU-8 negative-tone photoresist to produce an ~7 μm-thick layer. We spin-coat for 25 s at 1,000 rpm. Soft-bake the slides for 2 min at 65°C on a hot plate, and then for 2 min at 95°C.
2. Expose the coverslips to UV light through a mask (see Note 2). The mask contains the pattern which should be transferred to the SU-8. We additionally use a Schott BG12 filter and expose for 32 s.
3. Post-exposure bake the samples for 1 min at 65°C and subsequently for 2 min at 95°C.
4. Develop the coverslips with SU-8 developer by swirling them consecutively in three beakers:

   Developer: 90 s

   Developer to rinse: 20 s

   2-propanol to rinse
5. Blow-dry with N_2 and hard-bake for 10 min at 160°C on a hot plate.

3.2. Assays for Microscopy Experiments

3.2.1. Assay of MTs Growing Against Glass Barriers

With the glass barriers, we test the influence of MT pushing forces on MT dynamics (19, 23). For this assay, the glass barriers need to be high (~1.5 μm) and straight enough to prevent MTs from growing over them.

1. First, construct a flow cell by drawing two parallel lines of vacuum grease 4 mm apart on a clean microscope slide.

Mount a coverslip containing barriers on top. This forms a cavity, with the microscope slide and coverslip forming the bottom and top, respectively, and the two grease lines the side walls. The open sides are used as inlet and outlet. To ensure a well-defined volume, flow in 7 μl of solution (here PLL-PEG-biotin, see below) and press the coverslip evenly down.

2. In between each of the following incubation steps, rinse the sample with twice the flow cell volume of MRB80 by applying the solution at the inlet and at the same time absorbing solution with a filter paper at the outlet of the flow cell.
3. Block the glass surfaces with 0.2 mg/ml PLL-PEG-biotin by introducing it into the flow cell and incubating upside down for 5 min. Rinse with 100 μl of MRB80. Form another blocking layer by incubation with 1 mg/ml κ-casein.
4. Activate the surface with a streptavidin mix (1 mg/ml streptavidin, 1 mg/ml κ-casein). Incubate the stabilized MT seeds for 5 min upside down to let them adhere specifically via the streptavidin–biotin linkers.
5. Finally, introduce the tubulin mix.

3.2.2. Assay of MTs Growing Against Gold Barriers

To mimic the end-on interaction of MT ends with proteins at the cortex or kinetochore, we use gold barriers since gold allows the specific attachment of proteins. The assay starts by cleaning a microscope slide in chromosulfuric acid which, together with a coverslip with gold barriers, is used to construct a simple flow cell (see Subheading 3.2.1). A solution of centrosomes in MRB80 is flown in and incubated for 5 min in order to let the centrosomes nonspecifically adhere to the glass surfaces. Centrosomes that do not stick to the surface are washed out by rinsing with two flow cell volumes of MRB80. Subsequently, the barriers are functionalized with biotinylated dynein as described in Subheading 3.1.4. Rinse with MRB80, introduce the final tubulin mix, and seal the flow cell with candle wax.

3.3. Optical Trap Experiments

3.3.1. Optical Trap Setup

The optical trap setup (41, 42) consists of an infrared trapping laser (1,064 nm, Nd:YVO_4, Spectra-Physics, USA, CA), which is focused into the sample by a 100×/1.3 NA oil immersion objective. The laser beam is time-shared using acousto-optical deflectors (AODs) (IntraAction DTD-274HA6) to create a "keyhole" trap consisting of a point trap and a line trap, as previously described (24, 25). A low-power red laser (633 nm, HeNe, 1125P, Uniphase) is superimposed after the AODs on the IR beam. Before or after every experiment, the red laser light is focused on a bead trapped in the point trap and imaged onto a quadrant photodiode (QPD) for stiffness calibration. The stiffness of the point trap is then determined by analysis of the power spectrum of the thermal fluctuations of the bead. Typically, trap stiffnesses in the range of 0.03–0.12 pN/nm are used.

3.3.2. Optical Trap Assay

Experiments with optical tweezers offer high resolution and are, therefore, ideal for the determination of molecular details. Moreover, with these experiments, we can examine and exert pushing and pulling forces.

1. First, build a flow cell: Glue a piece of parafilm onto a microscope slide by heating both on a hot plate. Cut a channel into the parafilm and stick a clean coverslip with microfabricated SU-8 chambers over the channel. Leave one side of the channel open to allow injection of solutions during the experiment. With more parafilm, glue a piece of teflon tubing to the other end of the channel. The height of the tubing's end relative to the height of the flow cell's inlet determines the flow speed inside the flow cell and gives an easy and fast way to adjust the flow speed to the experiment.
2. Since proteins tend to nonspecifically adhere to SU-8, the surfaces of the flow cell have to be blocked thoroughly. Start with flowing in a solution of 0.2% agarose at 70°C. Pump out the agarose immediately by connecting a pump to the channel for a few minutes. Next, flow a 0.01% Triton X-100 solution to prevent bubble formation in the flow cell. Do a second blocking step by incubating the flow cell for 10 min with 0.1 mg/ml κ-casein (see Notes 4–5).
3. If MT force generation and regulation by dynein are being studied, introduce 20 nM of dynein subsequently and leave to incubate for 20 min (otherwise, leave this step out). Finally, rinse the flow cell with 100 μl of MRB80.

3.3.3. Performing the Experiments

1. Flow axonemes and beads into the flow cell. First, trap a bead in the point trap. Then, move the bead close to an axoneme to nonspecifically attach it and bring the axoneme into the line trap (25). Next, position this bead–axoneme construct in front of a barrier in one of the microfabricated chambers (Fig. 5a–c and see Notes 6–7).
2. Because the axoneme–bead construct is not infinitely stiff, the bead moves out of the trap center a smaller distance than the MT grows. Since we measure the bead displacement and deduct from it the MT length change, we need to know the conversion factor between bead and MT tip displacement.
3. Therefore, measure this conversion factor by repeatedly pushing the barrier against the bead–axoneme construct and plot the subsequent bead displacement as a function of barrier displacement. As an example, refer to Fig. 5d, where after a first

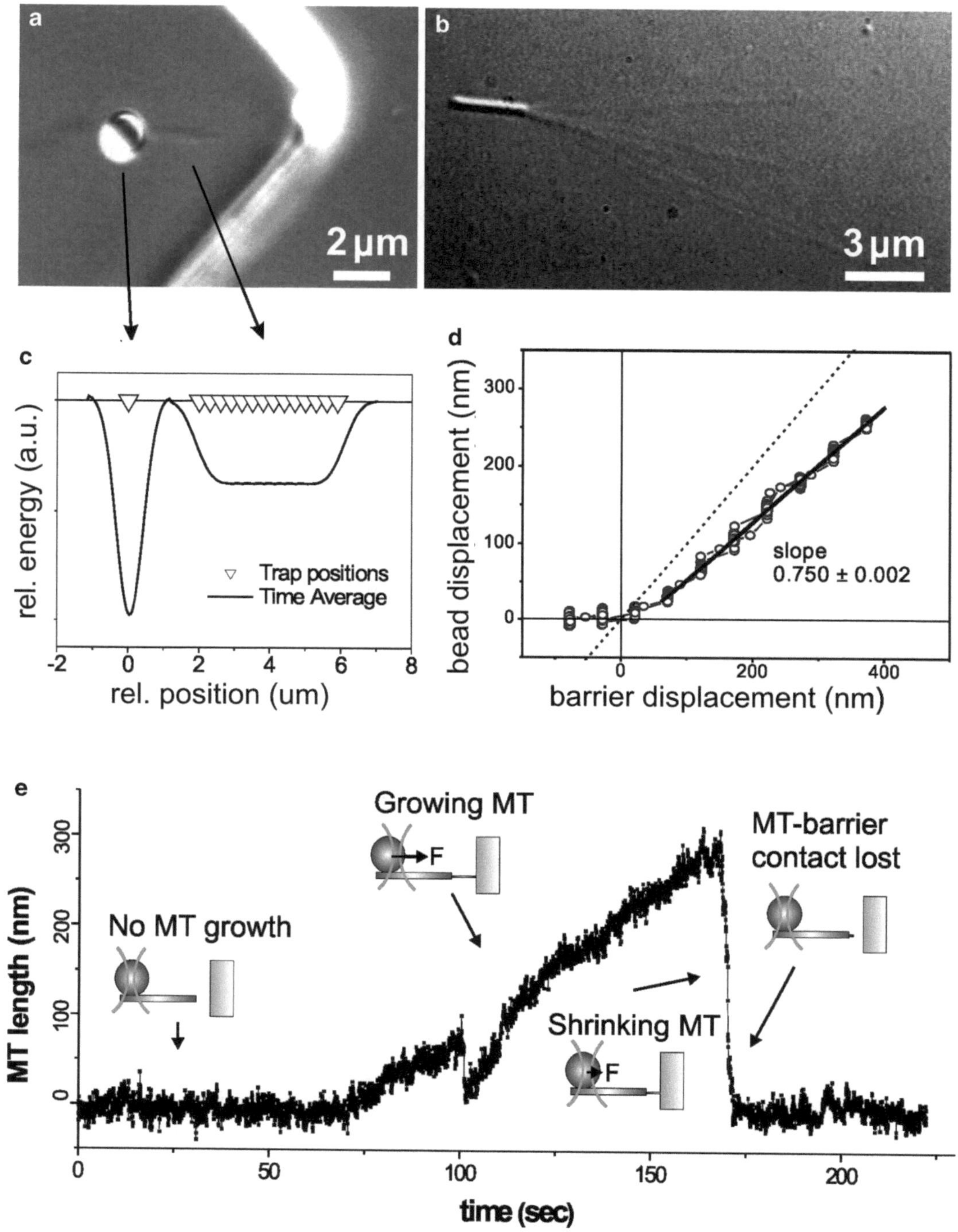

Fig. 5. The optical trap experiment. (**a**) Video-enhanced- differential interference contrast (VE-DIC) image of a construct consisting of a bead, held in the optical trap, connected to an axoneme, positioned in front of a barrier. (**b**) VE-DIC image of MTs growing from an axoneme. The number of MTs growing from the axoneme can be controlled by varying the tubulin concentration and the temperature (43). (**c**) Time-shared laser position (*open triangles*) and potential well of the keyhole-shaped optical trap. (**d**) Stiffness determination of the axoneme–bead construct. The wall displacement is plotted against the bead displacement. The axoneme–bead construct shows an initial soft regime of approximately 50 nm, followed by a linear regime. (**e**) Graph with a typical dataset of MT length against time. Cartoons insets show the bead position relative to the trap position for different phases of the experiment reproduced from (35), Copyright Elsevier 2010.

soft regime of approximately 50 nm, the conversion factor is constant over several hundreds of nanometers (25).

4. Rinse the chamber with MRB80 to remove leftover beads and axonemes (see Note 8). Next, position the tip of the axoneme approximately 100 nm away from the barrier and add the tubulin mix (variable tubulin concentration, 1 mM GTP, 0.1 mg/ml κ-casein in MRB80) to trigger MT growth (see Notes 9–10).
5. Figure 5e shows a typical dataset of MT growth. Initially, there is no MT growth so that the bead is positioned in the center of the point trap. Then, the MT starts to grow and pushes the bead out of the trap center. Eventually, the MT undergoes a catastrophe and while the MT shrinks, the bead moves back to the trap center (see Note 11).
6. We image axonemes and beads using VE-DIC microscopy. During the experiments, the image stream is digitized at 1 Hz and the position of the bead tracked online using a home-written cross-correlation routine for live monitoring. Afterward, the recorded images are digitized at a frame rate of 25 Hz. This image stream is used to track the bead offline. The bead tracking is automated using home-written IDL software. The position of the bead is obtained with a subpixel resolution of 2–3 nm (44).

3.3.4. Optical Trap with Force Feedback

With our optical tweezers, we trap a bead which is adhered to an axoneme. The tip is pointing toward the SU-8 barriers such that the MTs grow against the walls. A growing MT pushes the attached bead out of the trap center. The restoring force of the trap pushes the bead back, resulting in an increasing force on the MT. However, to elucidate the effect of force on MT dynamics, it is much more convenient to work with constant forces. This can be achieved with a force-feedback algorithm (26–29) such that upon displacement of the bead from the trap center, the force-feedback algorithm reduces the force onto the MT by mechanically increasing the space between trap center and barrier (Fig. 6). In our setup, this is achieved by translating the microscope stage, and with it the sample containing the barriers.

During an experiment with force feedback, we constantly monitor the bead position by focusing the red laser light on the bead trapped in the point trap and imaging it onto the QPD as described in Subheading 3.3.1. Before starting the algorithm, the user defines a "starting point" which specifies the position relative to the trap center where the bead should be during the experiment. The relative distance between the starting point and trap center defines the magnitude of the constant force and can be freely chosen. The force-feedback algorithm compares the bead

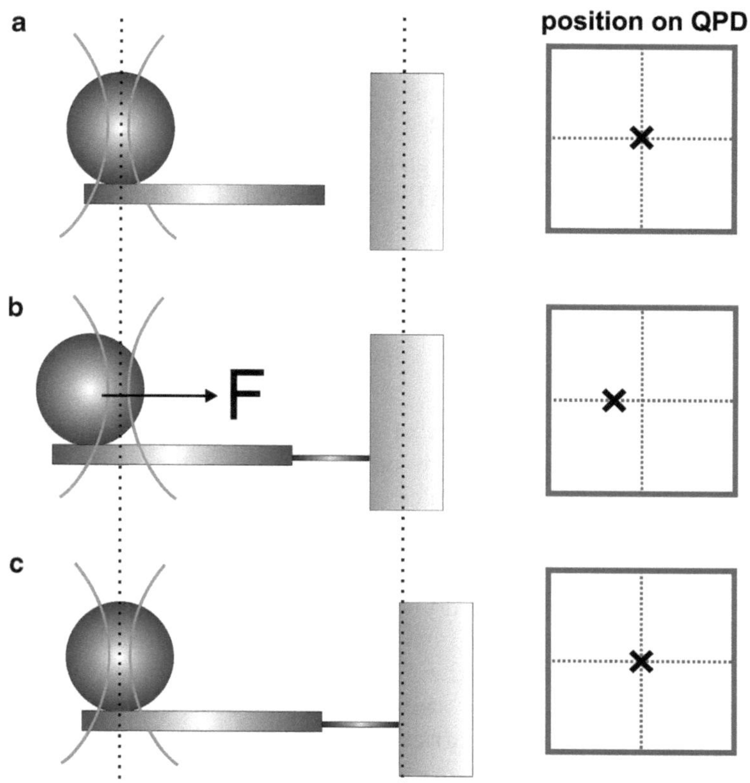

Fig. 6. Optical tweezers with force feedback. (**a**) No MT growth, the bead is at the chosen starting point. (**b**) Growing MT, the bead is pushed out of the trap center. This is measured on the QPD. (**c**) The force-feedback algorithm directs the microscope stage to move such that relative distance between the trap center and the barrier increases so that the bead is back at its starting position.

position calculated from the QPD signal to the starting point. If the bead is not at the starting point, then the software directs the piezo actuators of the microscope stage to move until the bead is located there. The software is written in LabView and functions with a 10 Hz iteration rate, which is fast enough to track MTs growing with typical speeds of 1–5 μm/min.

After making the construct and determining the conversion factor between MT tip displacement and bead displacement (see Subheading 3.3.3), the experiment with force clamp includes the following (see Note 12):

1. Conversion of the QPD signal in Volts to a signal in nanometers. For this, move the trapped construct a known distance (~100 nm) by translating the laser trap. Plot the subsequent trap displacement as a function of QPD read-off. From a fit to the resulting line, get the calibration factor.
2. Perform a trap stiffness calibration (see Subheading 3.3.1). This relates the QPD signal calibrated in nanometers to a force.

3. Choose the starting point by positioning the construct at the desired position.
4. Perform the experiment as described before (Subheading 3.3.3). Note that due to the repositioning of the stage, MT growth does not result in bead displacement. Instead, MT growth can be obtained from the stage position.

4. Notes

1. Evaporation of chromium, gold, or SiO should not exceed rates of a few nm/s to prevent the formation of big metal/glass granules and to avoid strain on the evaporation machinery.
2. If the contact between sample and mask is not tight enough, scattered UV-light exposes part of the barriers as well resulting in slanted walls. Therefore, ensure that the mask firmly touches the SU-8 or S1813 coating. Likewise, reflections from the sample holder below the coverslip can result in uneven barrier walls. We avoid this by placing the coverslip on black paper.
3. Timing of the wet etch process for gold barriers has to be done with extreme care. Etching a few seconds too long can result in complete liftoff of the lowest chromium layer and with it the barriers. We determine the timing by repeatedly etching for few seconds, drying the sample, and examining with a microscope.
4. Blocking of the SU-8 flow cell surfaces can be enhanced by repeating the agarose incubation step.
5. To reach constant and reproducible flow speeds, it is important that the flow cell does not contain even tiny air bubbles. Flowing 0.01% Triton X-100 removes air bubbles and makes the flow cell's surfaces hydrophilic. It can further help to degas the buffer before use by sonicating for 5 min.
6. Axoneme length should be carefully chosen. The axoneme tip has to be at least 1 μm away from the line trap to avoid light-induced MT depolymerization. Too long axonemes are not as stably trapped and can move out of the line trap during MT growth. Axonemes with length of 4–6 μm are ideal.
7. The optical tweezers experiment is performed with the axoneme tip relatively close to the barrier to keep the MTs short (less than 1 μm). As the force necessary to buckle these short MTs is much higher than the force that is maximally

applied with the optical trap, we can exclude MT buckling. Furthermore, the MTs are forced to stay in "end-on" contact with the barrier during the experiment.

8. Leftover beads and axonemes are removed with high flow speeds. However, the trapped construct can be lost due to viscous drag. Avoid this by hiding it in the "lee" of a microfabricated barrier.
9. Trapping of a second bead can distort the signal or knock the construct out of the trap. Avoid this by staying close to the bottom surface.
10. Flow speeds have to be adjusted such that beads, axonemes, and especially the tubulin mix reach the bottom of the flow cell. When flow speeds are too high, chemicals are flown out before they have diffused to the construct.
11. We could distinguish between plus- and minus-end growths because the plus end is known to grow faster and to have more catastrophes (8). Occasionally, we obtained measurements with much slower growth and no catastrophes which we attributed to minus ends.
12. When working with QPD position detection (see Subheading 3.3.4), it is best to start the experiment with the image in the center of the QPD to stay within the linear regime of the QPD.

Acknowledgments

We would like to thank Chris Rétif for his advice on microfabrication, Henk Bar for his help with microfabrication, Martijn van Duijn and Guillaume Romet-Lemonne for setting up the thiol chemistry assay and initiating the project, Jacob W. J. Kerssemakers, E. Laura Munteanu, and Julien Husson for developing and helping with the optical trap assay, Matt Footer (J. Theriot laboratory) for the axoneme purification, Sam Reck-Peterson for providing us with dynein molecules, and Claude Celati (M. Bornens laboratory) for help with the centrosome purification. We thank Magdalena Preciado Lopez for critical reading of the manuscript. Financial support from "Human Frontiers Science Program" and "Marie Curie Early Stage Training" is gratefully acknowledged. This work is part of the research program of the "Stichting voor Fundamenteel Onderzoek der Materie (FOM)," which is financially supported by the "Nederlandse Organisatie voor Wetenschappelijk Onderzoek (NWO)."

References

1. Inoue, S., and Salmon, E. D. (1995) Force Generation by Microtubule Assembly Disassembly in Mitosis and Related Movements. *Mol Biol Cell* 6, 1619–1640
2. Adames, N. R., and Cooper, J. A. (2000) Microtubule interactions with the cell cortex causing nuclear movements in Saccharomyces cerevisiae. *J Cell Biol* 149, 863–874
3. Dogterom, M., Kerssemakers, J. W., Romet-Lemonne, G., and Janson, M. E. (2005) Force generation by dynamic microtubules. *Curr Opin Cell Biol* 17, 67–74
4. Tolic-Norrelykke, I. M., Sacconi, L., Stringari, C., Raabe, I., and Pavone, F. S. (2005) Nuclear and division-plane positioning revealed by optical micromanipulation. *Current Biology* 15, 1212–1216
5. Tran, P., Marsh, L., Doye, V., Inoue, S., and Chang, F. (2001) A mechanism for nuclear positioning in fission yeast based on microtubule pushing. *J. Cell Biol.* 153, 397–411
6. Mitchison, J., and Kirschner, M. (1984) Dynamic instbility of microtubule growth. *Nature* 312, 237–242
7. Mitchison, J., and Kirschner, M. (1984) Microtubule assembly nucleated by isolated centrosomes. *Nature* 312, 232–237
8. Desai, A., and Mitchison, T. J. (1997) Microtubule polymerization dynamics. *Annu Rev Cell Dev Biol* 13, 83–117
9. Fygenson, D. K., Braun, E., and Libchaber, A. (1994) Phase-Diagram of Microtubules. *Physical Review E* 50, 1579–1588
10. Gadde, S., and Heald, R. (2004) Mechanisms and molecules of the mitotic spindle. *Current Biology* 14, R797–R805
11. Bieling, P., Kandels-Lewis, S., Telley, I. A., van Dijk, J., Janke, C., and Surrey, T. (2008) CLIP-170 tracks growing microtubule ends by dynamically recognizing composite EB1/tubulin-binding sites. *Journal of Cell Biology* 183, 1223–1233
12. Bieling, P., Laan, L., Schek, H., Munteanu, E. L., Sandblad, L., Dogterom, M., Brunner, D., and Surrey, T. (2007) Reconstitution of a microtubule plus-end tracking system in vitro. *Nature* 450, 1100–1105
13. Brouhard, G. J., Stear, J. H., Noetzel, T. L., Al-Bassam, J., Kinoshita, K., Harrison, S. C., Howard, J., and Hyman, A. A. (2008) XMAP215 is a processive microtubule polymerase. *Cell* 132, 79–88
14. Dixit, R., Barnett, B., Lazarus, J. E., Tokito, M., Goldman, Y. E., and Holzbaur, E. L. F. (2009) Microtubule plus-end tracking by CLIP-170 requires EB1. *Proc Natl Acad Sci USA* 106, 492–497
15. Honnappa, S., Gouveia, S. M., Weisbrich, A., Damberger, F. F., Bhavesh, N. S., Jawhari, H., Grigoriev, I., van Rijssel, F. J., Buey, R. M., Lawera, A., Jelesarov, I., Winkler, F. K., Wuthrich, K., Akhmanova, A., and Steinmetz, M. O. (2009) An EB1-binding motif acts as a microtubule tip localization signal. *Cell* 138, 366–76
16. Kinoshita, K., Arnal, I., Desai, A., Drechsel, D. N., and Hyman, A. A. (2001) Reconstitution of physiological microtubule dynamics using purified components. *Science* 294, 1340–1343
17. Kinoshita, K., Habermann, B., and Hyman, A. A. (2002) XMAP215: a key component of the dynamic microtubule cytoskeleton. *Trends in Cell Biology* 12, 267–273
18. Komarova, Y., De Groot, C. O., Grogoriev, I., Montenegro Gouveia, S., Munteanu, E. L., Schober, J. M., Honnappa, S., Buey, R. M., Hoogenraad, C. C., Dogterom, M., Borisy, G., Steinmetz, M., and Akhmanova, A. (2009) Mammalian end binding proteins control persistent microtubule growth. *J Cell Biol* 184, 691–706
19. Dogterom, M., and Yurke, B. (1997) Measurement of the force-velocity relation for growing microtubules. *Science* 278, 856–860
20. Fygenson, D. K., Marko, J. F., and Libchaber, A. (1997) Mechanics of microtubule-based membrane extension. *Phys. Rev. Lett.* 79, 4497–4500
21. Janson, M. E. (2002) Force Generation by Growing Microtubules. *PhD thesis*
22. Janson, M. E., and Dogterom, M. (2004) A bending mode analysis for growing microtubules: Evidence for a velocity-dependent rigidity. *Biophysical Journal* 87, 2723–2736
23. Janson, M. E., Dood, M. E. d., and Dogterom, M. (2003) Dynamic instability of microtubules is regulated by force. *The Journal of Cell Biology* 161, 1029–1034
24. Kerssemakers, J. W. J., Janson, M. E., Van der Horst, A., and Dogterom, M. (2003) Optical trap setup for measuring microtubule pushing forces. *Applied Physics Letters* 83, 4441–4443
25. Kerssemakers, J. W. J., Munteanu, E. L., Laan, L., Noetzel, T. L., Janson, M. E., and Dogterom, M. (2006) Assembly dynamics of microtubules at molecular resolution. *Nature* 442, 709–712

26. Finer, J. T., Simmons, R. M., and Spudich, J. A. (1994) Single myosin molecule mechanics: piconewton forces and nanometre steps. *Nature* 368, 113–119
27. Simmons, R. M., Finer, J. T., Chu, S., and Spudich, J. A. (1996) Quantitative measurements of force and displacement using an optical trap. *Biophys J* 70, 1813–1822
28. Visscher, K., Schnitzer, M. J., and Block, S. M. (1999) Single kinesin molecules studied with a molecular force clamp. *Nature* 400, 184–189
29. Schek, H. T., Gardner, M. K., Cheng, J., Odde, D. J., and Hunt, A. J. (2007) Microtubule assembly dynamics at the nanoscale. *Current Biology* 17, 1445–1455
30. Moudjou, M., and Bornens, M. (1994) Isolation of Centrosomes From Cultured Animal Cells. *Cell Biology: A Laboratory Handbook, ed. J.E.Celis Academic Press, New York* 595–604
31. Gibbons, I. R., and Frank, E. (1979) A latent adenosine triphosphatase form of dynein 1 from sea urchin sperm flagella. *J. Biol. Chem.* 254, 187–196
32. Tselutin, K., Seigneurin, F., and Blesbois, E. (1999) Comparison of cryoprotectants and methods of cryopreservation of fowl spermatozoa. *Poult. Sci.* 78, 586–590
33. Reck-Peterson, S. L., Yildiz, A., Carter, A. P., Gennerich, A., Zhang, N., and Vale, R. D. (2006) Single-molecule analysis of dynein processivity and stepping behavior. *Cell* 126, 335–348
34. Howard, J., and Hyman, A. A. (2007) Microtubule polymerases and depolymerases. *Curr Opin Cell Biol* 19, 31–35
35. The content of sections 1 to 3 as well as the one of Fig. 1–5 is based on Laan, L., and Dogterom, M. (2010) *In vitro* assays to study force generation at dynamic microtubule ends. *Methods in Cell Biology* in press, as well as on Laan, L. (2009) Force generation at microtubule ends: An *in vitro* approach to cortical interactions. *PhD thesis*
36. Laan, L., Husson, J., Van Duijn, M., Vale, R. D., Reck-Peterson, S. L., and Dogterom, M. (2009) Cortex-attached dynein regulates microtubule dynamics and pulls on shrinking microtubule ends in vitro. *submitted*
37. Schek, H. T., and Hunt, A. J. (2005) Micropatterned structures for studying the mechanics of biological polymers. *Biomed. Microdevices* 7, 41–46
38. Dogterom, M., Felix, M.A., Guet, C.C. and Leibler, S. (1996) Influence of M-phase chromatin on the anisotropy of microtubule asters. *J Cell Biol*, 133, 125–140
39. Romet-Lemonne, G., VanDuijn, M., and Dogterom, M. (2005) Three-dimensional control of protein patterning in microfabricated devices. *Nano Letters* 5, 2350–2354
40. Gittes, F., and Schmidt, C. F. (1998) Signals and noise in micromechanical measurements. *Methods in Cell Biology*, 55, 129–156
41. Ashkin, A. (1970) Acceleration and trapping of particles by radiation pressure. *Physical Review Letters* 24, 156–159
42. Svoboda, K., and Block, S. M. (1994) Force and Velocity Measured for Single Kinesin Molecules. *Cell* 77, 773–784
43. Laan, L., Husson, J., Munteanu, E.L., Kerssemakers, J.W.J. and Dogterom, M. (2008) Force-generation and dynamic instability of microtubule bundles. *Proc Natl Acad Sci USA* 105, 8920–8925
44. Visscher, K., Gross, S. P., and Block, S. M. (1996) Construction of Multiple-Beam Optical Traps with Nanometer-Resolution Position Sensing. *IEEE Journal of Selected Topics in Quantum Electronics* 2, 1066–1076

Chapter 12

Variations on the Single-Molecule Assay for Microtubule-Associated Proteins and Kinesins

Susanne Bechstedt, Michal Wieczorek, Michael Noujaim, and Gary J. Brouhard

Abstract

The direct observation of single kinesins and microtubule-associated proteins (MAPs) has become a core tool for cytoskeleton research. We outline several variations to the core experiment that allow the researcher to explore structural and biophysical mechanisms underlying kinesin motility and MAP function.

Key words: Microtubule, Microtubule-associated protein, Kinesin, Single molecule, Total internal reflection fluorescence

1. Introduction

Single-molecule assays based on total internal reflection fluorescence (TIRF) allow for the direct observation of single kinesins and microtubule-associated proteins (MAPs) as they interact with microtubules, as well as a reconstitution of microtubule growth and shrinkage on a single-filament basis. The advantages of these assays over bulk experiments have made them core tools for cytoskeleton research.

In addition to the core assay, where a single molecule of a kinesin or a MAP is directly observed, the assay has the potential to investigate additional characteristics that define the nature of the interaction between the MAP and the microtubule. These characteristics include the nucleotide state of the kinesin, structural domains involved in microtubule binding, the relative effect of electrostatic forces in mediating the interaction, as well

Anne Straube (ed.), *Microtubule Dynamics: Methods and Protocols*, Methods in Molecular Biology, vol. 777,
DOI 10.1007/978-1-61779-252-6_12, © Springer Science+Business Media, LLC 2011

as questions concerning kinetics and the behaviour of individual molecules alone versus ensemble conditions. The goal of this chapter is to provide protocols for these variations in the core single-molecule assay.

2. Materials

2.1. Cover Glass Preparation

2.1.1. Materials and Equipment

1. 18 × 18-mm and 22 × 22-mm cover glasses (e.g. Corning, No. 1½) or microscope slides.
2. Ceramic histology racks for holding cover glass (e.g. for 12 cover glasses per rack).
3. Rectangular histology staining dishes for submerging cover glass racks (e.g. 250-ml total volume for holding four cover glass racks).
4. Fume hood.
5. Gloves resistant to both strong acids and organic solvents (e.g. Viton brand).
6. Lab coat and safety glasses.
7. Chemical spill kit.
8. Acid-resistant tweezers.
9. 1-ml syringe with 5+-cm spinal needle.
10. Hotplate.
11. Graduated cylinders.
12. Sonicating water bath.

2.1.2. Reagents

1. Ultrapure water (>18 MΩ/m).
2. Acetone.
3. Ethanol.
4. 0.1 M potassium hydroxide (KOH) in H_2O.
5. Sulphuric acid (H_2SO_4) (97%).
6. Hydrogen peroxide (H_2O_2) (30%) (Sigma 21676-3, stored at 4°C).
7. Dichlorodimethylsilane (DDS; Aldrich, 440272).
8. Trichloroethylene (TCE; Sigma 251402).
9. Methanol.

2.2. Binding Microtubules to Surfaces

1. Tubulin (4 mg/ml in BRB80): Tubulin can be purified using established protocols (see Chapter 2) or purchased (e.g. Cytoskeleton, Inc.).
2. GMPCPP, 10 mM in H_2O (Jena Biosciences #NU-405).

3. Anti-beta-tubulin antibody (clone TUB2.1 Sigma T4026).
4. Anti-tetramethylrhodamine antibody (Invitrogen, Molecular Probes A6397).
5. BRB80 buffer: 80 mm PIPES/KOH pH 6.9, 1 mM EGTA, 1 mM $MgCl_2$. This buffer is commonly used to allow efficient polymerization of tubulin. KOH is used to set the pH. Buffer should be filtered (0.22-μm vacuum filter), degassed, and stored in aliquots at −20°C.
6. Pluronic F-127 (Sigma P2443), stock solution of 1% F-127 dissolved in BRB80 overnight, filtered (0.22-μm syringe filter), and stored at 4°C.
7. Double-sided scotch tape.

2.3. Spiking Experiments

2.3.1. Equipment

1. TIRF microscope suitable for single-molecule assays. The critical component is the use of an EMCCD camera (e.g. Andor iXon+). Commercial implementations of TIRF (e.g. Zeiss Laser TIRF III) are largely sufficient for these experiments.
2. Cover glass holders or silanized microscope slides.

2.3.2. Reagents

1. High-purity fluorescently tagged and untagged proteins (e.g. XMAP215 and XMAP215-GFP).
2. BRB80: 80 mM PIPES (Sigma, P6757), 1 mM $MgCl_2$, 1 mM EGTA in H_2O, pH 6.9 with KOH.
3. BRB20: 20 mM PIPES (Sigma, P6757), 1 mM $MgCl_2$, 1 mM EGTA in H_2O, pH 6.9 with KOH.
4. 1 M potassium chloride (KCl) in water.
5. D-Glucose (Sigma, G7528), 2 mg/ml in water and stored at −20°C.
6. Glucose oxidase (Sigma, G2133), 2 mg/ml in BRB80, snap-freeze 10-μl aliquots in liquid nitrogen, and stored at −20°C.
7. Catalase (Sigma, C9322), 0.8 mg/ml in BRB80 and 10-μl aliquots stored at −20°C.
8. β-Mercaptoethanol (Sigma, M3148).
9. Casein (Sigma, C0501), 10 mg/ml in BRB80 and 5-μl aliquots stored at −20°C.
10. BSA (Sigma A3059), 10 mg/ml in BRB80 and filtered.
11. Whatman filter paper (#1001-090).

2.4. Subtilisin Digestion

1. Subtilisin (Sigma, P5380), 1 mg/ml in BRB80.
2. Phenylmethanesulphonyl fluoride (PMSF) (Sigma, P7626), 2 M in ethanol.
3. BRB80 + 60% glycerol.

2.5. Nucleotide State of Kinesin

1. Adenosine-5′-triphosphate (ATP; magnesium salt) (Sigma, A9187), 100 mM in H_2O.
2. Adenosine-5′-diphosphate (ADP; sodium salt) (Sigma, A2754), 100 mM in H_2O.
3. Adenosine-5′-(β,γ-imido)triphosphate (AMP-PNP; tetralithium salt hydrate) (Sigma, A2647), 100 mM in H_2O.
4. Apyrase (Sigma, A6410), 100 mM in H_2O.

2.6. Flow-In Experiments

1. Soluble fluorophore with similar $\lambda_{excitation}$ and $\lambda_{emission}$ as fluorophore used for TIRF imaging of microtubules (i.e. TRITC, Alexa Fluor-546, etc.).
2. Steady hand and Whatman filter paper or a microfluidic device to flow in the imaging solution without perturbing the focal plane.

3. Methods

3.1. Cover Glass Preparation

Untreated glass is a sticky surface; fluorescent MAPs and kinesins adsorb to such a surface non-specifically, creating a background signal that swamps the experiment. The preparation of clean, silanized cover glass creates a surface that can be efficiently passivated by blocking agents (e.g. Pluronic F-127, see below), thus eliminating non-specific background signals.

3.1.1. Glass Cleaning

1. Prepare for the cleaning in a fume hood (see Note 1).
2. Place the cover glass into the ceramic histology racks.
3. Prepare the rectangular histology dishes by filling each dish with ~200 ml of the cleaning chemicals (described below).
4. Using tweezers to move the racks, submerge the cover glass in a histology dish containing acetone for 10 min.
5. Transfer the racks to ethanol and incubate for 10 min.
6. Transfer the racks to ultrapure water and rinse carefully by allowing running water to flow into the dish for 1 min.
7. Prepare a histology dish containing "Piranha" solution. Measure 175 ml of H_2SO_4 in a graduated cylinder. Pour the H_2SO_4 into the histology dish, and place the histology dish on a hotplate. Measure 75 ml of H_2O_2 in a graduate cylinder. Slowly add the H_2O_2 to the histology dish. Always add the peroxide to the acid; this method is preferred by most specialists, even though it goes against conventional rules. The solution becomes hot. Maintain the temperature at 60°C throughout the cleaning process using the hotplate.
8. Transfer the cover glass racks to the Piranha dish carefully and incubate for 1 h.

9. Transfer the racks to a dish with ultrapure water and incubate for 1 min. Repeat this procedure three times using fresh water.
10. Transfer the racks from the third ultrapure water bath to a 0.1 M KOH solution and incubate for 15 min.
11. Transfer the racks to a dish with ultrapure water and incubate for 1 min. Repeat this procedure two times using fresh water.
12. Dry the cover glass *thoroughly* by carefully blow drying the cover glass racks with clean nitrogen gas. Transfer the cover glass racks to a clean and dry histology dish.

3.1.2. Glass Silanization

Silanization is performed with a 0.05% solution of DDS in TCE. The clean glass surface is made hydrophobic by functionalization with silane.

1. Add sufficient TCE to a clean and dry histology dish.
2. Using a syringe fitted with a long spinal needle, puncture the DDS bottle and withdraw the appropriate volume. Avoid agitating the DDS bottle. Add DDS (to a final concentration of 0.05%) and stir while adding DDS to TCE.
3. Gently lower the cover glass racks into the TCE/DDS solution and incubate for 1 h.
4. Transfer the silanized glass into methanol, and place the histology dish in a sonicating water bath. Incubate for 5 min with the sonicating bath on maximum setting.
5. Repeat the methanol wash with two fresh dishes of methanol, incubating for 15 min and then 30 min in the sonicating bath on maximum setting.
6. Transfer the racks to a dish with ultrapure water and incubate for 1 min. Repeat this procedure two times using fresh water.
7. Transfer the cover glass racks to a clean and dry histology dish. Dry the cover glass by carefully blow drying the racks with clean nitrogen gas. The silanized cover glass can now be used to construct flow cells for single-molecule TIRF microscopy assays.

3.2. Binding Microtubules to Surfaces

Microtubules are attached to the silanized cover glass surface by antibodies. The antibodies adsorb to the silanized surface non-specifically, but a sufficient number bind in the correct orientation to interact with microtubules. The antibody molecules also introduce space between the microtubule and the glass surface. This space allows MAPs and kinesins, in principle, to interact with the microtubule around its complete circumference.

3.2.1. Selection of Binding Method

There are two broad categories of suitable antibodies: anti-tubulin and anti-fluorophore antibodies. There are several anti-tubulin antibodies available on the commercial market; our experience has favoured the TUB2.1 clone from Sigma. Anti-tubulin antibodies are effective with any fluorescently labelled microtubules, but the presence of soluble tubulin (e.g. in microtubule growth assays) causes dissociation of microtubules from the surface. Anti-fluorophore antibodies circumvent this problem as long as the soluble tubulin carries a different fluorescent label.

3.2.2. Preparation of GMPCPP Microtubules

1. Polymerize the microtubules: Add 2.5 μl tubulin, 5 μl GMPCPP, 0.5 μl $MgCl_2$, 42 μl BRB80 to a microcentrifuge tube.
2. Incubate on ice for 10 min.
3. Incubate at 37°C for 2 h.
4. Add 200 μl BRB80, prewarmed to 37°C.
5. Centrifuge at 150,000 × *g* for 5 min (e.g. in a Beckman Airfuge).
6. Discard the supernatant and resuspend the pellet in 200 μl BRB80.

3.2.3. Microtubule Binding and Surface Passivation

1. Construct a flow cell using silanized cover glass: Place four thin strips of double-stick tape onto a 22 × 22-mm cover glass or a silanized microscope slide. The strips should be approximately 2 mm in width and separated by approximately 4 mm. Place an 18 × 18-mm cover glass on top of the double-stick tape. This creates a thin "channel" between the strips of tape and the two pieces of cover glass. The channel has a volume of approximately 5 μl, and solution can be exchanged through the channel by applying vacuum to one side using a pipet tip inserted into vacuum tubing.
2. With the help of a vacuum, wet the channel by flowing through 20 μl of BRB80.
3. Flow through 20 μl of antibody solution. Incubate for 5 min. This step can be performed using Whatman paper or vacuum.
4. Flow through 40 μl BRB80 (to wash out unbound antibodies).
5. Flow through 20 μl F-127 stock solution and incubate for 10 min.
6. Flow through 80 μl BRB80 (to wash out excess F-127).
7. Flow through 20 μl polymerized microtubules in BRB80 and incubate for 5 min (see Note 2).
8. Wash with 20 μl of BRB80.
9. Introduce 20 μl of MAP/kinesin protein solution in imaging buffers (see below).

3.3. Spiking Experiments: Measuring Single Molecules at High Concentrations

The behaviour of MAPs and motors can be significantly influenced by the concentration of these molecules on a microtubule. For the observation of single-molecule behaviour in the context of larger amounts of protein (Fig. 1a), a mixture of labelled and unlabelled proteins can be used (1). These "spiking experiments" can also assay cooperative behaviour (2). The favourable ratio of labelled to unlabelled proteins has to be determined for each experiment, but lies typically between 1:50 and 1:10.

1. Prepare 50 μl of MAP/motor protein assay solution: 0.5 μl MAP/labelled-MAP mix, 0.5 μl casein, 1 μl glucose, 1 μl glucose oxidase, 1 μl catalase, 0.5 μl BME, 40.25 μl BRB20, and 0.75 μl 1 M KCl.

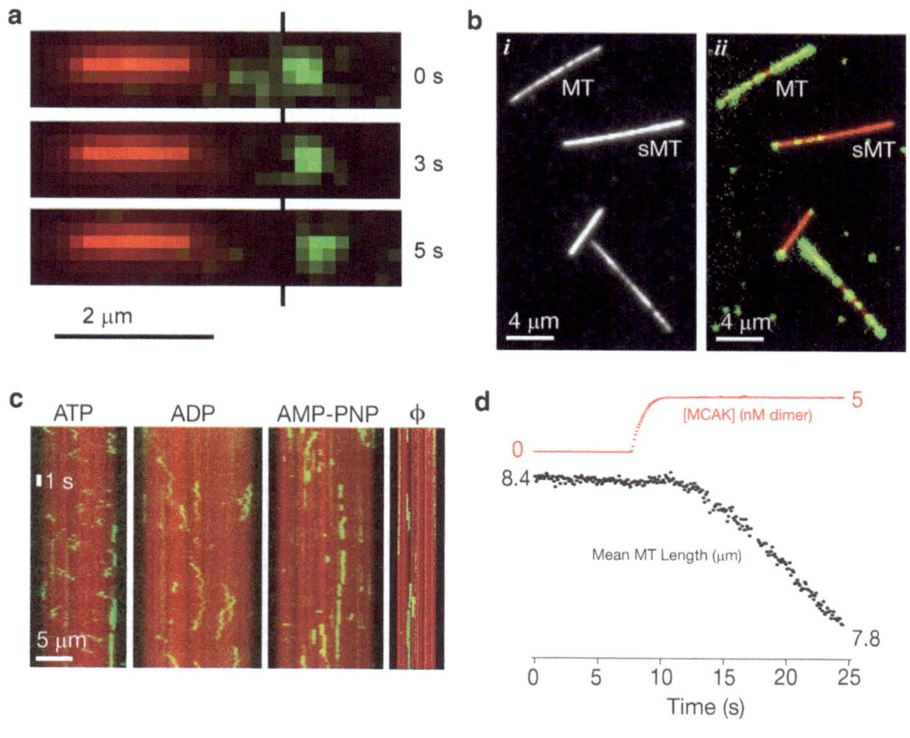

Fig. 1. Variations in the single-molecule assay. (**a**) Panels showing the observation of single molecules of XMAP215-GFP (*green*) in the presence of 200 nM total XMAP215 concentration. The XMAP215-GFP is seen to move away from the microtubule seed (*red*). Time points are indicated. Reprinted from (1), Copyright Elsevier 2008. (**b**) Images of XMAP215-GFP (*green*) interacting with undigested microtubules (MTs) and subtilisin-digested microtubules (sMTs). (i) Images of the microtubules. Different ratios of fluorescent label were used to distinguish the populations. (ii) XMAP215-GFP shows a clear preference for undigested microtubules. Reprinted from (1), Copyright Elsevier 2008. (**c**) Kymographs of MCAK in different nucleotide states. The nucleotide for each kymograph is noted at top. Diffusion occurs in ATP and ADP, but not in AMPPNP or the no-nucleotide state. Reprinted by permission from Macmillan Publishers Ltd: Nature (4), copyright 2006. (**d**) Flow in of MCAK and plot of microtubule length change over the time course of the experiment. The introduction of MCAK into the flow channel is shown in red. The average microtubule length across all microtubules is shown in black. Shortly after MCAK is introduced, the microtubules depolymerize at a steady rate. Reprinted by permission from Macmillan Publishers Ltd: Nature (4), copyright 2006.

2. Introduce the assay solution to the flow cell containing surface-immobilized microtubules using Whatman paper.
3. Observe the experiment under the TIRF microscope.

For a first trial, ratios of labelled/unlabelled proteins of 1:50, 1:20, and 1:10 should be assayed. The actual concentration and ratio of proteins used in this assay strongly depend on the affinity of the MAPs and motors to microtubules. It is important to confirm that the labelled protein is functionally equivalent to the unlabelled protein in overall enzymatic activity.

3.4. Subtilisin Digestion

The purpose of this experiment is to produce subtilisin-digested microtubules. Subtilisin removes the negatively charged "E-hook" located on the carboxy-terminus of tubulin (3). Along with possible dependencies on ionic strength, these data can provide important insight into the electrostatic effects responsible for MAP–microtubule interactions (4, 5) (Fig. 1b). Digestion can be achieved as follows:

1. Prepare GMPCPP-stabilized microtubules as described above.
2. Incubate microtubules with 10 μg/ml subtilisin for 20 min at 37°C.
3. Terminate the digestion by adding PMSF, 2 mM.
4. Place the reaction mixture on a 60% glycerol BRB80 cushion and centrifuge at 150,000 × *g* for 5 min.
5. Carefully aspirate the supernatant above the cushion (which contains subtilisin and the digested E-hooks). Wash the cushion interface, and then aspirate the remaining supernatant.
6. Resuspend the subtilisin-digested microtubules in an appropriate volume of BRB80 or a similar buffer.
7. Introduce the subtilisin-digested microtubules into the flow cell as described above. Undigested microtubules can be included as internal controls. The digested and undigested microtubules can be distinguished by using different fluorescent labels of different label brightness.
8. Introduce a MAP-imaging solution as described above.

3.5. Nucleotide State of a Kinesin

Kinesins are ATPases; their function depends on surrounding nucleotide (substrate) concentrations. Different nucleotide states, thus, result in different values for quantities, such as diffusion coefficients, k_{on}, etc. Studies of kinesin–microtubule interactions as a function of nucleotide state have produced elegant models for the enzymatic cycles of many kinesin family members (4–6). A typical TIRF assay for nucleotide state dependence of a kinesin (Fig. 1c) proceeds as follows:

1. Prepare rhodamine-labelled, GMPCPP-stabilized microtubules adhered to a silanized flow cell as described above.

2. Prepare assay solution as described above, substituting in the appropriate nucleotide (ATP, ADP, AMPPNP) at 1 mM (see Note 3). To obtain a nucleotide-free environment, add apyrase at a concentration of 0.15 U/ml to the imaging solution instead of nucleotide and incubate for 5 min prior to imaging.
3. After the microtubule incubation period, flush out excess microtubules with a BRB80 rinse, and flow in the imaging solution while acquiring a time-lapsed montage of single-molecule events.
4. Perform single-particle tracking analysis to obtain the desired biophysical parameters of the kinesin as described (4).

3.6. Flow-In Experiments

Conventional stopped-flow instruments rely on a basic principle: titrate a molecule into the system of interest at a certain concentration (e.g. urea to a protein solution) and observe the effect of that molecule on the system (e.g. increase in UV light absorbance due to protein unfolding). This technique has been used extensively to determine protein kinetics resulting in detailed biophysical models. A similar experiment can be performed in a TIRF assay. To perform a flow-in experiment to observe the effect of a MAP on microtubule length (Fig. 1d):

1. Prepare rhodamine-labelled, GMPCPP-stabilized microtubules adhered to a silanized flow cell as described above.
2. Add free fluorophore to imaging solution containing the MAP of interest at a known fluorophore:MAP molar ratio.
3. Begin the acquisition of a time-lapsed montage of the adhered microtubules at a frame rate of 0.1/s. Flow in the imaging solution without disturbing the current focal plane (see Note 4).
4. By tracking the average fluorescence intensity of a small region of the background as a function of time and assuming that the concentration of fluorophore is a maximum at steady state, back-calculate the concentration of the MAP as a function of time.
5. Track the length change of microtubules using tracking software; compare these data to MAP concentrations and extract kinetic parameters, such as k_{on} for the MAP (4).

4. Conclusion

The core single-molecule assay for kinesins and MAPs provides for the lifetime of interactions, the velocity of kinesins, and/or the diffusion coefficient of a MAP. By varying the assay, a great deal of additional information is accessible. We have outlined four variations in the assay that test the activity of single molecules under differing conditions. There is no doubt that further innovations will continue to develop these already powerful techniques.

5. Notes

1. Cover glass cleaning and silanization must be performed in a fume hood. The protocol uses chemicals that are toxic, flammable, corrosive, and explosive. Personal protective equipment, especially high-quality gloves, is essential. Read all Material Safety Data Sheets prior to starting the protocol. Have a chemical spill kit on-hand and prepare an action plan for spills or other problems.
2. For additional blocking of non-specific surface-binding sites, add 0.1 mg/ml BSA into your microtubule solution and throughout. BSA competes with tubulin and MAPs for non-specific binding sites and it does not fluoresce.
3. Note that nucleotide state analogues other than AMPPNP (i.e. ADP-VO_4, ADP-BeFx, etc.) may be used where appropriate.
4. This is where a bit of practice and a steady hand become important if you are using filter paper.

References

1. Brouhard, G., Stear, J., et al. (2008) XMAP215 Is a Processive Microtubule Polymerase. Cell 132(1):79–88.
2. Varga, V., Leduc, C., et al. (2009) Kinesin-8 Motors Act Cooperatively to Mediate Length-Dependent Microtubule Depolymerization. Cell 138(6):1174–1183.
3. Sackett, D., Bhattacharyya, B., Wolff, J. (1985) Tubulin subunit Carboxyl-Termini Determine Polymerization Efficiency. J Biol Chem 260(1): 43–35.
4. Helenius, J., Brouhard, G., et al. (2006) The depolymerizing kinesin MCAK uses lattice diffusion to rapidly target microtubule ends. Nature 441(7089):115–119.
5. Hirokawa, N., Nitta, R., Okada, Y. (2009) The mechanisms of kinesin motor motility: lessons from the monomeric motor KIF1A. Nature Rev Mol Cell Biol. 10(12):877–84.
6. Mori, T., Vale, R., Tomishige, M. (2007) How kinesin waits between steps. Nature 450 (7170):750–754.

Chapter 13

Analysing the ATP Turnover Cycle of Microtubule Motors

Claire T. Friel, Clive R. Bagshaw, and Jonathon Howard

Abstract

Proteins of the kinesin superfamily share a conserved motor domain, which both hydrolyses adenosine-5′-triphosphate (ATP) and binds microtubules. To determine the mechanism of action of a kinesin, it is necessary to relate the chemical cycle of ATP turnover to the mechanics of microtubule interaction. In this chapter, a number of methods are outlined by which the ATP turnover cycle of a kinesin can be analysed with a particular focus on the use of fluorescently labelled ATP and ADP analogues as a means of isolating individual steps in the cycle. By analysing the ATP turnover cycle of a kinesin, both in solution and in the presence of microtubules, the change in nucleotide state triggered upon microtubule binding can be determined. This provides information vital to understanding the coupling of the chemical and mechanical cycles that is integral to the action of members of the kinesin superfamily.

Key words: Kinesin, ATP hydrolysis, Enzyme kinetics, Mant-ATP, HPLC, Stopped flow

1. Introduction

Kinesins are a large family of proteins that use turnover of adenosine-5′-triphosphate (ATP) to regulate their interaction with the microtubule cytoskeleton (1). Changes in the nucleotide state of the kinesin motor domain are stimulated by interaction with the microtubule, and in turn the affinity of a kinesin motor domain for the microtubule is altered according to its nucleotide state. The exact nature of the nucleotide-dependent interaction with microtubules varies among members of the kinesin superfamily producing kinesins tailored to different functions, such as cargo transport (2), microtubule sliding (3), and microtubule depolymerisation (4, 5). However, for all kinesins, ATP turnover is closely coupled to the interaction with microtubules which stimulates the ATP hydrolysis rate of a kinesin often by several

Anne Straube (ed.), *Microtubule Dynamics: Methods and Protocols*, Methods in Molecular Biology, vol. 777,
DOI 10.1007/978-1-61779-252-6_13,

1,000-fold above the basal ATP turnover rate. The precise nature of the way in which a kinesin hydrolyses ATP (chemical cycle) as well as the nature of its interactions with microtubules (mechanical cycle) must be known in order to fully comprehend the action of an individual kinesin. In this chapter, a selection of protocols are described by which the ATP turnover cycle of a kinesin can be studied, particularly focusing on identifying the rate-limiting step in the basal ATP turnover cycle and how this is affected by interaction with microtubules.

2. Materials

1. 100 mM Mg-ATP: The disodium salt of ATP (Roche #10519979001) is made up to a concentration of 100 mM in 100 mM $MgCl_2$ and the pH adjusted to 6.9 with KOH. The concentration should be checked by absorption at 260 nm ($\varepsilon = 15{,}400$/M cm) (6) and the solution stored in aliquots at −20°C.
2. 2′/3′-(*N*-methylanthraniloyl)-adenosine-5′-triphosphate (mant-ATP) (Jena biosciences NU-202).
3. 2′/3′-(*N*-methylanthraniloyl)-adenosine-5′-diphosphate (mant-ADP) (Jena biosciences NU-201).
4. 3′-(*N*-methylanthraniloyl)-2′-deoxy-adenosine-5′-triphosphate (mant-dATP) (Jena biosciences NU-203).
5. G25 sephadex resin. It is convenient to use pre-packed columns, for example NAP-5 column (GE Healthcare #17-0853-01).
6. C-18 column. For example, the Luna 3 μ C18(2) 100Å (Phenomenex).
7. C-18 column running buffer: 100 mM potassium phosphate pH 6.5, 10 mM tetrabutylammonium bromide (TBABr) (Fluka #86860), plus acetonitrile, HPLC grade (Perbio #51101) as required, 10% (Subheading 3.1.1) or 30% (Subheading 3.2.2). Filter the buffer using 0.45-μm nylon filters prior to use.
8. BIOMOL green reagent (Enzo Life Sciences AK-111).
9. 0.6 M perchloric acid, pre-cool on ice (Sigma-Aldrich #24425-2).
10. 1 M Tris, 3 M KOH.
11. 2 M HCl.
12. 96-well plate (Greiner #655101).
13. Ethylenediaminetetraacetic acid (EDTA): 0.5 M stock solution in H_2O, pH 8.0.

14. Dithiothreitol (DTT): 1 M stock solution in H_2O, make fresh from powder for use on the same day.
15. BRB80 buffer: 80 mM PIPES, KOH pH 6.9, 1 mM $MgCl_2$, and 1 mM EGTA (see Note 4). This buffer is normally made up as a 5× stock solution (note that PIPES buffer is not very soluble in water until the pH is raised to ~6).

3. Methods

3.1. Analysis of Overall Cycle

The starting point for an analysis of the ATP turnover cycle (Fig. 1) of a kinesin is to determine the complete ATP turnover cycle rate: the rate at which ATP is hydrolysed under steady-state conditions. To do this, the rate of product production (either ADP or P_i) is determined under saturating substrate (ATP) concentrations. This rate is governed by the slowest step in the ATP turnover cycle and as such is a measure of the rate-limiting step in the cycle. When the steady-state rate is normalised for the kinesin concentration, e.g. μmole Pi/μmole kinesin/second, it becomes equivalent to a first order rate constant for the reaction. This value is referred to in the literature by a number of different terms (e.g. ATPase rate and turnover rate); this variation in nomenclature can cause confusion (see Note 1). This rate constant may also be determined directly by transient-state methods (see Subheading 3.2) allowing identification of the rate-limiting step in the cycle.

In order to measure the rate of ATP turnover under steady-state conditions, the substrate (ATP) must be in large excess over the concentration of kinesin motor domains (see Note 2) such that the reaction proceeds under saturating ATP concentrations. High concentrations of ATP are required for a linear initial rate not only to minimise the depletion of ATP, but also to ensure that ATP remains competitive as ADP concentrations build up.

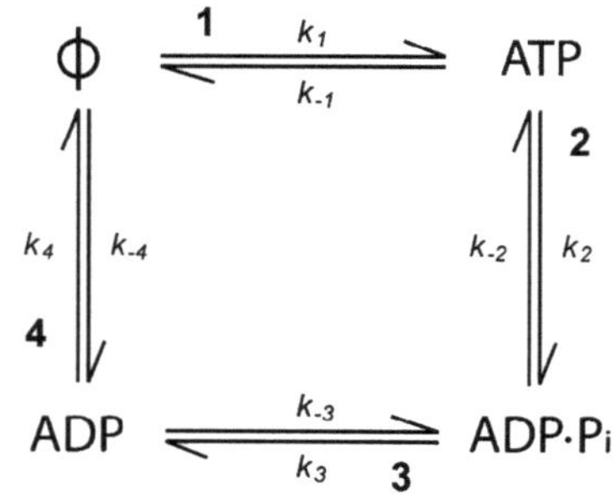

Fig. 1. The minimal ATP turnover cycle for an ATP-hydrolysing enzyme, such as a kinesin. Each kinesin motor domain can exist in four states with respect to nucleotide: nucleotide-free (ϕ), ATP-bound, ADP.P_i-bound, and ADP-bound.

In principle, to ensure that the ATP concentration is saturating, a titration of ATP should be carried out to determine the point at which the rate of ATP turnover no longer increases with increasing concentration of ATP. In practice, however, when working with micro-molar concentrations of kinesin, 2 mM ATP is generally found to be sufficient. Two methods are described by which the overall hydrolysis cycle rate can be determined: by following the production of ADP and the production of phosphate.

3.1.1. Measuring Production of ADP

There is no optical spectroscopic signal by which ADP can be distinguished from ATP. Therefore, in order to directly monitor the production of ADP, it is necessary to take samples as the reaction progresses and separate the product (ADP) from the substrate (ATP), for example by HPLC as described below (see Note 3).

1. Set up reaction mix with 2 mM Mg-ATP in the chosen buffer (see Note 4).
2. Start reaction by mixing kinesin (see Note 5) in chosen buffer with ATP-containing reaction mix (see Note 6).
3. Incubate reaction at chosen temperature (normally, either 25 or 37°C is used).
4. Take samples at selected time points (see Note 7), and quench reaction by mixing with an equal volume of ice-cold 0.6 M perchloric acid.
5. Neutralise samples by addition of the necessary volume of 1 M Tris, 3 M KOH, which is determined by trial reactions (e.g. if using 20 μl sample + 20 μl 0.6 M perchloric acid, 8 μl 1 M Tris, 3 M KOH is required). Neutralised samples can be stored at −80°C until all samples are ready for analysis.
6. Spin samples at 17,000 g 13,000 rpm, Heraeus Biofuge Microcentrifuge for 10 min to remove precipitate.
7. To separate ADP from ATP, run each reaction sample on a C-18 column (see Note 8) using isocratic flow with a 100 mM potassium phosphate, 10 mM TBABr, and 10% acetonitrile buffer (see Subheading 2, item 7). The flow rate used depends upon the pressure limits of the particular HPLC system and column used, but flow rates of ~1 ml/min are generally achievable.
8. The presence of both ADP and ATP is measured by absorbance at 260 nm. The integrated area under the respective peaks provides a measure of the quantity of each nucleotide present in the sample (Fig. 2a).
9. For each time point, calculate the proportion of ADP as a fraction of total nucleotide (area of ADP peak/[area of ADP + ATP peaks]).

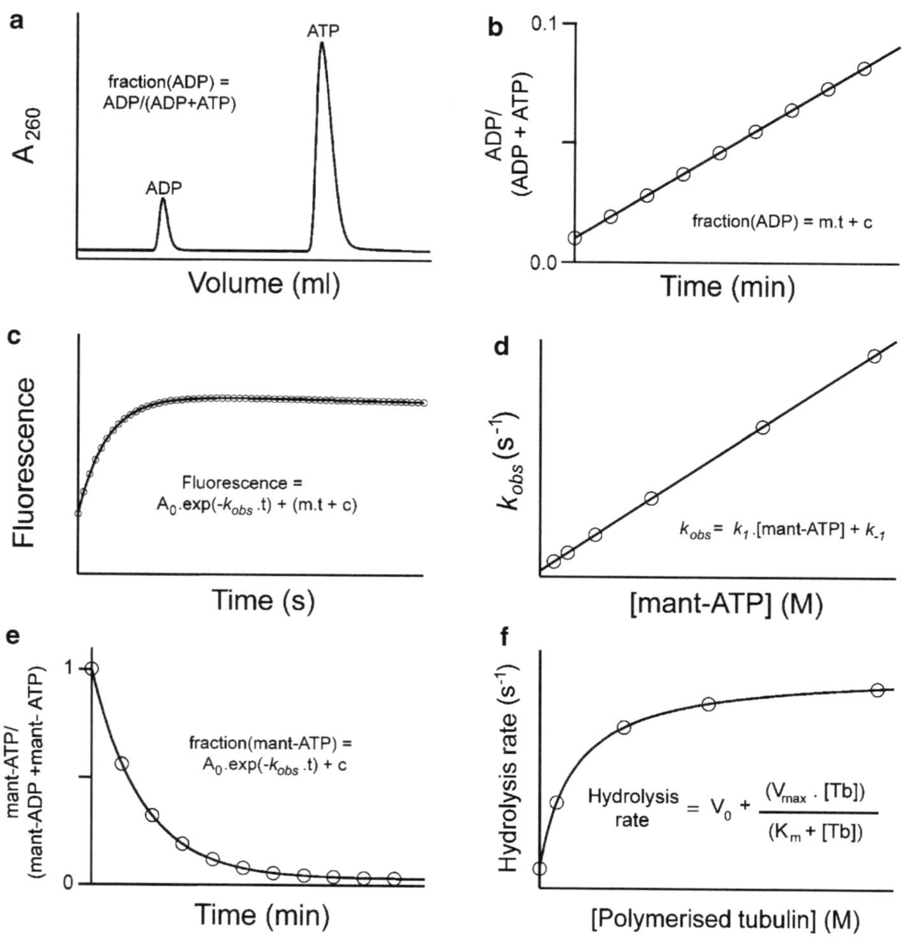

Fig. 2. (**a**) Example HPLC trace from a sample of the reaction of a kinesin with a saturating concentration of ATP. From this trace, ADP as a fraction of total nucleotide is calculated from area under the ADP peak divided by the sum of the areas under the ADP and ATP peaks. (**b**) Simulated data points (*open circles*), ADP as a fraction of total nucleotide [fraction(ADP)] versus time. The fraction(ADP) at each time point is determined from HPLC traces of the type shown in (**a**). The gradient (m) of a linear fit to these data gives the change in fraction(ADP) per unit time, from which the rate constant for ATP turnover can be calculated. (**c**) Simulated data points (*open circles*), fluorescence signal upon binding of mant-ATP to nucleotide-free kinesin. The fluorescence data is fit to an exponential function to obtain k_{obs} at each concentration of mant-ATP. A line of constant negative slope (m.t + c) is included, if necessary, in the fitting function to account for photo-bleaching of the mant group. (**d**) Simulated data points (*open circles*), k_{obs} [obtained from traces, such as (**c**)] versus the concentration of mant-ATP. Fitting this data to a linear function gives k_1 (gradient, association rate constant) and also k_{-1} (*y*-axis intercept, dissociation rate constant) (Fig. 1, step 1, when $k_{-1} > k_2$). (**e**) Simulated data points (*open circles*), mant-ATP as a fraction of total nucleotide [fraction(mant-ATP)] plotted against time. Data expected from the reaction of a kinesin with a sub-stoichiometric concentration of mant-ATP. The fraction(mant-ATP) is fitted to an exponential function to obtain k_{obs} for the hydrolysis of ATP. It is necessary to confirm whether this rate constant represents ATP binding or cleavage (Fig. 1, steps 1 and 2, see Note 18). (**f**) Simulated data points (*open circles*) showing the dependence of the complete ATP turnover cycle rate constant of a kinesin on the concentration of polymerised tubulin (microtubules). To obtain the ATP turnover rate (V_{max}) at saturating concentrations of polymerised tubulin ([Tb]), the data is fit to the function shown – V_0 is the ATP turnover rate in the absence of tubulin and K_m is the concentration of polymerised tubulin for which the half-maximal ATP turnover rate is observed.

10. Plot the fraction of ADP against time (Fig. 2b). The change in the fraction of ADP is linear if the reaction is in the steady state (see Note 9).
11. Fit the linear region of the data to a linear function, the slope of which gives the change in fraction of ADP per unit time.
12. To convert the change in fraction of ADP to change in concentration of ADP per unit time, multiply by the initial concentration of ATP (in this case, 2 mM).
13. Convert this into the rate constant for hydrolysis by dividing by the concentration of kinesin motor domains (see Note 10).
14. The value obtained has units of reciprocal time (normally quoted in s^{-1}) and represents the number of molecules of ATP hydrolysed per motor domain per second.

3.1.2. Measuring Production of Inorganic Phosphate

The progress of an ATP hydrolysis reaction can also be observed by monitoring the production of inorganic phosphate (P_i). As there is no signal by which phosphate can be directly observed, a secondary molecule is generally used via which P_i is detected. The most commonly used are molybdate/malachite green reagents that show an increase in absorbance at 620 nm when bound to inorganic phosphate. Such reagents can be prepared in the lab (7); however, there are a number of commercially available reagents (see Subheading 2, item 8).

1. Set up reaction mix with 2 mM Mg-ATP in the chosen buffer (see Note 4).
2. Start reaction by mixing kinesin (see Note 5) in chosen buffer with ATP-containing reaction mix (see Note 6).
3. Incubate reaction at chosen temperature (normally, either 25 or 37°C is used).
4. Take samples at suitable time points (see Note 7) and quench the reaction by addition to an equal volume of 2 M HCl.
5. Neutralise samples by addition of the necessary volume of 1 M Tris, 3 M KOH, which is determined by trial reactions (e.g. if using 25 μl sample + 25 μl 2 M HCl, 17.5 μl 1 M Tris, 3 M KOH is required).
6. When all samples are completed, spin at 17,000 g 13,000 rpm, Heraeus Biofuge Microcentrifuge for 10 min to remove precipitate.
7. To analyse sample using a plate reader, transfer 50 μl of the sample supernatant to a 96-well plate. Add 50 μl BIOMOL Green Pi detection solution.
8. Allow 20 min at room temperature for colour development (see Note 11).
9. Measure absorbance at 620 nm in a plate reader.

10. To convert the absorbance values to concentration of phosphate, a standard curve is required. Make a titration of an appropriate concentration range of phosphate standards (up to ~80 μM). Treat phosphate standards as the reaction samples (Subheading 3.1.2, steps 4–9).
11. Plot the concentration of phosphate against A_{620}, and fit a linear function to the data to obtain the relationship between A_{620} and phosphate concentration. Use this to convert absorbance data into phosphate concentration.
12. Plot phosphate concentration against time over the time course of the reaction.
13. Fit the linear region of the data to a linear function, the slope of which is the change in phosphate concentration per unit time (see Note 9).
14. Convert this into the rate constant for ATP turnover by dividing by the concentration of kinesin motor domains (see Note 10).
15. The value obtained has units of reciprocal time (normally quoted in s^{-1}) and represents the number of molecules of ATP hydrolysed per motor domain per second.

3.2. Analysis of Individual Cycle Steps

In order to determine which step in the cycle (Fig. 1) is responsible for the rate constant for ATP turnover of the complete cycle (Subheading 3.1), each step must be individually analysed. It is important to identify the rate-limiting step in the absence of microtubules, as this step must be accelerated by interaction with microtubules to produce the microtubule-stimulated increase in the ATP turnover rate generally observed for kinesins. Also, when the rate-limiting step in the cycle is known, the predominant intermediate is known, and therefore the predominant nucleotide-bound state in the absence of microtubules. This allows one to identify the change in nucleotide state of the kinesin that is triggered by interaction with microtubules.

One method used to determine rate constants for individual steps in the ATP turnover cycle (Fig. 1) is to take advantage of the properties of nucleotides labelled with the fluorescent group methylanthraniloyl (mant) (8). The mant group is most often used because being one of the smallest available fluorescent groups, when conjugated to the ribose it generally has little or no effect on the kinetics of ATP hydrolysis (9–11).

3.2.1. Association of ATP

The kinetics of ATP binding can be measured using mant-ATP taking advantage of the increase in fluorescence of the mant group that generally occurs upon nucleotide binding to the kinesin motor domain. To observe the kinetics of ATP binding, the ADP that remains trapped in the kinesin nucleotide-binding pocket in the absence of microtubules must be removed, allowing step 1 in the ATP turnover cycle to be observed in isolation (Fig. 1).

1. To a solution of kinesin in any appropriate storage buffer (preferably Mg^{2+}-free), add 1 mM (final concentration) EDTA pH 8 (see Note 12) and 1 mM (final concentration) DTT. The EDTA chelates Mg^{2+} ions causing the Mg^{2+} to be removed from the nucleotide-binding pocket. This in turn causes the nucleotide to be released (11).
2. Incubate at 25°C for 15 min.
3. Separate the free nucleotide from the protein by buffer exchanging into an appropriate Mg^{2+}-free buffer (see Note 13) using G25 sephadex resin (see Subheading 2, item 5).
4. Immediately after buffer exchange, add 1 mM $MgCl_2$ to the solution of nucleotide-free protein (see Note 14).
5. Make up a concentration series of mant-ATP in the appropriate reaction buffer. The range of concentrations of mant-ATP used depends on the concentration of kinesin available (see Note 15).
6. To simplify the observed kinetics, the reaction is carried out under pseudo-first-order conditions: the concentration of mant-ATP is in at least fivefold excess over the concentration of nucleotide-free protein. Therefore, it is assumed that the concentration of free mant-ATP remains constant over the course of the reaction; for further information on binding and dissociation kinetics, see (12).
7. Starting with the lowest mant-ATP concentration, rapidly mix the reactants (mant-ATP and nucleotide-free kinesin) in a 1:1 v/v ratio using a stopped-flow fluorimeter (see Note 16). The fluorescence of the mant-ATP is excited at 365 nm, and the emitted fluorescence above 400 nm is collected via a long-pass filter (GG400, Schott).
8. The observed rate constant (k_{obs}) at each mant-ATP concentration is determined by fitting the observed fluorescence increase to an exponential function (Fig. 2c) (see Note 17).
9. Plot the rate constants determined against the concentration of mant-ATP. There is a linear relationship between the observed rate constants and the concentration of mant-ATP, the gradient of which represents the association rate constant for mant-ATP (Fig. 1, k_1) with units of M^{-1}/s (Fig. 2d).

3.2.2. Cleavage of ATP

In order to determine whether the rate-limiting step is before or after the hydrolysis step, it is necessary to follow the kinetics of the ATP cleavage (hydrolysis) reaction (Fig. 1, step 2). This can be done using an excess of molar concentration of ATP over kinesin and determining if there is a burst of liberated phosphate when the reaction is terminated by an acid quench (which releases any bound phosphate) (13). It is also useful to determine the kinetics of ATP hydrolysis under single turnover conditions, i.e. mixing a molar concentration of motor domain in excess of the molar concentration of substrate (ATP). Multiple cycles of hydrolysis are prevented allowing transient intermediate species to be observed. However, in order that the ATP-binding phase is rapid compared with the rate-limiting step, it is necessary to use a kinesin concentration greater than the K_m for ATP, which may be difficult to achieve in practice (see Note 18).

As the reaction progresses, samples are quenched, denaturing the kinesin and releasing the protein-bound nucleotide. By separating the products using HPLC as described below, it is possible to observe the disappearance of the kinesin–ATP complex and thereby determine at least a lower limit for the rate constant for ATP cleavage (see Note 18).

When the reaction is carried out under single turnover conditions, the concentration of ATP used is limited by the concentration of kinesin available. Since most kinesins can be expressed and purified only at low micro-molar concentrations, it may be necessary to use mant-ATP and fluorescence detection to increase the sensitivity of the method. Alternatively, radioactive ATP with a labelled γ-phosphate can be used. However, using a more sensitive detection method will not overcome the problems associated with a kinesin, which has an inherently large K_m, because the observed kinetics will be limited by the binding reaction. Whether the observed reaction is limited by binding or cleavage should be checked (see Note 19).

1. Start reaction by mixing 1 μM nucleotide-free kinesin (Subheading 3.2.1, steps 1–4) motor domain (see Note 20) with 0.5 μM mant-deoxyATP (mant-dATP) (see Note 21).
2. Quench reaction at various suitable time points (see Note 22) by injection onto a C-18 HPLC column (see Note 23).
3. To separate mant-dATP from mant-dADP, run each sample over a C-18 column (see Note 24) using isocratic flow with a 100 mM potassium phosphate, 10 mM TBABr, and 30% acetonitrile buffer (see Subheading 2, item 7). The flow rate used depends upon the pressure limits of the particular HPLC system and column used, but flow rates of ~1 ml/min are generally achievable.
4. The presence of both mant-dADP and mant-dATP is measured using the fluorescence of the mant group (λ_{ex} = 355 nm, λ_{em} = 448 nm). The integrated area under the respective peaks provides a measure of the quantity of each nucleotide present in the sample.
5. For each time point, calculate the proportion of mant-dATP remaining as a fraction of total nucleotide (area of mant-dATP peak/[area of mant-dATP + mant-dADP peaks]).

6. Plot the fraction of mant-dATP/total nucleotide against time and fit the data to an exponential function in order to determine the rate constant for ATP cleavage (Fig. 2e).

3.2.3. Dissociation of Phosphate

For all kinesins studied to date, the dissociation of phosphate (Fig. 1, step 3) is faster than the cleavage of ATP (Fig. 1, step 2), so only a lower limit to the rate constant for the dissociation of inorganic phosphate can generally be determined. Detailed methods for the study of this step are, therefore, not described here. Information on observing P_i dissociation, by coupling the presence of P_i to the fluorescence signal of the phosphate-sensing protein, can be found in refs. 10, 14, 15.

3.2.4. Dissociation of ADP

To provide a signal by which the dissociation of ADP from a kinesin (Fig. 1, step 4) can be observed, ADP labelled with the mant fluorophore is used, again taking advantage of the signal difference generally observed between free mant-ADP and mant-ADP bound to the kinesin motor domain.

1. The unlabelled ADP bound to the kinesin motor domain in the absence of microtubules is exchanged for mant-ADP by the following method: to a solution of the kinesin of interest, add mant-ADP to a concentration in at least 25-fold molar excess over the concentration of kinesin motor domains.
2. Incubate at 25°C for 30 min.
3. Remove excess mant-ADP and other free nucleotides from the protein by buffer exchanging into the chosen reaction buffer using G25 sephadex resin (see Subheading 2, item 5).
4. The mant-ADP-loaded kinesin has a fluorescence signal, the loss of which is used to observe the dissociation of mant-ADP.
5. Rapidly mix the mant-ADP–kinesin complex (~1 μM) with a molar excess (50-fold or greater) of unlabelled ATP in a 1:1 v/v ratio using a stopped-flow fluorimeter (see Note 25).
6. The fluorescence of mant-ADP is excited at 365 nm, and the emitted fluorescence above 400 nm is collected via a long-pass filter (GG400, Schott).
7. The decrease in fluorescence, observed as the mant-ADP dissociates from the kinesin, is fit to an exponential function from which the rate constant for dissociation of ADP (Fig. 1, k_{-4}) is determined with units of s^{-1} (see Note 26).

3.3. Effect of Microtubules

To determine the effect of microtubules on the ATP turnover cycle, one must determine the rate constant of the complete ATP turnover cycle in the presence of microtubules. This can be done by either of the methods described in Subheading 3.1. The rate-limiting step in the microtubule-stimulated ATP turnover cycle can then be determined by analysing the individual steps in the cycle (Subheading 3.2) in the presence of microtubules.

1. Measure the rate constant of the complete cycle (as described in Subheading 3.1) at a range of concentrations of polymerised tubulin (microtubules) (see Note 27).
2. To determine the maximum ATP turnover rate constant at saturating concentrations of polymerised tubulin (microtubules), plot the rate constant for ATP turnover against the concentration of polymerised tubulin. Fit the data to a hyperbolic function which allows the determination of the V_{max} (i.e. the rate constant of the complete cycle for the kinesin under study in the presence of saturating polymerised tubulin) (Fig. 2f).
3. The rate constants for individual steps in the ATP turnover cycle (Fig. 1) can then be measured (according to the methods in Subheading 3.2) in the presence of a saturating concentration of polymerised tubulin (microtubules) in order to determine how the rate constants are affected by interaction of the kinesin with microtubules. At this point, the focus should be on steps in the cycle that are accelerated by the interaction with microtubules (i.e. have a rate constant in the absence of microtubules lower than the microtubule-stimulated ATP turnover rate).
4. The same procedure can be carried out to determine the effect of unpolymerised tubulin (free α/β-heterodimers) on ATP hydrolysis. This information can also be useful in uncovering the molecular mechanism of the action of a kinesin.

4. Notes

1. Much confusion in the literature has been caused by the tendency for the term "rate constant" to be shortened to "rate". These terms are not equivalent. A rate refers to a change in concentration with time and has units of concentration (or amount) per unit time. A rate constant is the coefficient that relates the rate to the concentration of reactants – effectively, it is the rate at unitary concentration(s) of the reactants (analogous to a molar absorption coefficient). The units of a rate constant depend on the molecularity of the reaction, e.g. s^{-1} for uni-molecular, M^{-1}/s for bi-molecular reactions. Rates can be positive or negative and generally increase with increasing concentrations (the law of mass action). Rate constants are positive coefficients, but observed rate constants can increase or decrease with increasing concentration.
2. In the context of studying the reaction of a kinesin with ATP, it is convenient to determine the molar concentration

as the concentration of ATP-binding domains (i.e. the monomer concentration). Kinesin molecules consist of different numbers of nucleotide-binding domains: monomeric, dimeric, and tetrameric.

3. It is possible to indirectly monitor the production of ADP by linking it to the spectroscopic signal of another molecule. This has the advantage of keeping the steady-state ADP concentration very low and preventing product inhibition of the kinesin, and also allowing the reaction to be continuously monitored (7).
4. The choice of reaction buffer is dependent on the kinesin under study. However, in general, a buffer is chosen that is suitable for microtubules, and therefore facilitates further studies in the presence of microtubules. One of the most commonly used buffers for this purpose is BRB80 (Subheading 2, item 15). Extra salt (e.g. KCl) is often added to BRB80 to modulate the interaction of the kinesin under study with microtubules. The solubility of many kinesins is limited in buffers suitable for microtubule stability. Therefore, for long-term storage, a more suitable buffer is generally used and the kinesin exchanged into the reaction buffer directly prior to use.
5. The concentration of kinesin used is chosen depending on its particular kinetics. If the rate of product production is slow, the concentration of kinesin used may be increased to produce detectable levels of product over a practical time period. It may be necessary to carry out some test reactions to determine the most suitable conditions, but make sure that the steady-state criteria remain satisfied and the reaction is monitored for a time corresponding to two or more turnovers.
6. The method of mixing used to start the reaction depends on the kinetics of the reaction. For a reaction under steady-state conditions, manual mixing is generally fast enough to allow observation of the subsequent reaction. If the reaction kinetics are too fast for manual mixing, a more rapid mixing technique, such as quench flow, may be used (16).
7. The time points selected depend on the rate of product production (see Note 5). Choose times points that give a good number of data points (>4) over the linear range of the reaction (see Note 9).
8. The Luna 3 μ C18(2) 100 Å, 100 × 4.6 mm (Phenomenex), gives excellent separation of ADP from ATP with relatively short run times (~10 min).
9. Depending on the kinetics of hydrolysis and the time points taken, it is possible that a non-linear phase may be observed at the start of the reaction as the system approaches the steady state. Also, as substrate is consumed, the kinetics of the reaction

may become non-linear due to both substrate depletion and product inhibition; this generally occurs when as little as 10% of the substrate has been consumed.

10. Make sure to match the units of kinesin motor domain concentration with the concentration of nucleotide.
11. If samples are to be measured separately, as when using a spectrophotometer, stagger the addition of Biomol green reagent such that all samples have the same time for colour development between addition of the reagent and measurement of absorbance.
12. The exact concentration of EDTA is not vital, but should be in at least 50-fold molar excess of the kinesin motor domain concentration. Also note that the concentration of EDTA must be in excess of any Mg^{2+} present in the kinesin storage buffer.
13. The kinesin should be exchanged into the buffer that is used for the reaction. If the chosen buffer contains Mg^{2+} (see Note 4), an Mg^{2+}-free version should be used at this stage and the Mg^{2+} added after the buffer exchange step. To reduce loss of kinesin on the resin during buffer exchange, add 0.05% Tween-20 to the reaction buffer.
14. Re-addition of Mg^{2+} helps to stabilise the nucleotide-free kinesin. However, the apo-motor domains are still relatively unstable and the nucleotide-free kinesin should be kept on ice and used within a few hours. There are reports of nucleotide-free kinesin remaining active for up to 2 weeks when stored frozen in a high-salt glycerol buffer (17).
15. The concentration of the kinesin is generally changed in accordance with the concentration of mant-ATP so that it remains at between five- and tenfold molar excess over the concentration of kinesin motor domains. This is so the signal change due to the binding of mant-ATP to kinesin is not swamped by the fluorescence from free mant-ATP. The maximum concentration of mant-ATP at which the reaction can be monitored is, therefore, limited by the concentration of protein available.
16. It is important to account for the dilution of the reactants that occurs upon mixing. The concentration in the syringes prior to mixing is twice that in the reaction. It is the reaction (post mixing) concentrations of mant-ATP that should be used to determine the ATP association rate constant. It is worth noting both concentrations in the lab book/file tag, e.g. 10/5 μM ATP so that in later reference it is clear what was used (it is easy to divide by two twice if one's memory fades!).
17. It may be necessary to include a linear component with a negative slope to the exponential fit function to account for

photo-bleaching of the mant-labelled nucleotide. The extent of photo-bleaching (and also mixing artefacts) can be determined by carrying out a control reaction in which mant-ATP is mixed with reaction buffer without kinesin. Also, temporary closing of the excitation shutter can be used to check if a drift is due to photo-bleaching or some other cause.

18. Single turnover conditions require that the binding step is rapid compared with the subsequent steps, if the latter are to be resolved. This is achieved by using concentrations of reactants which are greater than the K_m. For the scheme in Fig. 1, k_1. [kinesin] should be $>k_{-1}+k_2$, for the hydrolysis step to be resolved from the binding step.

19. One method of distinguishing between these possibilities is to monitor the reaction continuously via the fluorescence signal of the mant group in conjunction with the quenching-type assay described. If the rate constant for the increase in fluorescence associated with the binding of mant-ATP to nucleotide-free kinesin matches the rate constant determined by the quenching assay, then ATP binding is slow and therefore the rate constant determined for the dissipation of the ATP–kinesin complex represents binding rather than cleavage of ATP.

20. Since it is often difficult to determine the exact concentrations of active protein, a diagnostic test can be carried out to ensure that a reaction is proceeding under single turnover conditions. Double the concentration of protein and keep the concentration of ATP the same; if operating under single turnover conditions, the rate constants determined for steps following binding should be unaffected. This also provides a method to confirm which step the rate constant measured represents; if the rate constant measured represents the binding of ATP to the nucleotide-free kinesin, it will double upon doubling the protein concentration.

21. 3′-mant-2′-deoxyATP (mant-dATP) is used rather than regular mant-ATP which is a mixture of 2′/3′-mant-ATP, and therefore gives two peaks for both ATP and ADP when the nucleotides are separated on a C-18 column.

22. It is necessary to collect a number of time points (preferably ten or more) over a reasonable timescale to get a good description of the progress of the reaction. It may be necessary to carry out some test reactions to get an idea of the timescale over which the reaction proceeds.

23. As mant-labelled nucleotides can be sensitive to harsh quenching conditions, quenching can be carried out by injection of reaction samples directly onto the column. In this case, it is essential to use a pre-column/pre-filter to protect the resolving column from denatured protein and other precipitates.

24. The mant label means that nucleotides bind more strongly to the column and are more difficult to separate. Therefore, a longer column may be required than for unlabelled nucleotides [e.g. Luna 3 μ C18(2) 100 Å, 250×4.6 mm (Phenomenex)].
25. Excess unlabelled ATP is included to prevent rebinding of mant-ADP, therefore ensuring that only the dissociation of mant-ADP is observed. For further information on binding and dissociation kinetics, see (12).
26. It may be necessary to include a linear component with a negative slope to the exponential fit function to account for photo-bleaching of the mant-labelled nucleotide. In the case of the mant-ADP–kinesin complex, the extent of photo-bleaching can be determined by performing a control reaction in which mant-ADP–kinesin is mixed with mant-ADP–kinesin.
27. In these assays, microtubules stabilised using the non-hydrolysable GTP analogue, GMPCPP, or the drug taxol are used. For information on how to prepare stabilised microtubules, see refs. 18–20.

Acknowledgements

We acknowledge with thanks Roger Goody and Aymelt Itzen for assistance in developing the protocol described in Subheading 3.2.2, and also Christopher Gell for critical reading of the manuscript. This work was supported by the Alexander Von Humboldt Foundation and the Max Planck Society.

References

1. Miki, H., *et al.* (2005) Analysis of the kinesin superfamily: insights into structure and function *Trends Cell Biol* **15**, 467–476.
2. Hirokawa, N., *et al.* (2009) Kinesin superfamily motor proteins and intracellular transport *Nat Rev Mol Cell Biol* **10**, 682–696.
3. Peterman, E. J., and Scholey, J. M. (2009) Mitotic microtubule crosslinkers: insights from mechanistic studies *Curr Biol* **19**, R1089–1094.
4. Howard, J., and Hyman, A. A. (2007) Microtubule polymerases and depolymerases *Curr Opin Cell Biol* **19**, 31–35.
5. Wordeman, L. (2005) Microtubule-depolymerizing kinesins *Curr Opin Cell Biol* **17**, 82–88.
6. Burton, K. (1969) *in* "Data for Biochemical Research" (Dawson, R. M., *et al.*, Eds.), Oxford university Press.
7. Hackney, D. D., and Jiang, W. (2001) Assays for kinesin microtubule-stimulated ATPase activity *Methods Mol Biol* **164**, 65–71.
8. Bagshaw, C. (2001) ATP analogues at a glance *J Cell Sci* **114**, 459–460.
9. Lisal, J., and Tuma, R. (2005) Cooperative mechanism of RNA packaging motor *J Biol Chem* **280**, 23157–23164.
10. Gilbert, S. P., *et al.* (1995) Pathway of processive ATP hydrolysis by kinesin *Nature* **373**, 671–676.
11. Sadhu, A., and Taylor, E. W. (1992) A kinetic study of the kinesin ATPase *J Biol Chem* **267**, 11352–11359.

12. Goodrich, J. A., and Kugel, J. F. (2007) Binding and Kinetics for Molecular Biologists, Cold Spring Harbor Laboratory Press, New York.
13. Johnson, K. A. (1983) The pathway of ATP hydrolysis by dynein. Kinetics of a presteady state phosphate burst *J Biol Chem* **258**, 13825–13832.
14. Brune, M., *et al.* (1994) Direct, real-time measurement of rapid inorganic phosphate release using a novel fluorescent probe and its application to actomyosin subfragment 1 ATPase *Biochemistry* **33**, 8262–8271.
15. Cochran, J. C., *et al.* (2009) ATPase cycle of the nonmotile kinesin NOD allows microtubule end tracking and drives chromosome movement *Cell* **136**, 110–122.
16. Johnson, K. A. (1986) Rapid kinetic analysis of mechanochemical adenosinetriphosphatases *Methods Enzymol* **134**, 677–705.
17. Ma, Y. Z., and Taylor, E. W. (1997) Kinetic mechanism of a monomeric kinesin construct *J Biol Chem* **272**, 717–723.
18. Desai, A., and Walczak, C. E. (2001) Assays for microtubule-destabilizing kinesins *Methods Mol Biol* **164**, 109–121.
19. Hyman, A., *et al.* (1991) Preparation of modified tubulins *Methods Enzymol* **196**, 478–485.
20. Gell, C., *et al.* (2010) Microtubule Dynamics Reconstituted in vitro and Imaged by Single-Molecule Fluorescence Microscopy *Methods in Cell Biology* **95**, 221–245.

Chapter 14

Cryo-electron Tomography of Microtubules Assembled In Vitro from Purified Components

Frédéric M. Coquelle, Sophie Blestel, Claire Heichette, Isabelle Arnal, Charles Kervrann, and Denis Chrétien

This volume includes electronic supplemental material that will appear on the website http://extras.springer.com/straube.

Abstract

Cryo-electron tomography of vitrified specimens allows visualization of thin biological samples in three-dimensions. This method can be applied to study the interaction of proteins that show disorder and/or bind in a nonregular fashion to microtubules. Here, we describe the protocols we use to observe microtubules assembled in vitro in the presence of XMAP215, a large and flexible protein that binds to discrete sites on the microtubule lattice. Gold particles are added to the mix before vitrification to facilitate image acquisition in low-dose mode and their subsequent alignment before tomographic reconstruction. Three-dimensional reconstructions are performed using the IMOD software, processed with ImageJ and visualized in UCSF Chimera. Extraction of features of interest is performed using a patch-based algorithm (CryoSeg) developed in the laboratory. All the software used in this procedure is freely available or can be obtained on request, and run on most operating systems.

Key words: Microtubule, Tubulin, Microtubule-associated proteins, XMAP215, Cryo-electron tomography, Fiducial markers, Three-dimensional reconstruction, Segmentation

1. Introduction

Three-dimensional reconstructions of microtubule assemblies have long been obtained from two-dimensional electron microscope images by exploiting the helical symmetry of the tubulin molecules in the microtubule wall. Indeed, this method can provide high-resolution information on microtubules alone (1) or in association with accessory proteins (2). Yet, limitations may arise from the averaging that is necessary to perform these approaches, or from the requirement that the accessory proteins are arranged

Anne Straube (ed.), *Microtubule Dynamics: Methods and Protocols*, Methods in Molecular Biology, vol. 777, DOI 10.1007/978-1-61779-252-6_14, © Springer Science+Business Media, LLC 2011

periodically onto the microtubule lattice (3). Discrete structural features, such as proteins, that bind to microtubule extremities, or in a nonregular fashion to their wall, cannot be analyzed using helical reconstruction methods. Moreover, when considering the tubulin heterodimer as the basic repeating unit, only a limited number of microtubule types are fully helical. This excludes, for instance, the 13 and 14 protofilament microtubules, even though they are the most commonly found architectures in vitro and in vivo (4).

During the last years, improvements came from electron microscope automation and the use of sensitive CCD cameras to record images. Using this technology, it is now possible to obtain series of tilted images of fragile samples, such as vitrified biological specimens, and compute three-dimensional reconstructions using back-projection algorithms. Here, we detail the protocol we use to obtain three-dimensional reconstructions of XMAP215 in association with Taxol-stabilized microtubules, which includes specimen preparation, image acquisition, and computer processing. Although the present procedure is described for microtubule assemblies, most of the image acquisition and processing can be applied to other types of macromolecular assemblages.

2. Materials

2.1. Taxol-Stabilized Microtubules and XMAP215

1. Tubulin (see Note 1) at a concentration of about 300 μM in BRB80: 80 mM Pipes (Sigma, P6757), 1 mM $MgCl_2$ (ICN Biomedicals Inc., 191421), 1 mM EGTA (Sigma, E3889), pH 6.9 with KOH. Store at –80°C.
2. BRB80 stock solution (5×): 400 mM Pipes, 5 mM $MgCl_2$, 5 mM EGTA, pH 6.9 with KOH. Store at 4°C.
3. Guanosine 5′-Triphosphate (GTP) (Sigma, G8877). Powder dissolved in BRB80 buffer at 20 mM. Store at –20°C.
4. Paclitaxel (Taxol, see Note 2) from Pacific Yew Tree (ICN Biomedicals Inc., 193532). Powder dissolved in dimethyl sulfoxide (Sigma, D8418) at 10 mM. Store at –20°C.
5. XMAP215 (see Note 3) at a concentration of about 23 μM in BRB80, 20% glycerol. Store at –80°C.

2.2. Electron Microscopy

1. Electron microscope grids covered with a holey carbon film (see Note 4).
2. 10-nm gold particles functionalized with cationic bovine serum albumine (BSA, see Note 5) supplied in phosphate

buffered saline (PBS), with 15 mM NaN_3 (Aurion Gold Tracers, 210111).

3. Nanosep 10 K device (Pall Corporation, OD010C34).
4. Filter paper (see Note 6) Whatman no. 1.
5. Environmental guillotine device (see Note 7).
6. Cryo-stage: Single tilt liquid nitrogen cryo-transfer holder from Gatan allowing acquisition of tilt series (e.g., Gatan model 626).
7. Cryo-electron microscope equipped for tomography: 200 kV Tecnai G^2 Sphera, with LaB_6 filament, Gatan Digital Micrograph and FEI (TEM User version 3.1.1) software, 2k×2k Gatan CCD camera model Ultrascan 1000.

2.3. Image Analysis

1. IMOD software (Boulder Laboratory): Freely available at http://bio3d.colorado.edu/imod/. Runs on Mac OS X, Linux, and Windows platforms.
2. ImageJ software (NIH): Freely available at http://rsbweb.nih.gov/ij/. Platform independent.
3. MRC and Spider reader/writer plugin for ImageJ: Freely available at http://u759.curie.u-psud.fr/softwaresu759.html.
4. UCSF Chimera software: Freely available at http://www.cgl.ucsf.edu/chimera/. Runs on Mac OS X, Linux, and Windows platforms.

3. Methods

3.1. Preparation of Taxol Stabilized Microtubules

1. Prepare a 20 μl Taxol solution at 50 μM in BRB80 buffer and warm it up at 36°C in a water bath.
2. Prepare a 60 μl Taxol solution at 20 μM in BRB80 and keep it at room temperature.
3. Prepare 20 μl tubulin at 150 μM with 1 mM GTP in BRB80 at 4°C and incubate for 10–15 min on ice.
4. Induce tubulin polymerization by incubating the tubulin mix at 36°C for 1 h.
5. Add one volume (20 μl) of 50 μM Taxol solution to the microtubule suspension, mix by pipetting gently up and down with a truncated micropipet tip (see Note 8). Incubate for 1 h at 36°C.
6. Centrifuge the microtubule suspension at 25,000 × *g*, 15 min at room temperature.

7. Resuspend the microtubule pellet in two volumes (40 μl) of 20 μM Taxol solution and centrifuge at 25,000 ×*g*, 15 min at room temperature.
8. Resuspend in one volume (20 μl) of the 20 μM Taxol solution. The Taxol-stabilized microtubules can be kept at room temperature.
9. Estimate the protein concentration: Dilute a 1-μl aliquot 1:10 in a 1 M $CaCl_2$ solution. Incubate for 10 min at 4°C and evaluate the protein concentration using a Bradford assay (Sigma, B6916). The protein concentration of the undiluted Taxol-stabilized microtubules should be around 30 μM.

3.2. Preparation of Fiducial Markers

1. Resuspend the stock solution of cationic beads by tube inversions and deposit 150 μl in a Nanosep 10 K device.
2. Centrifuge 5 min at 5,000 ×*g*, 4°C.
3. Rotate the tube by 180° and centrifuge at 5,000 ×*g*, 5 min at 4°C.
4. Resuspend the beads in 40 μl of cold BRB80. Keep the beads at 4°C until the freezing procedure.

3.3. Preparation of the Mix Between Taxol-Stabilized Microtubules and XMAP215

1. Dilute Taxol-stabilized microtubules in BRB80 in order to reach a final tubulin concentration of about 5 μM.
2. Mix gently microtubules (1 μM final concentration) with XMAP215 (3 μM final concentration) in cold BRB80. Keep the mix at 4°C until the freezing procedure.
3. Just before freezing, mix gently but thoroughly at room temperature 4 μl of the previous mix with 4 μl of gold beads prepared before in BRB80 (see Subheading 3.2, step 4).

3.4. Specimen Vitrification

The aim of this step is to freeze rapidly at very low temperature (about −180°C) a thin layer of the suspension. The rate of freezing is such that water molecules have no time to organize into crystals, and thus remain in a vitreous state. In addition, the suspension layer must be sufficiently thin before freezing to allow water vitrification. Moreover, grids must be rendered slightly hydrophilic to facilitate the spreading of the suspension over the carbon membrane. For this purpose, we use an Edwards E306 unit.

1. Place the grids on a wire frame between electrodes.
2. Pump vacuum to ~0.1 atm.
3. Raise the current until a light glow discharge appears between electrodes and wait 30 s.
4. Close the valves and recover the charged grids.
5. Cut rectangular pieces of Whatman filter no. 1 of about 1 cm × 3 cm and fold them at right angle to facilitate their handling.

6. Prepare the guillotine device.
7. Take a grid with tweezers and attach to the guillotine device.
8. Pipet 4 μl of the suspension from step 3 in Subheading 3.3 and deposit them onto the grid.
9. Blot with filter paper and plunge into liquid ethane (see Notes 9 and 10).
10. Store grids in liquid nitrogen until observation (see Notes 11 and 12).

3.5. Electron Microscopy

The detailed steps concerning image acquisition vary depending on the instrumentation available. Here, we describe our procedure on an FEI Tecnai G^2 Sphera 200 kV electron microscope. Vitrified specimens are very sensitive to the electron beam. The main concern is to minimize radiation damage as much as possible when localizing suitable areas to be imaged and during acquisition. Movements of the cryo-stage during tilting must be corrected during acquisition, without additional electron dose. For this purpose, the mechanical displacements of the cryo-stage must have been pre-calibrated (see Fig. 1). Acquisition is performed using the post-tracking mode of the acquisition software. During this procedure, the displacement between two successive images is calculated by cross-correlation and is applied to the following acquisition. Cross-correlation is facilitated by the presence of the gold particles that provide a signal even at very low electron doses (see Fig. 2a). The present procedure assumes that your acquisition software allows acquisition of images in low-electron dose mode, with a search position at low magnification (e.g., 5,000×), a focus position, and an exposure position that is used to take pictures. Blanking of the electron beam must be performed whenever it is possible.

1. Load the specimen on the cryo-stage and insert it into the microscope. Let settle at least 15 min before removing the anti-contamination blade and opening the valve of the microscope. Make sure you reset the position of the cryo-stage so that you start at the grid center.
2. Localize a suitable area at low magnification in search mode and in very low electron dose conditions (see Note 13). The beam can be spread so that it illuminates a whole grid square. Use a binning of 4 on the camera to increase its sensitivity, and high defocus values to increase contrast (e.g., 300 μm).
3. Adjust eucentrism by tilting clockwise and anticlockwise the cryo-stage using the wobbler. Determine the maximum tilt angles on the specific area you have chosen.
4. Adjust focus (focus mode at a magnification of 29,000×) on an area adjacent to the one you are planning to image (see Note 14).

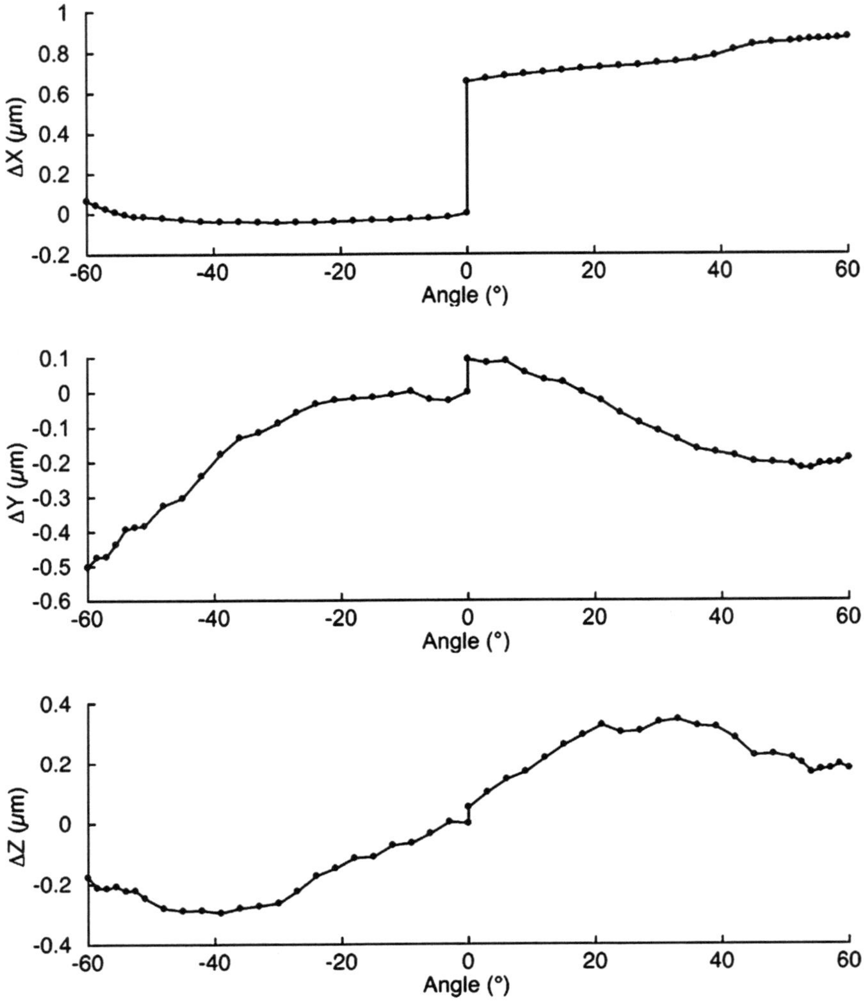

Fig. 1. Sample-holder calibration curves. Sample holders display mechanical imperfections characterized by small movements during rotation. The three curves describe the grid displacements in *X*, *Y*, and *Z* (focus) directions as a function of the rotation angle. The tomography software uses these curves to compensate the small movements during tilt series acquisition so that objects of interest are kept in the illumination field. Note the discontinuity at 0° since two half series are taken, one from 0° to –60° and a second from 0° to +60°.

5. Switch back to search mode and move the stage to the area to be imaged. Blank the beam.
6. Adjust acquisition parameters in the tomography component: enter the maximum angular ranges (see Note 15), e.g., from –60° to +60°. Select a Saxton scheme (see Note 16) and an angular increment according to the angular range chosen before so that you can record ~80 images. Set the starting angle at 0° (see Note 17). Fix a 5 s interval between each rotation of the stage and 30 s after readjustment at 0° tilt angle. Use the post-tracking mode (see Note 18).

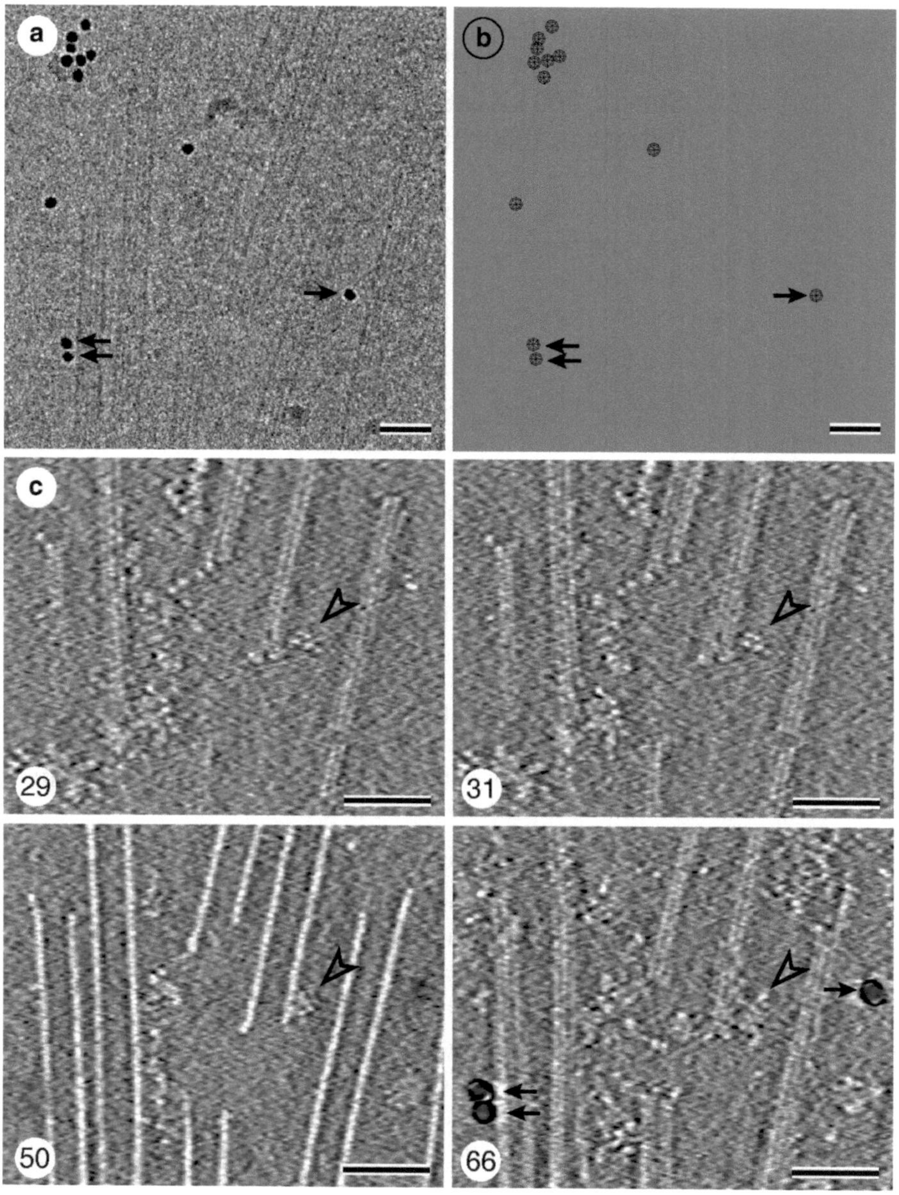

Fig. 2. (**a**) Crop of a single image (at 4.6°) from the tilt series (see Movie 1). Note the low contrast of the microtubules (protein is imaged dark with respect to the background), and the 10 nm gold particles (*arrows*) that have been added to the suspension before vitrification. Scale bar, 50 nm. (**b**) Bead model calculated from the position of the gold particles. Scale bar, 50 nm. (**c**) Sections through the final tomogram (see Movie 4). *Arrowheads* point to XMAP215 molecules attached to the extremity of a microtubule (section numbers are indicated in the *bottom left* corner of each image). *Arrows* in section 66 points to the gold particles marked in (**a**) and (**b**). Note the strong reinforcement of the protein contrast (inverted with respect to (**a**)) after 3D reconstruction. Scale bars, 50 nm.

Set magnification to 29,000×. Set the camera parameters as shown in Table 1 (see Note 19).

7. Start acquisition and save image series in Molecular Research Cambridge (MRC) format.

Table 1
Tomography acquisition parameters on the FEI Tecnai G^2 Sphera, 200 kV LaB_6, equipped with a 2k × 2k CCD camera (Ultrascan 1000, Gatan)

Mode	Search	Focus	Exposure	Tracking
Binning	4	1	2	2
Exposure time (s)	0.5	0.5	1	1
Top-left pixel	0/0	512/512	0/0	0/0
Bottom-right pixel	512/512	1536/1536	1024/1024	1024/1024

3.6. Image Processing

The types of files obtained may depend on the instrumentation and software used to acquire and record the images. Yet, the minimal requirements are the image series, obtained usually as stacks, the tilt angles, and the angle of rotation of the stage with respect to the CCD camera. On the Tecnai G^2 Sphera, we obtain four files: the stack of images (file.mrc), a text file with the angles (angles.rawtlt), and two other text files with all the acquisition parameters. Images are processed using the eTomo procedure of the free IMOD package from Boulder Laboratory (http://bio3d.colorado.edu/imod/, see Note 20).

1. Change the image file extension from .mrc to .st (for single tilt).
2. Launch eTomo and open the .st file. Specify "single tilt," and "tilt angles existing from a .rawtlt file." Scan the header of the image series to retrieve the image rotation angle and the pixel size. Enter a value of 10 nm for the fiducial diameter.
3. Check the quality of the tilt series by clicking "view raw image stack" and press on "create com scripts" (see Note 21).
4. Remove extreme values due to X-rays using default parameters and create a fixed stack (see Fig. 2a and Movie 1). Click "use fixed stack" and move on to the next procedure by clicking "done".
5. Create a coarsely aligned stack by cross-correlation. Uncheck "convert to bytes" if you wish to keep the whole dynamic gray level range of the images (see Note 22). Generate a coarse-aligned stack and check the result in 3dmod. If jumps appear between images (see Note 23), adjust the filters in the "tiltxcorr program" (low- and high-frequency rolloff sigma and the high-frequency cutoff radius) to optimize the cross-correlation. Once the images have been coarsely aligned, move on to the next procedure.
6. Fine alignments require first the make of a fiducial model based on the gold particles that are spread out over the specimen.

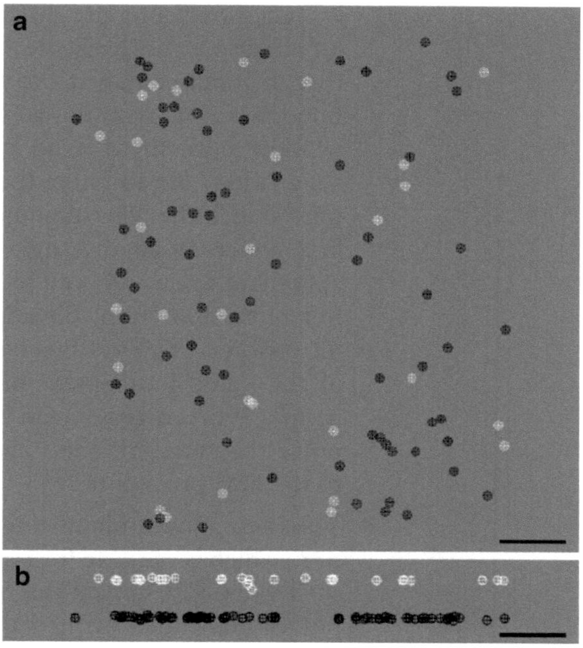

Fig. 3. Example of a fiducial seed model showing beads adsorbed at the two air–water interfaces before vitrification (see Movie 2). (**a**) *Top view*. (**b**) *Side view*. The ice layer is ~70 nm in thickness. Beads attached on opposite surfaces are depicted in *black* and *white*. Scale bar, 100 nm.

For this purpose, check the repartition of the gold particles in the specimen volume by examining the stack. Particles located on opposite sides of the ice layer should move in opposite directions. Select beads evenly dispersed on the surface of the tomogram, and whenever possible from both sides (see Fig. 3 and Note 24). Once the beads have been selected on one tilted view, run the "beadtrack program" to find automatically all the beads on all other views by pressing on "track seed model". The fiducial model that is generated will be the basis for fine alignments and distortion corrections. This model is usually not perfect (beads are not always selected on certain views and/or selections of beads are not well positioned) and has to be optimized either manually or using the iterative procedure in the bead fixer window.

7. Start the "tiltalign program" to perform fine alignments. This procedure solves for the displacements, rotations, tilts, and magnification differences in the tilted views. The 3D model (see Figs. 2b and 3 and Movie 2) may help you to choose the subsequent parameters to further improve the alignment. Thus, you will verify the presence of one or two bead layers (Fig. 3) and chose the calculation options accordingly in the "analysis of surface angles" box.

Click on "compute alignment". In the bead fixer window, we suggest to uncheck the "examine points once" option so that all points are taken into account in each round of calculation. Make the required adjustments through the bead fixer window until there is no big residual anymore. Open the "aligna.log" file and note the "total tilt angle change". Enter this value in the "fine alignment volume position parameters" box under "total tilt angle offset" and then compute the alignment again. Proceed in an iterative manner until you do not refine the "total tilt angle change" anymore. Start the distortion corrections by choosing the option "full solution" in the "global variables" menu. Click on "compute alignment". Proceed again in an iterative manner until you do not refine the "total tilt angle change" anymore. Then, move on to the next procedure.

8. This step allows minimizing the volume of the reconstruction, which is valuable for future analyses with other software, such as UCSF Chimera (see Subheading 3.7). Take the *X*-axis tilt recommended in the "aligna.log" file from the previous procedure and enter the value in the appropriate box. Given the low signal-to-noise ratio in the cryo-electron microscope images, the option "use whole tomogram" must be chosen. Indeed, by contrast to resin embedded specimens, it is almost impossible to visualize the ice borders in the cryo-tomograms. Then, start "create whole tomogram". Press on "create boundary model" and check if the tomogram reorientation and thickness are suitable. If not, adjust the parameters (essentially the *X*-axis tilt and tomogram thickness) accordingly, and repeat the whole procedure. In the case where two bead layers are present (Fig. 3), take the "*Z*-shift" recommended from the "aligna.log" file, enter the value in the appropriate box, then press on "create final alignment", and move on to the next procedure.
9. At this stage, all the adjustments that have been calculated before are pulled together in order to generate a final aligned stack of tilted views. Press the button "create full aligned stack". We recommend masking completely the gold particles since they may perturb future image analyses (see Movie 3). For that purpose, in the submenu "erase golds", transform the fiducial model and then erase the beads with a fiducial diameter slightly larger than proposed. Move on to the next procedure.
10. The three-dimensional reconstruction can be performed at this stage by pressing the button "generate tomogram". You may adjust different parameters available in the control panel, with a particular interest for the "radial filter cutoff and fall-off" that will optimize the overall contrast (see Note 25) (Fig. 2c and Movie 4). Move on to the post-processing procedures by clicking "Done".

11. In this final step, you may trim and bytes scale the volume to reduce the file size as much as possible, making easier its manipulation with diverse software, such as UCSF Chimera.

3.7. Volume Visualization

Volumes can be directly visualized in UCSF Chimera. However, the volume will be originally displayed in inverted contrast since UCSF Chimera interprets protein density as white. To overcome this problem, we usually use the following procedure:

1. Open a terminal window and move to your working directory.
2. Change the image mode of the file from 16 to 32 bits by typing the following Imod command: newstack-mode 2 file.rec newfile.rec.
3. Launch ImageJ and open the file: File/Import/MRC Reader.
4. Invert the contrast: Edit/Invert.
5. Save the file in Spider format: Save as/Spider writer. At this stage, the volume can be further processed using the diverse tools present in ImageJ. Crops of particular regions can be performed to facilitate their visualization at high resolution in UCSF Chimera.
6. Visualize the volume in UCSF Chimera: launch UCSF Chimera and open the Spider (name.spi) file. In the volume data window, select solid (densities) or surface representation (see Fig. 4 and Movie 5).

3.8. Segmentation

Cryo-electron tomograms of vitrified biological samples remain very noisy (Fig. 4a). The surface representation uses a simple threshold that can cut off important features close to the noise level. To extract the structures of interest in a semiautomatic manner, we have developed a patch-based recognition algorithm (CryoSeg) that can be obtained from Charles.Kervrann@inria.fr (see Note 26):

1. Open the volume in ImageJ and save it in Tiff format.
2. Launch CryoSeg and select the volume in the "tomogram path".
3. Select the "output path" in the lower box and give the segmented tomogram a name (see Note 27).
4. Press on "display tomogram" so that you will be able to explore the volume through three different directions ((x,y), (z,y), and (z,x)). Then, move on to the "parameters" menu.
5. As a first attempt, we do suggest to leave the default values for each segmentation parameter (patch width of nine pixels and regularization coefficient of 0.05) and, as a next step, it will

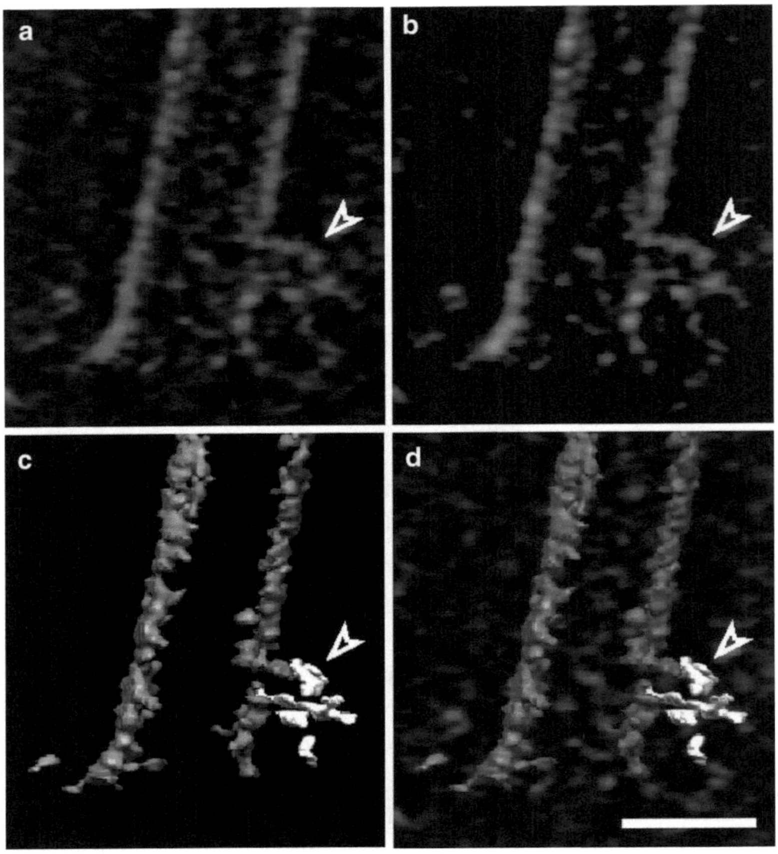

Fig. 4. Volume visualization using UCSF Chimera (see Movie 5). (**a**) Raw volume in solid mode representation showing protein densities in *white*. The *arrowhead* points to the same XMAP215 molecules as depicted in Fig. 2c. (**b**) Volume after segmentation and extraction of densities using the CryoSeg software. (**c**) Surface mode representation of the segmented volume in (**b**). The microtubule and XMAP215 molecules are shown in *gray* and *white*, respectively. (**d**) Superimposition of the raw (**a**) and segmented (**c**) volumes. Scale bar, 25 nm.

be worth adjusting those parameters to improve the segmentation precision (in Fig. 4b, c and Movie 5), we have used a patch width of five pixels and a regularization coefficient of 0.5 (see Note 28).

6. In the "select slices" box, the number of initial patches can be chosen according to the number of directions observed in the objects of interest. Most of the time, microtubules are roughly in one or two different direction(s), and thus two patches are sufficient. Before starting a time-costly full segmentation of the whole tomogram, it is preferable to perform partial segmentations on a few slides that have to be selected carefully as representative of the whole volume. These preliminary tests will be helpful to determine the optimum segmentation parameters.
7. Press the "run" button and select two points to define each patch along protofilaments that can be chosen on different

slices. Press on any letter key to validate each set of two points; otherwise, press on the space bar to redo the selection. Once all the initial patches are determined, the segmentation starts and can be followed slice by slice on the screen (press "ctrl-c" in the terminal window to kill the process).

8. The software generates three files. Among them, the "file.intensities.tif" contains the segmented volume that can be directly opened in UCSF Chimera (Fig. 4).

4. Notes

1. Tubulin can be prepared from calf or pig brains according to published protocols (5) or purchased (Cytoskeleton Inc., Denver, USA).
2. Paclitaxel is irritant, toxic to reproduction category 2, carcinogen category 2, and mutagen category 2. Do not breathe vapors and wear suitable protective clothing, gloves, and eye/face protection.
3. XMAP215 can be purified according to published procedures (6).
4. Holey carbon grids can be prepared according to published protocols (7), or are commercially available (Lacey carbon type-A, 300 M, copper, 01890-F, Ted Pella, Inc.).
5. Three types of gold-BSA particles from Aurion are available: negatively charged (ref. 210122), positively charged (ref. 210111), or neutral (ref. 210133). Negatively charged particles can be used with tubulin alone to avoid unwanted interactions.
6. The grade of the filter paper can be adapted to the viscosity of the suspension (high concentrations of microtubules may necessitate Whatman no. 4).
7. It is important to keep a high humidity around the specimen to avoid evaporation, which otherwise may induce solute concentration and temperature loss. A homemade apparatus has been described (8) and more sophisticated ones are commercially available (Vitrobot™ Mark IV from FEI, EM GP from Leica, Cryoplunge™ 3 from Gatan).
8. In all steps that necessitate resuspending microtubules, it is advisable to use truncated tips to avoid breakage of microtubules.
9. The duration of blotting along with the grade of the Whatman filters are very sensitive parameters in this procedure and have to be adapted by the experimenter according to the sample viscosity. They determine the ice thickness and quality.

We currently blot for 2–4 s by pressing slightly on the grid with the filter paper.

10. Ethane is a flammable gas; risk of asphyxiation and dermal contact with a rapidly evaporating liquid could result in freezing of the tissues or frostbite. Ensure adequate ventilation and wear suitable protective clothing, gloves, and eye/face protection.
11. Solid ethane tends to disappear upon storage.
12. Risk of asphyxiation and dermal contact with a rapidly evaporating liquid could result in freezing of the tissues or frostbite. Ensure adequate ventilation and wear suitable protective clothing, gloves, and eye/face protection.
13. It is preferable to choose microtubules oriented along the tilt axis to get a better visualization of the protofilaments in the final tomograms.
14. We currently use a defocus value between −3 and −5 μm.
15. The angular range varies depending on the type of cryostage (some models allow up to ±80°) and also on the position on the grid. It is important to check before acquisition the maximal values allowed at a specific position on the grid.
16. Using this scheme, the incremental angle is adjusted so that the space is filled evenly in Fourier space. As an example, an incremental angle of 1.9° with an angular range from −60° to +60° generates 81 pictures.
17. Starting at 0° minimizes movement of the specimen as a consequence of its initial exposure to the electron beam, since its apparent thickness is minimal.
18. It is important that the stage has been pre-calibrated to anticipate the mechanical movements that will occur during acquisition (see an example of calibration curves in Fig. 1). In addition, corrections must be performed without additional irradiation of the specimen. For this purpose, we use the post-tracking mode that compares two successive images and apply their displacement to the next. The cross-correlation will perform better if a filter is applied to the images. This must have been determined before on test specimens with gold particles. We currently use radius and sigma values of 50 and 1 nm, respectively, for the cut-off of long wavelengths, and 8 and 0.1 nm, respectively, for the cut-off of short wavelengths.
19. To determine the electron dose and exposure time that the specimen can sustain, we have used the following procedure. The resolution that can be currently achieved by cryo-electron tomography is on the order of a few nm. The Fourier transform of the microtubule images are characterized by layer

lines, which reflect the arrangement of the tubulin molecules in the microtubule lattice. One major layer line is located at 4 nm^{-1}, corresponding to the repetition of the tubulin subunits along the protofilaments. The strategy is thus to use an electron dose and a total exposure time sufficiently low enough so that this layer line is still visible after irradiation. For an 80 images tilt series of 1 s each, expose an area of the specimen 80 s and check for the presence of the 4 nm^{-1} layer line after exposure. Adjust illuminating conditions accordingly.

20. We advice experimenters to perform, before starting image processing, the IMOD tutorial (http://bio3d.colorado.edu/imod/doc/etomoTutorial.html) and to read carefully the IMOD user's guide. Here, we propose a particular usage of these procedures with respect to the specificities of the cryo-electron tomography approach.
21. Bad images can be unselected at this stage, but they might also be removed using ImageJ, for instance, while editing the angle file accordingly. Note that the images may appear unsuitable until X-rays have been removed. Proceed to the next step even if the image series appears unsuitable.
22. It might be preferable to convert to bytes (256 gray values) depending on the computer used. The 16 bits (65,535 gray values) of the original .mrc files provide a larger dynamic range at the cost of longer calculation times.
23. Jumps often appear at 0° due to the cryo-stage resetting between the two half-tilt series. If this is not improved enough, the "seed tracking" procedure might not work correctly.
24. Beads that have been selected at this step can be erased later before the 3D reconstruction step.
25. We usually use a radial filter cutoff around 0.05 and a radial filter falloff around 0.1.
26. In this statistical method, a small number of patches (images of a few pixels width) are semiautomatically selected by the user to form a dictionary. The features present in this dictionary are then detected and extracted inside the volume.
27. We suggest mentioning in the file name, the two main parameter values (patch width and regularization coefficient) since numerous attempts have to be performed in order to optimize the segmentation process.
28. Small patches enable to be less sensible to orientation but detect more noise. The regularization coefficient prevents the software from selecting isolated pixels; a high coefficient, such as 0.5 induces high regularization.

Acknowledgments

This work was supported by grants from the French National Agency for Research (ANR PCV06_142769 and PCV07_190830) and from the Federative Research Institute of Rennes IFR140 Functional genomics, Agronomy and Health.

References

1. Li H, DeRosier DJ, Nicholson WV, et al (2002) Microtubule structure at 8 Å resolution. Structure (Camb) **10**: 1317–1328.
2. Sindelar C, Downing KH (2010) An atomic-level mechanism for activation of the kinesin molecular motors. Proc Natl Acad Sci USA **107**: 4111–4116.
3. Sosa H, Hoenger A, Milligan RA (1997) Three different approaches for calculating the three-dimensional structure of microtubules decorated with kinesin motor domains. J Struct Biol **118**: 149–158.
4. Chrétien D, Wade RH (1991) New data on the microtubule surface lattice. Biol Cell **71**: 161–174.
5. Ashford AJ, Hyman AA, Julio EC (2006) Preparation of tubulin from porcine brain. In: Celis JE (ed) Cell Biology, 3rd edn. Academic Press, Burlington.
6. Tournebize R, Popov A, Kinoshita K, et al (2000) Control of microtubule dynamics by the antagonistic activities of XMAP215 and XKCM1 in Xenopus egg extracts. Nat Cell Biol **2**: 13–19.
7. Fumaki A, Adachi K (1965) A new method of preparation of a self-perforated micro plastic grid and its application. J Electron Microsc **14**: 112–118.
8. Arnal I, Malleter M, Chrétien D (2008) Vitrification of dynamic microtubules. In: Cavalier A, Spehner D, Humbel BM (eds) Handbook of cryo-preparation methods for electron microscopy. CRC Press, Boca Raton.

Chapter 15

Kinetochore Microtubules, Analyzed by Correlated Light and Immunoelectron Microscopy

Andreas Merdes and Jan De Mey

Abstract

The movement of chromosomes in mitosis requires spindle microtubules, as well as a set of specific motor proteins located at the kinetochores of the chromosomes. The exact mechanisms of chromosome movement have remained ambiguous for many years. Cumulating evidence indicates that chromosome movement in early mitosis occurs by lateral sliding of kinetochores along the surface of microtubules. We provide here the protocol for an immunological staining method of microtubules that allows electron microscopic analysis of spindle microtubules over a long distance and that has helped clarifying this biological question. The technique involves the use of ultra-small immunogold, enhanced by silver.

Key words: Mitosis, Spindle microtubules, Kinetochore, Chromosome movement, Immunoelectron microscopy, Immunogold

1. Introduction

During mitosis, equal numbers of chromosomes need to be segregated to the newly forming daughter cells. To achieve this, pairs of sister chromosomes are aligned at the equatorial plane and subsequently split and separated (for a recent review, see (1)). Alignment of chromosomes is a time-consuming process that involves chromosome movement towards and away from the poles during prometaphase, until a stable position is reached. This process is also termed "congression." Once the equatorial alignment is completed, the chromosome pairs are usually bi-oriented, with their two kinetochores attaching to spindle microtubules of opposite poles. Electron microscopy of mitotic cells has revealed that microtubules contact the kinetochores with their plus-ends

Anne Straube (ed.), *Microtubule Dynamics: Methods and Protocols*, Methods in Molecular Biology, vol. 777,
DOI 10.1007/978-1-61779-252-6_15,

(2–4). For this reason, it has been suggested that the movement of chromosomes is coupled to polymerization and depolymerization of the microtubule plus-ends. Experiments with isolated chromosomes and microtubules in vitro have confirmed that microtubule depolymerization alone is indeed capable of driving chromosome movement (5–10). This process requires no additional energy in the form of nucleoside triphosphates, but is likely powered by conformational changes of tubulin at the depolymerizing microtubule ends (5, 8, 11). In addition to end-on contacts between microtubules and kinetochores, lateral interactions between kinetochores and the microtubule surface have also been reported (12–15). Such lateral interactions have been found in prometaphase cells of various species and throughout mitosis in spindles of diatoms. In particular, at the onset of prometaphase in vertebrate cells, single microtubules have been found to penetrate the disintegrating nuclear envelope and to serve as tracks for fast poleward chromosome motion (13, 14, 16). In these cases, chromosome movement occurs along the surface of microtubules, independently of microtubule plus-end dynamics. To study the interaction between microtubules and kinetochores in detail, various techniques of light microscopy have been exploited during the last few decades, including differential interference contrast and advanced methods of fluorescence microscopy, such as confocal, spinning disc, and deconvolution microscopy (13, 15, 17–20). Nevertheless, the analysis has often been restricted by the limited resolution of the light microscopical techniques and has been complemented by electron microscopy to resolve the exact position of individual microtubules. However, the necessity of serial ultra-thin sectioning for electron microscopy, to reconstruct the entire length of microtubules, posed additional technical difficulties. Here, we describe a method of decorating microtubules with small immunogold particles of approximately 1 nm size that are further enhanced by silver lactate treatment. With our technique, microtubules are densely labelled and can be investigated over a long distance in semi-thin sections. This improves the visibility of microtubules and facilitates the reconstruction of spindle microtubules.

2. Materials

2.1. Cell Culture

1. PtK2 rat kangaroo kidney cells.
2. Modified Eagle's medium (Gibco/Invitrogen), supplemented with 10% foetal calf serum (Sigma-Aldrich), 1% glutamine, 1% non-essential amino acids, and 1% sodium pyruvate (Gibco/Invitrogen).

3. 20-mm square glass coverslips (Menzel, Braunschweig, Germany).
4. 10-cm plastic cell culture dishes (Nunc/Thermo Fisher Scientific).

2.2. Video Microscopy

1. Leitz Orthoplan microscope (Leica Microsystems, Wetzlar, Germany), equipped with a 63×/1.3NA phaco 3 objective lens, and a zoom lens 5–12.5× at the camera port.
2. Video camera Dage-MTI, model 81 (Dage-MTI, Michigan).

2.3. Electron Microscopy

1. PHEM: 60 mM PIPES, pH 6.9, 25 mM HEPES, 1 mM EGTA, and 2 mM magnesium acetate (all Sigma-Adrich).
2. Fixation solution: 0.06% glutaraldehyde and 3% formaldehyde (Polysciences, Eppelheim, Germany) in PHEM. Make fresh just before use.
3. 0.5% Triton X-100 (Sigma-Aldrich) in PHEM.
4. 1 mg/ml sodium borohydride (Sigma-Aldrich) in phosphate-buffered saline (Dulbecco's PBS), adjusted to pH 8.0. Make fresh just before use.
5. Blocking solution: 0.8% bovine serum albumin, 0.1% fish skin gelatine, 5% normal goat serum (all from Sigma-Aldrich), in PBS, pH 7.4.
6. Primary antibodies were a mixture of monoclonal mouse anti-alpha tubulin and anti-beta-tubulin (GE Healthcare), diluted 1:500 each, in PBS, pH 7.4, containing 0.8% bovine serum albumin, 0.1% fish skin gelatine, and 1% normal goat serum (Sigma-Aldrich).
7. Washing buffer: 0.8% bovine serum albumin and 0.1% fish skin gelatine in PBS, pH 7.4.
8. Secondary antibodies were goat anti-mouse, coupled with ultra-small immunogold (0.8 nm; Aurion, Wageningen, The Netherlands), diluted 1:20 in PBS containing 0.8% bovine serum albumin, 0.1% fish skin gelatine, and 1% normal goat serum (Sigma-Aldrich).
9. Postfixation solution: 2% glutaraldehyde in PBS.
10. 0.2% tannic acid in PBS, pH 7.4 (Mallinckrodt Baker).
11. Silver enhancement solution, according to (21), freshly mixed from stocks of 3-ml gum arabic (25% in distilled water), 3-ml distilled water, 85 mg hydroquinone in 1.5-ml distilled water, 255 mg citric acid monohydrate in 1-ml distilled water, 235 mg trisodium citrate dihydrate in 1-ml distilled water, and 11 mg silver lactate in 1.5-ml distilled water. All solutions were prepared in advance, aliquoted, and stored at −20°C.

12. 0.5% uranyl acetate and 1% phosphotungstic acid in 70% ethanol (Sigma-Aldrich).
13. 100% pure ethanol, stored over molecular sieves, to remove any traces of water (Sigma-Aldrich).
14. Araldite CY212 embedding kit (Agar Scientific, Stansted, UK).
15. Hydrofluoric acid, 40% (Sigma-Aldrich; very toxic and corrosive!).
16. Rotating diamond scribe (Leica Microsystems, Wetzlar, Germany).
17. Cyanoacrylate glue (Loctite Super Glue, Henkel, Duesseldorf, Germany).
18. Formvar/carbon-coated slot grids (Agar Scientific, Stansted, UK).
19. Diamond knife (Diatome, Biel, Switzerland), and ultramicrotome (Reichert OM U3, Leica Microsystems).
20. Lead citrate solution, according to (22): 80 mM lead nitrate, 120 mM trisodium citrate, and 160 mM sodium hydroxide. The solution is prepared freshly, by dissolving lead nitrate and trisodium citrate in triple distilled water that has been boiled and cooled down, to remove carbon dioxide (thus avoiding lead carbonate precipitates). Sodium hydroxide is added last, from a fresh 1 M stock solution; its addition will clear the solution.
21. Electron microscope Philips EM 400, equipped with a tilting stage (Philips, Eindhoven, The Netherlands).

3. Methods

3.1. Cell Culture

Culture of PtK2 cells was performed according to standard procedures. Prior to use, glass coverslips were marked with a diamond pen, to define reference points for the microscope stage coordinates. Cells were seeded onto glass coverslips and grown until they reached approximately 50% confluence. This led to improved cell spreading and flattening, and therefore improved light microscopic observation. Moreover, the lower cell density facilitated the relocation of observed cells after video microscopy for subsequent ultra-sectioning.

3.2. Video Microscopy

The coverslips were mounted onto a live observation chamber of our own design, permitting the use of microscope optics of high numerical aperture and short working distance. Alternatively, coverslips can be observed with long distance lenses in a standard Petri dish filled with culture medium. The stage coordinates of the microscope were noted for the reference points marked on the coverslip surface. Cells in late prophase or early prometaphase were identified with the help of phase contrast, and their respective stage coordinates were noted. Time-lapse recordings were made with a video recorder. Once chromosome movement was noted, cells were immediately fixed on the microscope stage. For this, we had connected a peristaltic pump to our observation chamber that helped to remove the culture medium and to add fixation solution simultaneously. Alternatively, removal of medium and addition of fixation solution can be performed with a pipette within a few seconds.

3.3. Electron Microscopy

1. Fixation was carried out for 1 min in the microscope observation chamber, and after transfer of the coverslip to a petri dish of 3.5 cm diameter, fixation was continued for nine additional minutes in 2 ml of fixation solution.
2. After three quick rinses in PHEM buffer, extraction with 0.5% Triton X-100 in PHEM was performed for 15 min.
3. The coverslips were subsequently rinsed three times in PBS and treated twice for 15 min with freshly made sodium borohydride solution (2 ml/petri dish). This served to reduce excessive, free aldehyde groups, to prevent unspecific binding of antibody in the following labelling steps. Small bubbles of hydrogen were often seen during the reduction step. This step may be performed on a rocking or shaking table to ensure that samples are covered with buffer and to prevent floating and drying of coverslips.
4. After three rinses with PBS, cells were treated with 1 ml blocking solution per coverslip for 30 min.
5. The incubation of the primary antibodies was performed for 1 h at 37°C using 40 μl of diluted antibody per coverslip. Dilutions may vary, depending on the type of antibody. Incubation was performed in a closed humid chamber to prevent evaporation of the antibody solution. Humid chambers were assembled by placing a wet paper towel onto the bottom of a plastic box and covering the box with a lid. The petri dishes containing the coverslips were placed inside the box for incubation.
6. Primary antibody was rinsed off, and coverslips were incubated three times for 10 min with washing buffer.
7. Incubation with secondary antibody was performed in a humid chamber, as described for the primary antibody. Because of the conjugation of the secondary antibodies to

immunogold, the diffusion was found to be slower. To obtain dense labelling, we compensated slow diffusion by performing two incubation steps: one for 50 min at 37°C and a second step with a fresh aliquot of antibody overnight at room temperature.

8. Secondary antibody was removed by incubating the coverslips three times for 10 min in washing buffer, followed by three short rinses with PBS.
9. Postfixation was carried out for 5 min with glutaraldehyde to immobilize the secondary antibodies/immunogold conjugates, followed by a treatment with tannic acid solution for 30 min to increase the contrast of microtubules.
10. Cells were rinsed in excess distilled water (at least three times 5 min), and subsequently incubated with 2 ml silver enhancement solution per coverslip in a dark room, equipped with a red safelight. The silver enhancement was followed by eye; a brown colouration of the cells was usually visible after 6–9 min (see Note 1). Development was stopped by rinsing the specimens with excess distilled water (at least five changes of water). Any treatment with osmium tetroxide after silver enhancement should be avoided (see Note 2).
11. Cells were subsequently dehydrated by incubating the coverslips three times for 3 min with 70% ethanol, and contrasted with a mixture of 0.5% uranyl acetate and 1% phosphotungstic acid in 70% ethanol for 15 min.
12. The contrasting solution was rinsed off, and dehydration was continued by incubating four times for 3 min with 90% ethanol, and four times for 3 min with 100% ethanol (desiccated over molecular sieves).
13. Cells were infiltrated with epoxy resin (Araldite CY212), diluted 50% in desiccated ethanol for 30 min, and then with pure resin overnight at room temperature (see Note 3). A drop of pure resin was placed onto a sheet of parafilm, and coverslips were mounted with cells facing down.
14. Flat embedding in epoxy resin was performed by mounting the coverslips onto microscope slides with the cells facing down (Fig. 1a). A defined thickness of the resin layer was obtained by using halves of new coverslips as spacers. Polymerization of the resin was performed in an oven at 60°C for 16–24 h (Fig. 1b).

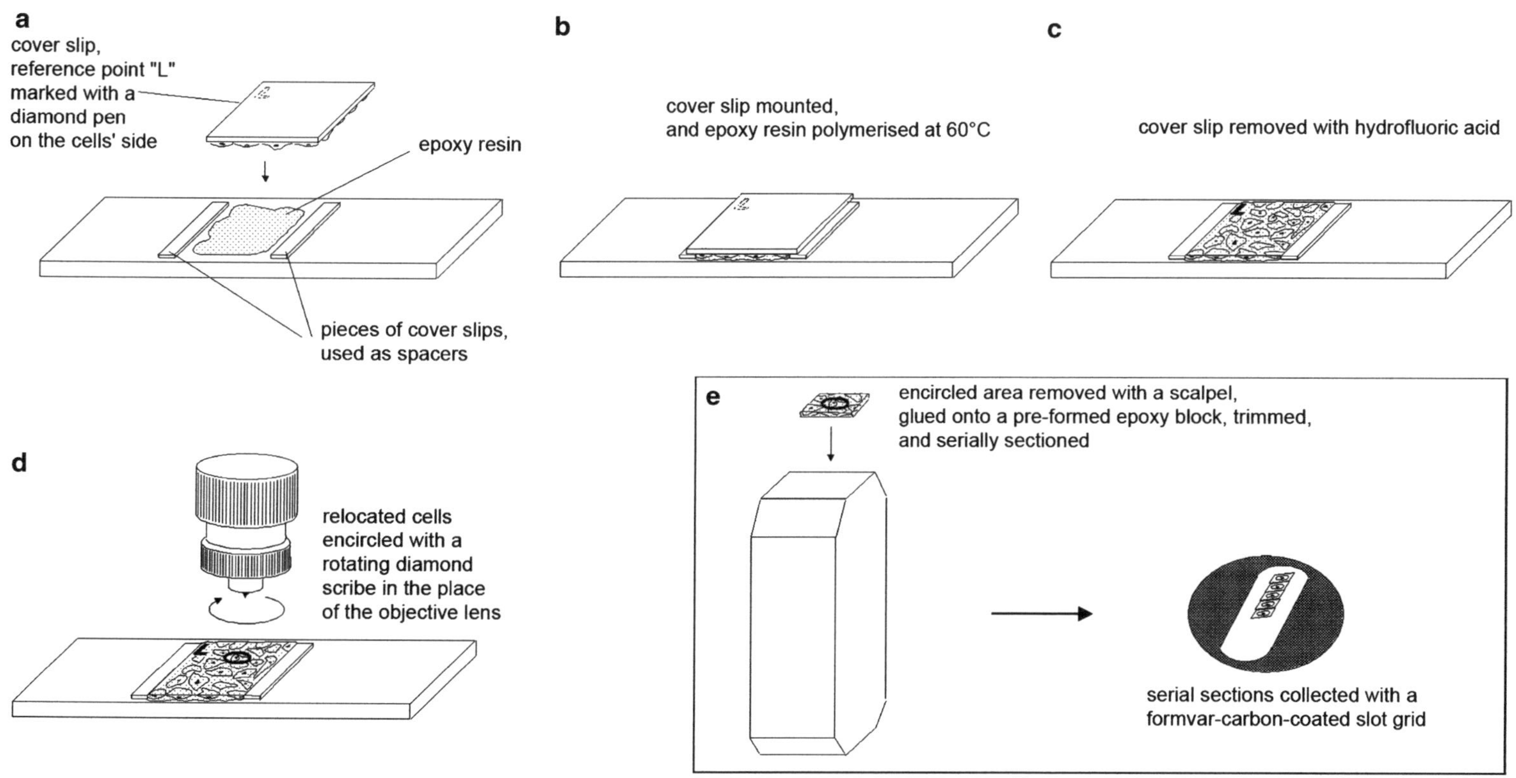

Fig. 1. Scheme, explaining sample preparation for correlated light and electron microscopy. (**a**) Cells were grown on coverslips that were marked on the cells' side with a diamond pen ("L") to provide a reference point for the microscope stage coordinates. Cells were observed by time-lapse microscopy, and both coordinates of the cells and of the reference point were noted. After fixation and processing, cells were embedded on a glass slide in a drop of epoxy resin, using pieces of coverslips as spacers. (**b**) The epoxy resin was polymerized at 60°C overnight. (**c**) The glass coverslip was dissolved with several drops of 40% hydrofluoric acid. The cells were protected in the underlying layer of epoxy resin. A replica of the marked reference point was visible. (**d**) After relocation of the previously observed cells, these were encircled with a rotating diamond scribe that was mounted in the place of the microscope objective lens. (**e**) The encircled area was removed with the help of a scalpel and glued on top of a preformed block of epoxy resin. After trimming, semi-thin serial sections were cut and collected on a Formvar/carbon-coated slot grid.

15. The coverslips were removed by dissolving the glass with several drops of hydrofluoric acid (Fig. 1c; to be carried out with protective clothing, under a fume hood! see Note 4). Excess acid was rinsed off with distilled water.
16. The flat embedded cells were reinvestigated by light microscopy. The previously recorded cells were relocated by adjusting the microscope stage to the relative distance from the stage coordinates of the reference points.
17. A rotating diamond scribe was placed in the position of the objective lens, and the cells of interest were encircled at a diameter of approximately 1 mm (Fig. 1d).
18. With the help of a scalpel, a larger area of the resin outside the circle was cut. Using cyanoacrylate glue, this piece of resin was glued onto a preformed resin block that had been polymerized in a flat embedding mould (cells facing up; Fig. 1e).
19. Under a dissection microscope, excess resin was trimmed away from the cells of interest.
20. Serial sections of approximately 200–250 nm thickness (blue/green interference colour) were cut on an ultramicrotome with a diamond knife and recovered with Formvar/carbon-coated slot grids (Fig. 1e).
21. The sections were contrasted for 1 min on a drop of lead citrate, followed by three rinses on drops of bi-distilled water. After drying, the specimens were viewed in an electron microscope.

The method provides homogeneous and dense immunogold/silver labelling that permits tracking and high resolution of microtubules all along their length. The method is suited for both light and electron microscopy (Fig. 2). The flat embedding technique on glass slides allows correlated analysis of cells by time-lapse and electron microscopy (Figs. 3 and 4). The high contrast of the immunogold/silver staining allows tracking of single microtubules over a long distance, even in single semi-thin sections, thus avoiding difficult reconstruction of microtubules that would otherwise be necessary in series of conventional ultra-thin sections.

4. Notes

1. The exact time of silver lactate development may vary; longer development times lead to a higher contrast of the labelled structures due to the formation of larger silver grains, but this occurs at the risk of elevated background staining.

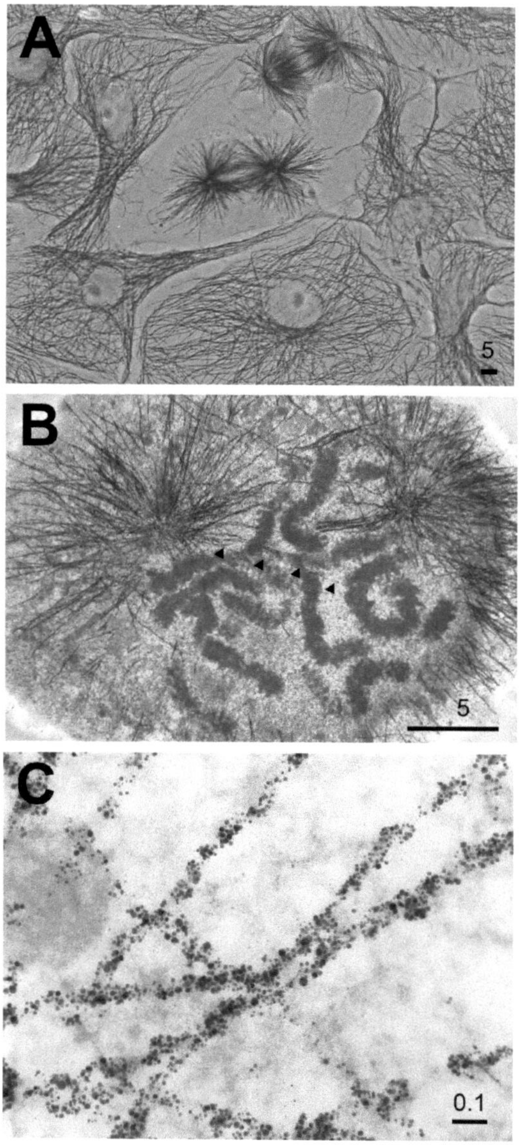

Fig. 2. The microtubule network of PtK2 cells, visualized with ultra-small immunogold and silver enhancement. (**a**) Light microscopic image (phase contrast) of cells at different stages of the cell cycle. *Bar*: 5 μm. (**b**) Electron microscopic image of a cell in early prometaphase. A single microtubule reaching from the left spindle pole to the equatorial region can be seen, running laterally past the chromosomes (*black arrowheads*). Most microtubules, at this early stage, are astral microtubules, radiating from the centrosome towards the cell periphery. These microtubules are densely labelled and can be tracked over a long distance. *Bar*: 5 μm. (**c**) High magnification of microtubules, revealing the dense, homogeneous labelling with ultra-small immunogold, after enhancement with silver lactate. *Bar*: 0.1 μm.

2. Postfixation of the samples with osmium tetroxide after the silver enhancement step is not recommended, since it can oxidize and dissolve the silver grains.

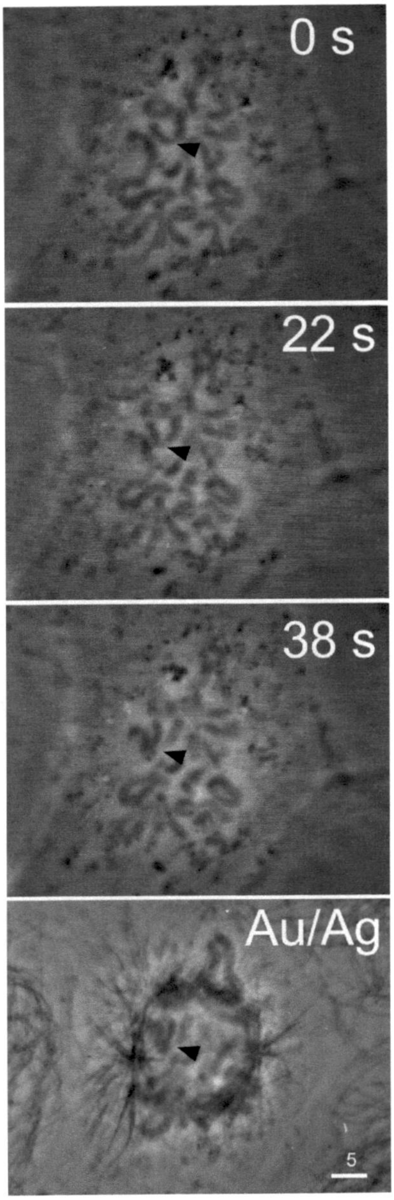

Fig. 3. Phase contrast images of an early prometaphase cell, recorded immediately after nuclear envelope breakdown (0 s). The time-lapse sequence of a chromosome moving towards the left spindle pole is shown at 22 and 38 s, respectively. The position of the moving chromosome is indicated with a *black arrowhead*. The cell was fixed immediately after 38 s. The same cell is shown after fixation and processing for immunogold/silver staining of the microtubules (Au/Ag). The *arrowhead* indicates the previously moving chromosome. *Bar*: 5 μm.

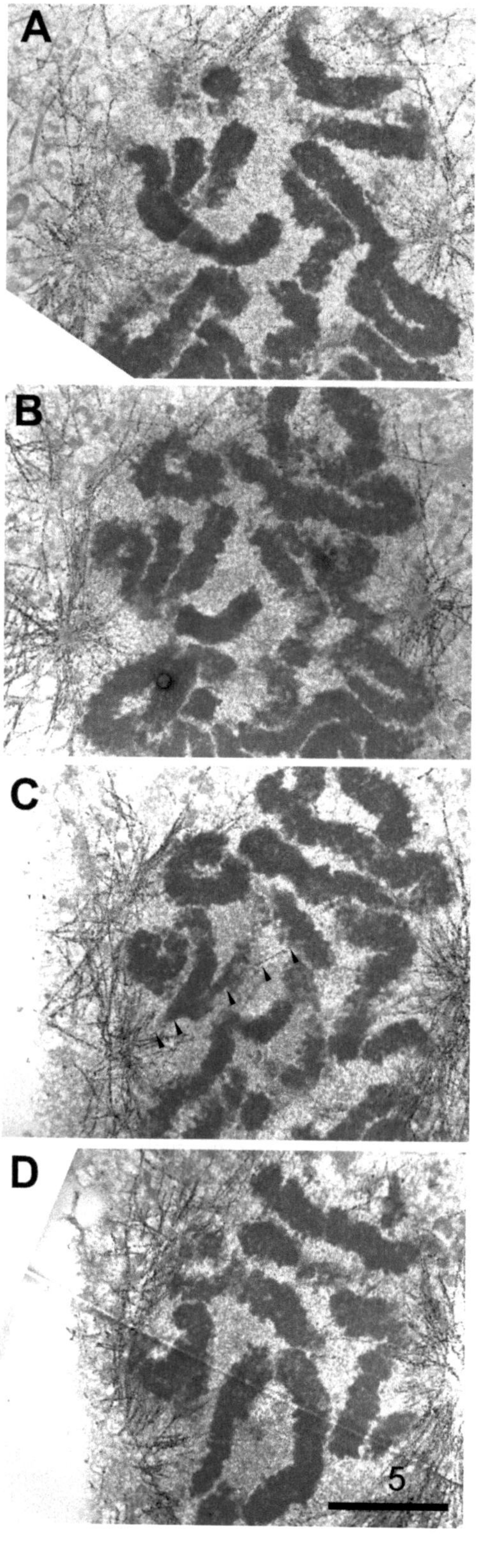

Fig. 4. Serial semi-thin sections of the same cell as in Fig. 3. The microtubules are labelled with immunogold/silver. The previously moving chromosome can be seen in sections (**a**)–(**d**). A single microtubule is indicated in (**c**) (*arrowheads*), running from the left spindle pole past the region of the leading kinetochore, and terminating several microns behind the moving chromosome. The orientation of the microtubule coincides with the direction of movement of this chromosome, suggesting that this chromosome was running laterally along the microtubule. *Bar* in (**d**): 5 μm.

3. Alternative to the described infiltration procedure with ethanol/epoxy resin, infiltration may be performed with a 1:1 mixture of propylene oxide/epoxy resin, following incubation with pure propylene oxide after the last ethanol dehydration step. Since propylene oxide residues are more volatile than ethanol, this may improve subsequent polymerization of the epoxy resin.
4. Hydrofluoric acid should be handled with care, wearing eye protection and gloves. Since hydrofluoric acid dissolves glass, the acid must not be stored in glass bottles.

References

1. Tanaka TU, Stark MJR, Tanaka K (2005a) Kinetochore capture and bi-orientation on the mitotic spindle. Nat Rev Mol Cell Biol **6**:929–942.
2. Roos UP (1973b) Light and electron microscopy of rat kangaroo cells in mitosis. II. Kinetochore structure and function. Chromosoma **41**:195–220.
3. Euteneuer U, McIntosh JR (1981) Structural polarity of kinetochore microtubules in PtK1 cells. J Cell Biol **89**:338–345.
4. McEwen BF, Heagle AB, Cassels GO, Buttle KF, Rieder CL. (1997) Kinetochore fiber maturation in PtK1 cells and its implications for the mechanisms of chromosome congression and anaphase onset. J Cell Biol **137**: 1567–1580.
5. Koshland DE, Mitchison TJ, Kirschner MW (1988) Polewards chromosome movement driven by microtubule depolymerization in vitro. Nature **331**:499–504.
6. Hyman AA, Mitchison TJ (1990) Modulation of microtubule stability by kinetochores in vitro. J Cell Biol **110**:1607–1616.
7. Hyman AA, Mitchison TJ (1991) Two different microtubule-based motor activities with opposite polarities in kinetochores. Nature **351**:206–211.
8. Coue M, Lombillo VA, McIntosh JR (1991) Microtubule depolymerization promotes particle and chromosome movement in vitro. J Cell Biol **112**:1165–1175.
9. Lombillo VA, Nislow C, Yen TJ, Gelfand VI, McIntosh JR (1995) Antibodies to the kinesin motor domain and CENP-E inhibit microtubule depolymerization-dependent motion of chromosomes in vitro. J Cell Biol **128**: 107–115.
10. Hunt AJ, McIntosh JR (1998) The dynamic behavior of individual microtubules associated with chromosomes in vitro. Mol Biol Cell **9**: 2857–2871.
11. Grishchuk EL, Molodtsov MI, Ataullakhanov FI, McIntosh JR (2005) Force production by disassembling microtubules. Nature **438**:384–388.
12. Tippit DH, Pickett-Heaps JD, Leslie R (1980) Cell division in two large pennate diatoms Hantzschia and Nitzschia III. A new proposal for kinetochore function during prometaphase. J Cell Biol **86**:402–416.
13. Rieder CL, Alexander SP (1990) Kinetochores are transported poleward along a single astral microtubule during chromosome attachment to the spindle in newt lung cells. J Cell Biol **110**:81–95.
14. Merdes A, De Mey J (1990) The mechanism of kinetochore-spindle attachment and polewards movement analyzed in PtK2 cells at the prophase-prometaphase transition. Eur J Cell Biol **53**:313–325.
15. Tanaka K, Mukae N, Dewar H, van Breugel M, James EK, Prescott AR, Antony C, Tanaka TU (2005b) Molecular mechanisms of kinetochore capture by spindle microtubules. Nature **434**:987–994.
16. Roos UP (1973a) Light and electron microscopy of rat kangaroo cells in mitosis. I. Formation and breakdown of the mitotic apparatus. Chromosoma **40**:43–82.
17. Hayden JH, Bowser SS, Rieder CL (1990) Kinetochores capture astral microtubules during chromosome attachment to the mitotic spindle: direct visualization in live newt lung cells. J Cell Biol **111**:1039–1045.
18. Merdes A, Stelzer EH, De Mey J (1991) The three-dimensional architecture of the mitotic spindle, analyzed by confocal fluorescence and electron microscopy. J Electron Microsc Tech **18**:61–73.
19. Skibbens RV, Skeen VP, Salmon ED (1993) Directional instability of kinetochore motility during chromosome congression and segregation in mitotic newt lung cells: a push-pull mechanism. J Cell Biol **122**:859–875.

20. Kapoor TM, Lampson MA, Hergert P, Cameron L, Cimini D, Salmon ED, McEwen BF, Khodjakov A (2006) Chromosomes can congress to the metaphase plate before biorientation. Science **311**:388–391.
21. Danscher G (1981) Histochemical demonstration of heavy metals. A revised version of the sulphide silver method suitable for both light and electron microscopy. Histochemistry **71**:1–16.
22. Reynolds ES (1963) The use of lead citrate at a high pH as an electron opaque stain in electron microscopy. J Cell Biol **17**: 208–212.

Chapter 16

Photoactivatable-GFP-α-Tubulin as a Tool to Study Microtubule Plus-End Turnover in Living Human Cells

Catarina P. Samora and Andrew D. McAinsh

Abstract

The development of photactivatable (PA) variants of Green fluorescent protein (GFP) has allowed the dynamics of spatially restricted protein pools within living cells to be determined. Over the last 5 years, experiments utilizing PA-GFP fused to α-tubulin have provided important insights into the mechanisms that control microtubule dynamics in living cells. In this chapter, we describe the methodology required to generate stable cell lines expressing photoactivatable-GFP-α-tubulin and to derive quantitative measurements of tubulin turnover at microtubules plus-ends in living cells.

Key words: Microtubule, α-Tubulin, Photoactivatable-GFP, Plus-end turnover, Dynamics, Kinetochore

1. Introduction

The mitotic spindle is constituted by a bipolar array of parallel microtubules each of which is assembled from parallel microtubules with their minus-ends nucleated at the spindle poles and the plus-ends projecting inwards. Spindle microtubules can either have their plus-end embedded in kinetochores, the attachment sites on chromosomes, can overlap with microtubules emanating from the opposite pole at the spindle midzone forming anti-parallel overlaps, or remain "free" within the spindle. Although the mitotic spindle is able to maintain its overall architecture at steady state, the microtubules are highly dynamic exhibiting non-equilibrium dynamic behaviour termed dynamic instability (1). Dynamic instability results in microtubules alternating between periods of growth and shrinkage with the transitions to growth

Anne Straube (ed.), *Microtubule Dynamics: Methods and Protocols*, Methods in Molecular Biology, vol. 777,
DOI 10.1007/978-1-61779-252-6_16,

and shrinkage called rescue and catastrophe, respectively. Within the spindle, tubulin has been observed to undergo continuous movement towards the spindle pole. In human cells, this poleward "flux" is driven primarily by the "treadmilling" of tubulin within the lattice of kinetochore–microtubules due to the balanced addition of tubulin at plus-ends and its removal at the minus-end (reviewed in (2)). In other organisms, such as flies and frogs, microtubule flux is primarily driven by the sliding of microtubules towards the pole (3–11). These findings were possible due to the development of a number of methods, including microinjection of labelled tubulin (12), fluorescence speckle microscopy (13), and expression of **photoactivatable-green fluorescent protein** fused to α-tubulin (PA-GFP-α-tubulin) (14). The latter technique takes advantage of an engineered variant of the GFP protein, which only emits fluorescence after irradiation with ~413 nm light, allowing precise activation of specific pools of protein in a cell (14). Importantly, the dissipation of the PA-GFP-α-tubulin fluorescence signal as a function of time provides information (expressed as a half-life) on the rate of tubulin turnover at microtubules plus-ends (15). This half-life is an indirect read-out of microtubule dynamic instability in the spindle and is related to the catastrophe frequencies of microtubule populations. This assay has led to a number of important insights into how spindle proteins, including molecular motors and kinetochore components, regulate kinetochore–microtubule dynamics (16–20). In this chapter, we provide a step-by-step guide to perform quantitative measurements of tubulin turnover at microtubules plus-ends in living human cells. We describe how to produce a cell line that stably expresses photoactivatable-GFP-α-tubulin and how to specifically photoactivate the tagged protein and analyse the behaviour of dynamic microtubules in live cells.

2. Materials

2.1. Generation and Culture of Stable Cell Lines

1. Wild-type strain of cell line (see Note 1).
2. DNA with the gene fusion of interest (in this case, PA-GFP-α-tubulin and H2B-mRFP).
3. Dulbecco's modified Eagle's medium (DMEM) supplemented with 10% foetal bovine serum (FBS) and 1% PenStrep (all Invitrogen).
4. 0.05% Trypsin–EDTA (Invitrogen).
5. FuGENE 6 (Roche – Applied Science).
6. Serum-free media Optimem (see Note 2; Invitrogen).
7. Sterile phosphate-buffered saline (PBS).

8. Puromycin ready-made solution 10 mg/ml, store at –20°C (Sigma-Aldrich).
9. Geneticin/G418 Selective antibiotic 50 mg/ml, store at 4°C (Invitrogen).
10. Chambered coverslips (e.g. Lab-Tek or 35-mm dishes, Nunc, or Thermo-Fisher).
11. 3-mm cloning discs (Sigma-Aldrich).
12. CO_2-independent media, for example, Leibovitz-15 medium (L-15) supplemented with 10% FBS and 1% PenStrep (all Invitrogen).
13. Paclitaxel, stock at 100 mM in DMSO and store at –20°C (Taxol®, Tocris Biosciences).

2.2. Movies Acquisition and Analysis

1. DeltaVision RT microscope equipped with a solid-state 405-nm laser (Applied Precision, LLC), or confocal or widefield microscope system equipped with a solid-state 405-nm laser.
2. Sedat quad filter set, which incorporates single band-pass excitation and emission filters for DAPI, FITC, Cy3/TRITC/Texas Red and Cy5 (Chroma).
3. Image analysis software. For example, SoftWorx (Applied Precision, LLC).
4. Graph plotting and curve-fitting software. For example, KaleidaGraph (Synergy Software).

3. Methods

3.1. Production of Histone2B-mRFP, PA-GFP-α-Tubulin Stable Cell Line

The methodology described here uses a cell line stably expressing both Histone2B-mRFP and PA-GFP-α-tubulin (20). Stable cell lines ensure a high reproducibility between experiments that cannot be provided by transient transfection due to the variability in expression levels of the proteins in the latter case. The inclusion of a Histone2B-mRFP marker provides a number of advantages (1) the mitotic stage (prophase/prometaphase/metaphase/anaphase) of a rounded up cell can be determined and (2) the chromosomes can be used as a reference for positioning the laser focus (i.e. at the metaphase plate).

1. Transfect cells (see Note 3) with the Histone2B-mRFP construct and change the media to new media containing the selective agent 24 h after transfection (see Note 4).
2. 24 h after adding the selective agent, transfer the cells to 25-cm dishes (follow steps 1–7 in Subheading 3.2, but transfer different dilutions for each dish; for example, 1:50, 1:100, and 1:1,000).

3. Change the media (DMEM supplemented with 10% FBS and 1% PenStrep and the selective agent) on the cells every 3 days to ensure that the cells always have fresh media (see Note 5).
4. In a 10-cm dish, place 5 ml of warm trypsin, and add the same number of cloning discs as the colonies to select. At least 30 colonies must be selected, as it gives a good range of the different levels of protein expression between the clones.
5. Remove the media from the cells and wash twice with PBS.
6. Place the cloning discs on the surviving individual colonies of cells and leave for 3–5 min.
7. Remove the cloning discs and place each one into a separate well of a 12-well dish with media (DMEM supplemented with 10% FBS and 1% PenStrep and the selective agent).
8. Change media (DMEM supplemented with 10% FBS and 1% PenStrep and the selective agent) in each well of the 12-well dishes every day and wait until the cells reach 70–80% confluency.
9. Transfer the cells in each well of the 12-well dish onto a separate well of a 6-well dish with media (DMEM supplemented with 10% FBS and 1% PenStrep and the selective agent) (follow steps 1–7 in Subheading 3.2, but transfer all the cells) and wait until the cells reach 70–80% confluency.
10. View each individual well under a fluorescence microscope (e.g. in this chapter, we use a DeltaVision microscope with a DAPI/FITC/RD/Cy5 filter set) and confirm the integration of the construct. If the population is heterogeneous, it is useful to FACS sort the cells based on the expression level of the integrated plasmid. This way you get a more homogeneous population.
11. Expand positive clones by transferring each clone to a 10-cm dish. Positive clones should be frozen down in liquid nitrogen (see Note 6).
12. Once the Histone2B-RFP cell line is established, use this cell line and repeat steps 1–8 with PA-GFP-α-tubulin plasmid (see Note 7).
13. Once the new cell line is established, it is critical to confirm that the insertion of new plasmids did not alter cell proliferation by measuring mitotic index and calculate the doubling time of the new cell line compared to wild-type cells.

3.2. Setting Up a Live-Cell Imaging Experiment

1. Culture the cells in a 10-cm dish until they reach 80–90% confluency.
2. Aspirate off media from the cells and wash with 10 ml PBS.
3. Aspirate off PBS and repeat PBS wash.

4. Add 1 ml 0.05% trypsin–EDTA and incubate at 37°C for 3–4 min, until the cells start to detach from the dish.
5. Meanwhile, add 9 ml of warm media in a new 10-cm dish (if splitting 1:10), and 1.5 ml in each well of a 2-well Lab-Tek.
6. Add 9 ml of fresh DMEM supplemented with 10% FBS and 1% PenStrep and the selective agents to the trypsinized cells (see Note 8).
7. Add 1 ml of trypsinized cells into the new 10-cm dish and incubate at 37°C, this is the stock culture.
8. Add 80 μl of trypsinized cells into each well of the Lab-Tek (see Note 9).
9. Approximately 1 h before start imaging, change the media on the Lab-Tek to 1.5 ml of warm L-15, supplemented with 10% FBS and 1% PenStrep.

3.3. Imaging Microtubule Turnover Using PA-GFP-Tubulin

1. Approximately 2 h before the start of imaging set the microscope environmental chamber to 37°C and make sure that the temperature is stable before starting the experiment.
2. The experiments described here were performed on a DeltaVision RT microscope with a quantifiable laser module (Applied Precision, LCC). 512×512 images were acquired with a 100× oil-immersion objective with a numerical aperture of 1.4. The light source for collecting widefield images was mercury–halide using the Sedat quad filter set, which incorporates single band-pass excitation and emission filters and a multiband beamsplitter (Chroma). Emitted photons were detected with a Coolsnap HQ2 (photometrics) cooled CCD camera, which gives 65% quantum efficiency (in the FITC channel) and has a pixel size of 6.45×6.45 μm. Other microscope–filter–camera combinations could be used equally as well.
3. Find rounded up mitotic cells using the transmission light and then confirm that they are in the right stage of mitosis using the Histone2B-mRFP (using the Cy3/TRITC/Texas Red excitation/emission filters). Only look briefly with these filters so that the cells are not too exposed to the fluorescence light. Since the PA-GFP-α-tubulin is photoactivatable, you will not be able to see anything when using excitation with 488-nm light.
4. Centre the field of view on the centre of the metaphase plate (this is where the laser pulse will be located), using the Histone2B-mRFP as a reference, and acquire an image of the Histone2B-RFP (see Note 10 and Fig. 1a).
5. Capture three images using the FITC channel prior to photoactivation (step 7). This step ensures that no tagged protein is active before the laser pulse. As a control, quantify the

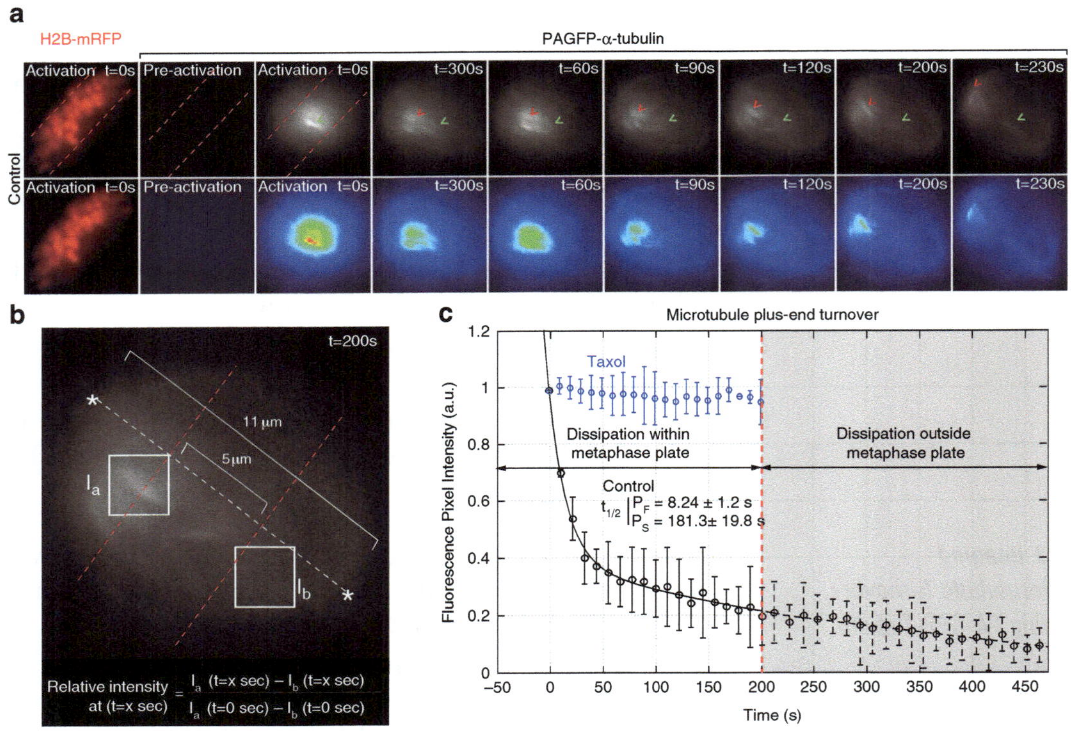

Fig. 1. Example of a live-cell photoactivation experiment. (**a**) Representative frames of a PA-GFP-α-tubulin photoactivation experiment. Photoactivatable GFP-α-tubulin fluorescence was activated in a circular region near the chromosome mass in metaphase cells (detected by the H2B–mRFP signal and represented as *red dotted lines*). An H2B–mRFP (DNA) frame is shown for the first time point of the live-cell movie after activation. *Green arrowheads* indicate the initial position of the photoactivated spot; *red arrowheads* show the final position of the activated spot (*upper panels*). Heat maps (*lower panels*) show the microtubule poleward flux and the dissipation of the photoactivated spot due to microtubule plus-end turnover. (**b**) Scheme for the quantification of the intensity of the activated spot at the metaphase plate. Intensities were normalized to $t=0$ s after photoactivation. Metaphase plate is represented as red dotted lines. (**c**) Quantification of fluorescence intensity decay of the activated regions over time. Indicated is the half-lives of the fast and slow MT populations. *Error bars* represent SD.

intensity of the spindle before photoactivation, and it should be identical to background levels.

6. Activate the PA-GFP-α-tubulin with a 50-ms pulse from the 406-nm laser (20%) (see Note 11).
7. Capture widefield images in the FITC channel every 10 s for 600 s, using 100% of the transmitted light and a 1-s exposure (see Note 11).
8. Representative frames of such a time-lapse sequence are shown in Fig. 1a.
9. For each condition tested, image 15–20 cells per experiment that meet the requirements described in Subheading 3.4, step 1, and repeat at least three times every experiment.

3.4. Quantification of Plus-End Turnover in Dynamic Microtubules

Different molecules ensure the control of MT dynamics within the spindle. Combining the method described in this chapter with siRNA or small molecule inhibitors can provide powerful insights to the roles of different proteins in the regulation of microtubule dynamics. From the photoactivation time-lapse sequences, it is possible to determine two key parameters (1) poleward MT flux, as the movement of the marks towards the poles and (2) MT plus-end turnover (expressed as half-life). Microtubule plus-ends in a wild-type mitotic spindle follow either fast or slow turnover kinetics. The first population (fast) is believed to be free spindle MTs, while the second population (slow) represents the more stable kinetochore–microtubules, which form the k-fibres.

1. Choose all movies, where you can see a clear activation spot against the background and discard the movies, where the cell enters anaphase (see Note 12).
2. Open the movie file (see Note 13) and place a 20×20 pixel box over the activated spot.
3. Measure the intensity and copy the value to an excel spreadsheet.
4. On the half spindle opposite the one the activated spot is moving, take a measurement of the background with the same pixel square (view Fig. 1b), and subtract that value from the intensity of the activated spot.
5. Normalize the background corrected intensity of the activated spot at $t = 0$ s to 1.
6. Follow the activated spot through time and repeat steps 3–5 for each time point until at least $t = 200$ s (see Note 14).
7. Normalize the fluorescence pixel intensity measured for each time point to the intensity at $t = 0$ s.
8. Repeat the measurements for each movie and average the normalized values for all time points.
9. On the appropriate software (see Note 15), plot the normalized fluorescence intensity values as a function of time on a scatter plot.
10. Fit a double-exponential curve of the type $I = P_f \exp(-k_f t) + P_s \exp(-k_s t)$, where I is the proportion of the initial fluorescence intensity, P is the proportion of fluorescence decay due to the fast (f) or slow (s) process, k is the time constant (τ) for fluorescence decay of the fast (f) or slow (s) process, and t is time.
11. Set the initial parameters on the double exponential curve P_f, k_f, P_s, and k_s to 1, 0.1, 0.2, and 3.3×10^{-3}, respectively (15).

Table 1
Previously reported microtubule turnover half-lives

		MT turnover half-live			
Cell line	**Organism**	**Fast (s)**	**Slow (s)**	**Method**	**Lab (references)**
Ptk1	Rat kangaroo	40.8 ± 24	318 ± 48	Caged fluorescein tubulin	Borisy (15)
Ptk1	Rat kangaroo	20.0 ± 1.2	607.0 ± 78	Recombinant PA-GFP-α-tubulin	Salmon (16)
Ptk2	Rat kangaroo	7.3	307.18	Recombinant PA-GFP-α-tubulin	Walczak (22)
U2OS	Human	16.6 ± 1.2	155.2 ± 13.9	Recombinant PA-GFP-α-tubulin	Maiato (21)
HeLa	Human	8.9 ± 0.9	209.1 ± 21.1	Recombinant PA-GFP-α-tubulin	McAinsh (20)

12. Calculate the half-time ($t_{1/2}$) from the time constant (τ) as $\ln 2/k$ for each fast and slow process. An example of microtubule plus-end turnover plot is shown in Fig. 1c and a comparison of our data to previously reported microtubule plus-end turnover rates from different systems and using different imaging set-ups are shown in Table 1.

3.5. Correcting for Photobleaching

Photobleaching is a common problem when using fluorescent molecules, and it may eliminate potentially useful information. Is the result of the decomposition of fluorophores in the excited state because of their interaction with molecular oxygen before emission. Since photobleaching cannot be prevented, several techniques have been developed to slow it down or correct the alterations due to this process. In this chapter, we use the microtubule drug taxol to correct for photobleaching. Taxol has been known for stopping all microtubule dynamics and, consequently, every decrease in the intensity of the photoactivated tubulin in the presence of taxol is exclusively due to photobleaching.

1. Follow the instructions in Subheading 3.2 to have cells ready to image in the same conditions as the previous experiments.
2. Approximately 10 min before start imaging, add 10 μM paclitaxel to the cells.
3. Follow the instructions in Subheading 3.3 (see Note 16) and analyse the movies as described in Subheading 3.4 until step 8.
4. Set the intensity measured at time = 0 s after photoactivation as 100% and determine the percentage of fluorescence loss for each frame during the first 200 s (see Note 17).

5. Add, in each frame of the experiments performed previously, the percentage of fluorescence loss determined in the cells treated with taxol.

4. Notes

1. Here, we use HeLa E1 cells as an example, but other cell lines can be used. For example, hTERT-RPE1 cells, U2OS cells, or PTK1 cells.
2. The pH of Optimem changes over time, so you need to change the bottle every month.
3. For transfection, plate 1×10^5 cells in a 35-mm culture dish (or a 6-well plate) with 2 ml of DMEM supplemented with 10% FBS and 1% PenStrep. 24 h later, prepare a transfection reaction at three different ratios (1) 97 μl Optimem + 3 μl FuGENE 6 + 1 μg DNA, (2) 96 μl Optimem + 3 μl FuGENE 6 + 2 μg DNA, (3) 94 μl Optimem + 6 μl FuGENE 6 + 1 μg DNA. Dilute FuGENE® 6 Transfection Reagent with serum-free medium (without antibiotics or fungicides). Pipette the FuGENE® 6 Transfection Reagent directly into the medium *without allowing contact with the walls of the plastic.* Add the DNA and incubate for 30 min at room temperature. Add the transfection reaction directly to the cells.
4. The plasmid used for transfection must contain a selective resistance. The concentration of the selective agent can be determined by a kill curve. As an example, for HeLa E1 cells, we use 200 μg/ml for G418 (Geneticin) and 1 μg/ml for puromycin.
5. Two days after addition of the selective reagent the majority of cells will begin to die and colonies of cells that are now resistant to the drug start to grow. These are the cells that we are interested in and need to be selected.
6. Before freezing the cells, prepare an ice bucket and make 5-ml freezing solution (77% media + 13% foetal calf serum + 10% molecular biology grade DMSO) on ice. Aspirate off the medium, wash the cells 1× with PBS, and add 1.5 ml trypsin to the dish. Incubate at 37°C for 2–3 min. Re-suspend the dish with 10 ml DMEM, and transfer the total into a falcon tube. Centrifuge 5 min at 16.1×*g*. In the mean time, label five freezing vials (with screw cap) and unscrew the cap. Aspirate off the supernatant and re-suspend the cell pellet in 5 ml of freezing solution. Aliquot the 5 ml of cells into the five vials (1 ml per vial) and transfer the vials to an isopropanol freezing container. Freeze down the cells directly to −80°C

and the day after transfer cells to liquid nitrogen. 1 or 2 days later, thaw one vial and check how they survive.

7. Make sure the plasmid you use with the PA-GFP-α-tubulin has a different drug resistance from the plasmid used for the Histone2B-mRFP transfection.
8. Pipette gently but vigorously up and down until no more lumps can be seen.
9. Depending on the conditions (siRNA or drug treatment), the amount of cells need to be adjusted. We use 80 μl if the experiment is to be performed 24 h after seeding the cells, 60 μl for 48 h and 40 μl for 72 h.
10. This image will give you an idea of the stage in which the metaphase cell is. You can also acquire an image at the end of the experiment.
11. The settings for the laser and acquisition are needed to be standardized for each microscope and cell line. The settings described here are for a DeltaVision RT microscope and HeLa cell line. Be careful not to use too high laser power as this will bleach the PA-GFP and you will not see a signal. Decrease and increase laser power to find the correct conditions for photoactivation.
12. By discarding all the cells that enter anaphase during the movie, you ensure that all measurements are taken from cells in late prometaphase/metaphase, as it is known that microtubules in anaphase have a different (even slower) turnover time (15).
13. It is easier to follow the activated spot if you use the greyscale mode.
14. The photoactivated spot will move in each time point due to poleward MT flux, so you will need to move the 20 × 20 pixel square in each time point to adjust for that movement.
15. Here, we use KaleidaGraph as an example.
16. We imaged each well of the Lab-Tek for approximately 30–40 min after the addition of taxol to the cells. We found no difference between the cells in this time interval. If you wish to acquire a significant number of movies, you can prepare multiple Lab-Teks and add taxol at different times.
17. We found that, for the conditions described in this chapter, the percentage of signal photobleaching is approximately 6%.

References

1. Mitchison, T. & Kirschner, M. (1984) Dynamic instability of microtubule growth. *Nature* 312:237–242.
2. Kueh HY, Mitchison TJ. (2009) Stuctural plasticity in actin and tubulin polymer dynamics. *Science.* 325.5943:960–3.
3. Brust-Mascher I, et al (2009) Kinesin-5–dependent Poleward Flux and Spindle Length Control in *Drosophila* Embryo Mitosis *Molecular Biology of the Cell* Vol. 20, 1749–1762.
4. Maddox P, et al (2002) Poleward microtubule flux is a major component of spindle dynamics and anaphase a in mitotic Drosophila embryos. *Curr Biol.* 12(19):1670–4.
5. Brust-Mascher I, Scholey JM. (2002) Microtubule flux and sliding in mitotic spindles of Drosophila embryos. *Mol Biol Cell.* 3967–75.
6. Buster DW, Zhang D, Sharp DJ. (2007) Poleward tubulin flux in spindles: regulation and function in mitotic cells. *Mol Biol Cell.* 18(8):3094–104.
7. Matos I, et al (2009) Synchronizing chromosome segregation by flux-dependent force equalization at kinetochores. *J Cell Biol.* 186(1):11–26.
8. Uteng M, et al (2008) Poleward transport of Eg5 by dynein-dynactin in Xenopus laevis egg extract spindles. *J Cell Biol.* 182(4):715–26.
9. Mitchison TJ. (2005) Mechanism and function of poleward flux in Xenopus extract meiotic spindles. Philos Trans R Soc Lond B Biol Sci. 360(1455):623–9.
10. Miyamoto DT, et al (2004) The kinesin Eg5 drives poleward microtubule flux in Xenopus laevis egg extract spindles. *J Cell Biol.* 167(5):813–8.
11. Shirasu-Hiza M, et al (2004) Eg5 causes elongation of meiotic spindles when flux-associated microtubule depolymerization is blocked. *Curr Biol.* 14(21):1941–5.
12. Mitchison TJ. (1989) Polewards microtubule flux in the mitotic spindle: evidence from photoactivation of fluorescence. *J Cell Biol.* 109(2):637–52.
13. Vallotton P, et al (2003) Recovery, visualization, and analysis of actin and tubulin polymer flow in live cells: a fluorescent speckle microscopy study. *Biophys J.* 85(2):1289–306.
14. G.H. Patterson, J. Lippincott-Schwartz (2002) A photoactivatable GFP for selective photolabeling of proteins and cells. *Science.* 297:1873–1877.
15. Zhai, Y., Kronebusch, P.J. & Borisy, G.G. (1995) Kinetochore microtubule dynamics and the metaphase-anaphase transition. *J Cell Biol.* 131:721–734.
16. DeLuca, J.G. *et al.* (2006) Kinetochore microtubule dynamics and attachment stability are regulated by Hec1. *Cell.* 127:969–982.
17. Ganem, N.J., Upton, K. & Compton, D.A. (2005) Efficient mitosis in human cells lacking poleward microtubule flux. *Curr Biol.* 15:1827–1832.
18. Cimini D, Wan X, Hirel CB, Salmon ED. (2006) Aurora kinase promotes turnover of kinetochore microtubules to reduce chromosome segregation errors. *Curr Biol.* 16(17):1711–8.
19. Cameron LA, et al. (2006) Kinesin 5-independent poleward flux of kinetochore microtubules in PtK1 cells. *J Cell Biol.*173(2):173–9.
20. Amaro AC, et al (2010) Molecular control of kinetochore-microtubule dynamics and chromosome oscillations *Nat Cell Biol.* 12(4): 319–29.
21. Maffini S, et al (2009) Motor-independent targeting of CLASPs to kinetochores by CENP-E promotes microtubule turnover and poleward flux. *Current Biology* 19(18):1566–72.
22. Rizk RS, et al (2009) MCAK and paclitaxel have differential effects on spindle microtubule organization and dynamics. *Mol Biol Cell.* 20(6):1639–51.

Chapter 17

Quantification of Asymmetric Microtubule Nucleation at Subcellular Structures

Xiaodong Zhu and Irina Kaverina

Abstract

Cell polarization is important for multiple physiological processes. In polarized cells, microtubules (MTs) are organized into a spatially polarized array. Generally, in nondifferentiated cells, it is assumed that MTs are symmetrically nucleated exclusively from centrosome [microtubule organizing center (MTOC)] and then reorganized into the asymmetric array. We have recently identified the Golgi complex as an additional MTOC that asymmetrically nucleates MTs toward one side of the cell. Methods used for alternative MTOC identification include microtubule regrowth after complete drug-induced depolymerization and tracking of growing microtubules using fluorescently labeled MT +TIP binding proteins in living cells. These approaches can be used for quantification of MT nucleation sites at diverse subcellular structures.

Key words: Microtubules, MTOC, Centrosome, Golgi, Cell polarity, Cell motility

1. Introduction

The asymmetric distribution of MT network contributes to the establishment or maintenance of cell polarity during a number of biological processes, such as cell migration, cell division, and embryonic development. How could MTs achieve asymmetric distribution pattern? In the canonical model, MTs are exclusively nucleated from the centrosome with the plus ends extending toward the cell periphery creating a radial distribution pattern. As both in vivo (1) and in vitro (2) experiments have demonstrated that MTs nucleated from the centrosome are symmetric, it is assumed that nucleation does not contribute to the asymmetric organization of MTs. Instead, other mechanisms are responsible for it, which include the local stabilization of MTs at a specific region (3), relocating the centrosome away from the cell center (4), or severing and releasing

Anne Straube (ed.), *Microtubule Dynamics: Methods and Protocols*, Methods in Molecular Biology, vol. 777,
DOI 10.1007/978-1-61779-252-6_17, © Springer Science+Business Media, LLC 2011

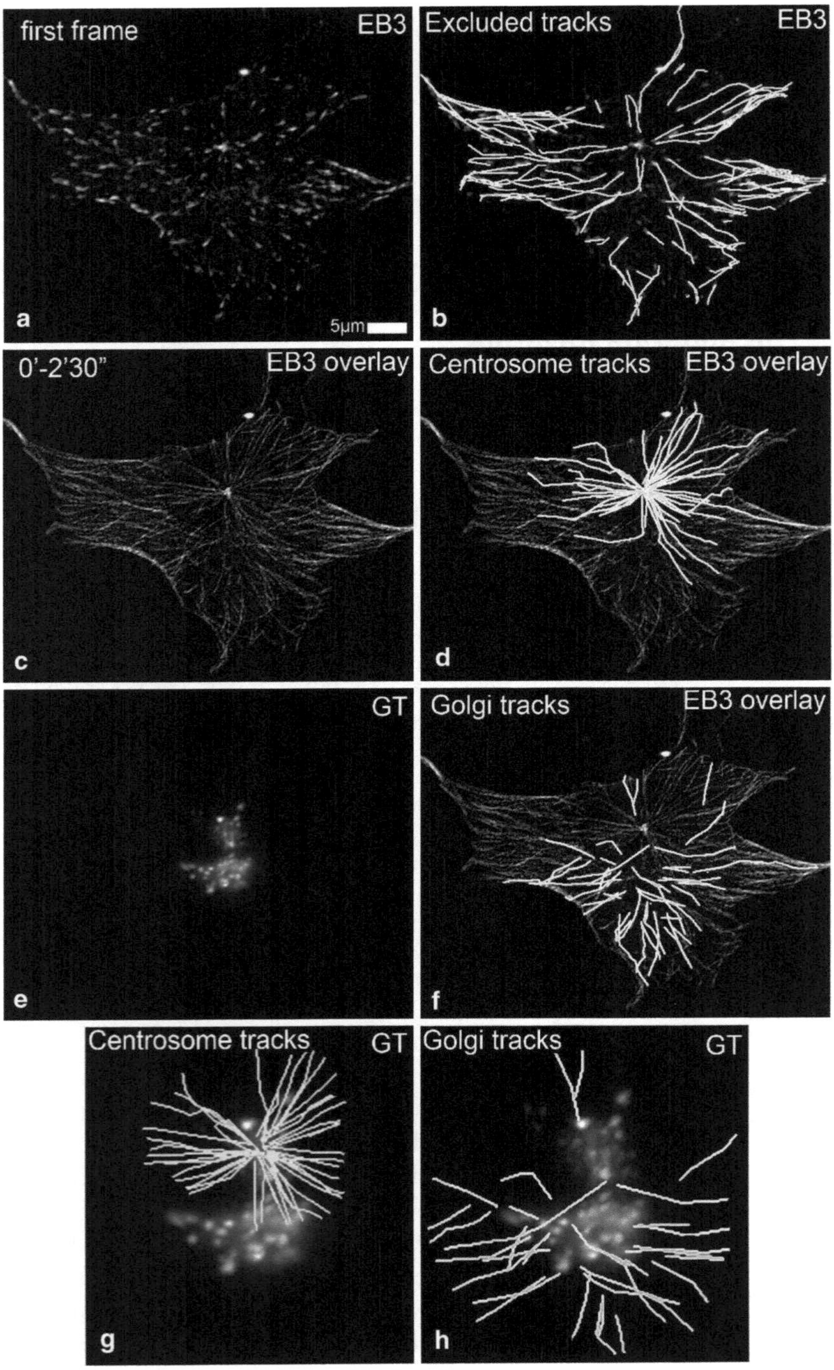

Fig. 1. Detection of Golgi-originated MTs in time-lapse recording of GFP-EB3- and mCherry-GT-expressing RPE1 cells (5 s/frame). (**a**) GFP-EB3 in the first frame of the video. Currently growing MTs are detected as EB3 dots. (**b**) MT tracks starting at EB3 dot at the first frame. These tracks are excluded from the analysis. (**c**) Overlaid GFP-EB3 images within 2.5 min mark all MT growth paths. (**d**) MT tracks starting from a single site are identified as centrosomal tracks. (**e**) mCherry-GT identifies Golgi complex localization. (**f**) Noncentrosomal MT tracks (**g**) Centrosomal MT tracks and their relation to the Golgi position. (**h**) Noncentrosomal (Golgi-derived) MT tracks in the cell center and their relation to the Golgi position. Modified for grayscale from (12) with permission from Elsevier Limited.

MTs from the centrosome (5, 6). However, a number of observations have suggested that MTs may be nucleated independently of the centrosome. For example, during cell division, MTs are nucleated from kinetochores (7). In interphase cells, noncentrosome-based MT organization centers (MTOCs) have been identified in several specialized cells (8, 9). These observations have raised two questions. First, are noncentrosomal MTs also nucleated in nondifferentiated cells? Second, can the noncentrosomal MTs contribute to the asymmetric pattern of MT array?

Technically, it is difficult to directly quantify MT nucleation at MTOC. The high density of the existing MTs around the MTOC makes it difficult to detect the newly nucleated MTs emerging from the MTOC. Thus, MT nucleation from centrosome has been studied by monitoring the fluorescently labeled +TIP proteins, such as CLIP170, EB1, etc. (10, 11). +TIPs are proteins which only localized to the plus end of the growing MTs. In the region around MTOC, +TIPs are associated only with newly nucleated MTs and move away from the MTOC as an MT polymerizes. Thus, the emergence of one +TIP dot from the MTOC indicates one nucleation event.

We have adopted this approach to monitor the MT nucleation by tracking the GFP-EB3 in interphase RPE1 cells in the steady state (Fig. 1). A complementary approach involves complete disassembly of the existing MTs by nocodazole and monitoring the initiation of MT nucleation after nocodazole washout. In nondifferentiated RPE1 cells, these approaches allow the identification of the Golgi apparatus as an alternative MT organizing-center (12). MT nucleation at the Golgi is asymmetric, with the majority of MTs oriented toward the leading edge in motile cells (Fig. 2). This observation indicates that MT nucleation can be an alternative way to form the asymmetric MT array.

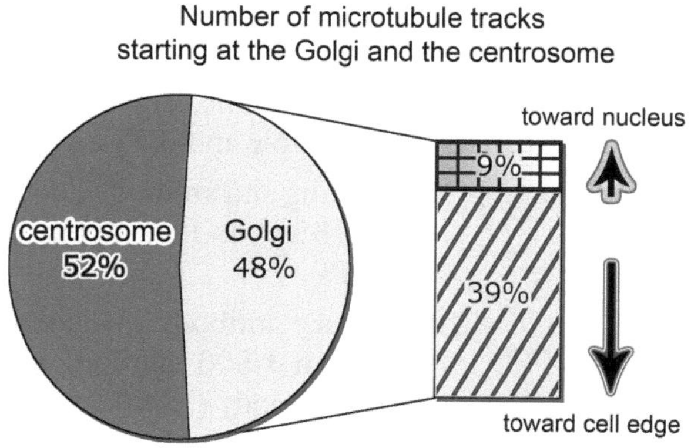

Fig. 2. Centrosome-nucleated MTs are symmetric and Golgi-nucleated MTs are asymmetric. Percentage and directionality of Golgi-associated tracks (583 tracks in ten cells, analyzed as in Fig. 1). Modified for grayscale from (12) with permission from Elsevier Limited.

2. Materials

2.1. Cell Culture and Transfection

1. Dulbecco's Modification of Eagle's Medium/Ham's F-12 50/50 Mix (DMEM/F12K) from Mediatech, supplemented with 1.5 mM $NaHCO_3$ solution (Thermo Scientific HyClone), 10% fetal bovine serum (Maverick Biosciences), and 1 mM L-glutamine (Mediatech).
2. Trypsin: 0.25% trypsin solution with EDTA, without sodium bicarbonate, calcium, and magnesium from Mediatech.
3. Fugene 6 transfection reagent from Roche.
4. Plasmid: +TIP marked with EGFP-EB3 (kind gift of Dr. J.V. Small, Vienna); and the centrosome marked with GFP-centrin (kind gift of Dr. M. Bornens, Paris); Golgi apparatus marked with mCherry-GT (modified from Clontech).

2.2. Live Cell Imaging and Immunofluorescence

1. Cell culture dishes for live cell imaging: 35-mm glass bottom dish with No. 1.5 coverslips from MatTek Corporation.
2. No. 1.5 microscope coverslips and microscope slides for immunofluorescence from Fisher Scientific.
3. Human Fibronectin (BD Biosciences) stock: 2.27 μM (1 mg/ml) in 1 M urea.
4. Phosphate-buffered saline (PBS): Prepare 10× stock solution: 1.37 M NaCl, 27 mM KCl, 80 mM Na_2HPO_4, 20 mM KH_2PO_4. Autoclave and dilute to 1× working solution before use.
5. Fixative: Methanol (EMD Chemicals Inc.).
6. Nocodazole (Sigma) stock: 16.6 mM in DMSO; store at −20°C.
7. Taxol stock: 5 μM in DMSO; store at −20°C.
8. Extracting buffer: 60 mM PIPES, 25 mM HEPES, 10 mM EGTA, 2 mM $MgCl_2$, 0.1% Saponin; adjust pH to 6.9. Store at 4°C. Immediate before use, supplement with 0.25 nM nocodazole and 0.25 nM taxol (13).
9. Blocking and antibody dilution buffer: 1% *b*ovine *s*erum *a*lbumin (BSA)/in PBS supplemented with 5% *d*onor *h*orse *s*erum (DHS).
10. Primary antibody: Monoclonal anti-alpha tubulin antibody (used at 1:500 dilution) from Sigma. Guinea pig anti-GCC 185 antibody (1:800) was developed by Covance Inc.
11. Secondary antibody: Alexa Fluor 568 anti-guinea pig IgG, highly cross absorbed Alexa Fluor 568 anti-mouse IgG (Invitrogen).
12. Mounting medium: ProLong Gold antifade reagent (Invitrogen). Store at −20°C.

2.3. Confocal Microscope

1. Microscope: Nikon TE2000E microscope with a built-in Z motor is equipped with a heated stage (Warner Instruments) and a PerfectFocus automated focusing device.
2. Optics: A Yokogawa QLC-100/CSU-10 spinning disk head (Visitec assembled by Vashaw) is attached to the microscope. Samples are illuminated by a krypton–argon laser (75 mW 488/568; Melles Griot) with AOTF and visualized with CFI PLAN APO VC 100× oil lens, double dichroic mirror and filters (Chroma) in a filter wheel (Ludl) were used for emission light.
3. Camera: Back illuminated EM-CCD camera Cascade 512B (Photometrics) is driven by IPLab software (Scanalytics).
4. Software packages, such as IPLab and NIH ImageJ, can be used for image analysis.

3. Methods

Here, we suggest three experiments allowing the identification of noncentrosomal nucleation sites, with the Golgi as an example.

First experiment involves transfection of cells with a +TIP marker and a Golgi marker with subsequent live cell imaging of MT tip movement in steady state (Fig. 1). This method allows the identification of nucleation sites in cells with unchanged MT organization.

However, monitoring MTs nucleated at the Golgi as compared to the centrosome may be challenging due to the fact that Golgi apparatus is localized adjacent to the centrosome. This can be avoided by monitoring those cells, where Golgi and centrosome are physically separated.

Another way to solve this difficulty is by monitoring MT repolymerization after nocodazole washout. When MTs are completely dissembled by nocodazole, Golgi is also dissembled into ministacks distributed all over the cell and imaging cells shortly after nocodazole treatment can clearly identify those MTs which are nucleated from the Golgi. Thus, *the second experiment* is live cell imaging of cells expressing a +TIP marker and a Golgi marker in the course of nocodazole washout. This experiment is conclusive but time consuming. For the purpose of statistical evaluation of MT number organized at the Golgi, we suggest *the third experiment* that involves nocodazole washout approach in nontransfected cells with fixation at specific time points and immunostaining to visualize MTs and the Golgi. Below, detailed techniques required for these three experiments are described.

3.1. Live Cell Imaging of MT Nucleation in Steady State

1. Cell maintenance and preparation for transfection: RPE1 cells are maintained in 5-ml tissue culture flask and passaged when reaching 80% confluence. To passage, cells were treated with trypsin for 1–2 min at room temperature and split into a new 5-ml flask for maintenance at 1:4 or 1:8 dilution. For transfection, cells are plated at 1:16 dilution in 35-mm tissue culture dishes. 2 h incubation before transfection is necessary for cells to attach to the culture dish and achieve 40–60% confluence (see Note 1).
2. Transfection: The cells are transfected with 0.5 μg of PEGFP-EB3 using Fugene 6 according to manufacturer's instructions. Briefly, in a 1.5-ml tube, 3 μl Fugene 6 is added into 100 μl plain DMEM/F12k medium, then 0.5 μg GFP-EB3 is added to form the transfection mixture. After incubated at room temperature for 30 min, 1.4-ml complete medium is added to stop the reaction. The final mixture is added to cells in the 35-mm culture dish. Cells are then incubated with the transfection mixture overnight (see Note 2).
3. Glass coating: Fibronectin stock solution is diluted with 1× PBS at 1:100. 150 μl diluted fibronectin is added onto the center of the glass coverslip in the MatTek dish and then incubated at 37°C for 30 min or room temperature for 1 h.
4. Plating transfected cells on glass: At least 4 h prior to live cell imaging, fibronectin is removed from the glass surface by aspiration. Transfected cells are plated on the glass bottom culture and incubated to allow cells attached to the glass surface.
5. Before placing the cells on the stage, microscope stage is warmed up to 37°C.
6. For imaging, cell culture medium is overlaid with a layer of mineral oil to avoid evaporation of medium and then placed on the microscope stage.
7. In order to monitor all the MTs nucleation toward all directions, Z-stack images of the entire cells with 0.2 μm Z-interval are captured at 5 s/frame. All the images are captured with laser power 100–150 mW and 200–500 ms exposure time.

3.2. Live Cell Imaging of MT Nucleation After Nocodazole Washout

1. Cells are maintained and transfected as the above procedure (see Subheading 3.1, steps 1–4).
2. Nocodazole treatment: To dissemble MT cytoskeleton, nocodazole stock is diluted with complete medium at 1:2,000 dilution. Then, the solution is added to the cells and incubated at 37°C for 2 h.
3. Nocodazole washout: Cells incubated with nocodazole were directly placed on the microscope stage and imaged, then washed with cold medium on the stage. Image recording is resumed immediately after washing.

4. To monitor MT nucleation after nocodazole washout, images are captured at 5 s/frame with laser power 100–150 mW and 200–500 ms exposure time. Three Z-slices with 0.2 μm interval is taken at each time point to capture the Golgi mini-stack (see Note 3).

3.3. Immuno-fluorescence Analysis of MT Nucleation After Nocodazole Washout

1. Cells are maintained as the above procedure (see Subheading 3.1, step 1).
2. Preparation of coverslips: Place a pack of coverslips in glass petri dish and autoclave; the sterilized coverslips can be kept in the tissue culture hood.
3. Glass coating: Use the forceps to lift the coverslip from the petri dish and place inside the 35-mm plastic culture dish. 1 mg/ml fibronectin stock solution is then diluted with 1× PBS at 1:100 dilution. 60–100 μl diluted fibronectin is added onto the center of the coverslip and then incubated at 37°C for 30 min or room temperature for 1 h.
4. Plating cells on glass: At least 4 h prior to nocodazole treatment, fibronectin is removed from the glass surface by aspiration. Cells are plated on 35-mm tissue culture dish containing coated coverslips and incubated to allow cells attach to the glass surface.
5. Nocodazole treatment: To dissemble MT cytoskeleton, nocodazole stock is diluted with complete medium at 1:2,000 dilution. Then, solution is added to the cells and incubated at 37°C for 2 h.
6. Preparation of fixative: During the nocodazole incubation, 100% methanol is added into 12-well plate with 1–2 ml/well and chilled at −20°C.
7. Nocodazole washout for immunofluorescence: Place two 60-mm tissue culture dishes containing 5-ml complete medium and extraction buffer, respectively, on a floating raft in the 37°C water bath. Directly remove the 35-mm culture dish containing nocodazole treated cells from the incubator and place it on ice. Wash the cells with ice-cold plain medium five times on ice (see Note 4), then quickly lift the coverslip and immerse it into the complete medium in the water bath, gently swirl the dish for 45 s to allow the regrowth of the MT (see Note 5). Then, immediately remove the coverslip from the medium and immerse it into the extraction buffer and gently swirl the dish for 40 s to extract the free tubulin from the cytoplasm. Cells were then subjected to fixation.
8. Fixation: Carefully pick up the coverslip from the culture dish with the forceps and quickly submerge the coverslip into methanol prechilled in step 6 with the cell side facing up.

9. Cells are incubated with methanol at −20°C for 5 min. Then, washed with PBS and kept in PBS for at least 1 h to rehydrate.
10. Cut a piece of parafilm and place it into a plastic box. Attach a piece of wet Kimwipe inside the cover of the box to keep the humidity.
11. Add 18 μl drops of blocking solution onto the parafilm and carefully place each coverslip onto a solution drop with the cell side facing the solution. Let the coverslip rest on the solution so that the cell side is in contact with the solution uniformly. Do not press the coverslip down! Incubate at room temperature for 1 h.
12. In a separate plastic box, prepare a similar humidity chamber as in step 10. Place 18 μl drops of primary antibody solution on parafilm. Gently lift the coverslip from the blocking solution using forceps and place the coverslip on primary antibody drops by the same procedure as described above. Incubate at room temperature for 1 h.
13. Immerse the coverslip into PBS in 12-well plate with the cell side up. Wash with three changes of PBS each for 5 min.
14. Repeat steps 12 and 13 for secondary antibody.
15. Add a drop of mounting medium on the microscopic slide and invert the coverslip on to it with the cell side facing down. Then, keep the slide at room temperature overnight or 37°C for 1 h to dry. Seal the edge of the coverslip with the nail polish.
16. Once the nail polish has dried, carefully clean the surface of the coverslip with ethanol to remove the precipitates from PBS, and samples are ready for microscopic examination. For long-term storage, samples should be kept at −20°C.

3.4. Image Processing for Plus-Tip Tracking

1. IPLab, ImageJ, Metamorph, or any other appropriate software package can be used for processing. To subtract background, time-lapse images are processed by rolling ball background correction. Brightness and contrast, gamma settings were then adjusted to make both the major structure and the minor structures visible.
2. Z-stack images at each time point and fluorescent channel is projected by Maxi-Projection algorithm individually. In order to clearly follow MT tip dislocation, image sequence is processed to create a rolling average sequence over four frames.
3. All MT plus-tip tracks are determined by a tracking algorithm, such as "Manual Tracking" plugin of ImageJ software. As all the plus-tip signals existing in the first frame of the sequence present already existing MTs nucleated before the observation,

those MTs tracks have to be excluded from the further analysis. All the other tracks initiated during the imaging have to be tracked back to the sites of origin and classified according to the origin (Fig. 1).

4. Identify centrosomal MTs: As +TIP binding proteins bind to the centrosome, and MT tracks continually emanating from centrosome, the position of the centrosome can be identified as a single spot with high fluorescence intensity. This spot normally is less than 1 μm in diameter. All the MTs tracks emanating from this common spot can be considered centrosomal MTs.
5. To identify Golgi-derived MTs, sites of MT origin have to be co-localized with the Golgi. Similarly, specific markers can be used for any subcellular structure under consideration.

4. Notes

1. Transfection can also be performed directly with the cells growing on the glass surface. As it is easy to image when cells are very well spread and flat, a dense culture should be avoided. We find that it is easy to adjust the cell confluence when transfecting cells in 35-mm plastic dishes and then splitting onto the glass surface on the day before imaging.
2. The exact amount of plasmid used and the incubation time after transfection need to be adjusted according to the desired expression level for imaging. Increasing the amount of plasmid and incubation time normally leads to a higher expression level. It is easy to obtain high-quality images with a strong fluorescence signal. However, when overexpressed, +TIPs decorate the entire MT network and affect MT dynamics. On the other hand, weak fluorescence signal requires high laser intensity and/or long exposure time to image, which may induce photo damage to the cells. Therefore, it is crucial to select cells with appropriate expression levels. Unfortunately, different imaging systems have different sensitivity, so no fixed criteria can be applied to select the appropriate cells. It is necessary to determine the best condition through trial and error. Alternatively, a stable cell line expressing +TIPs at the appropriate level can be established first, and then used for imaging.
3. Both the time interval and Z interval need to be adjusted according to the cell line and the imaging system used. Nyquist sampling rate can be calculated according to the individual image system and used as a starting point to test

for Z interval. Apart from photodamage, oversampling is better than undersampling.

4. It is important to wash on ice to inhibit MT repolymerization during the washing step. To completely wash away any trace amount of nocodazole, wash media should fill the whole volume of the cell culture dish. For one 35-mm culture dish, 5-ml medium is necessary per wash.
5. Short regrowth times should be considered. Because MTs are very dynamic structures, a small variation in time may generate significant difference in MT organization. In general, MT nucleation and growth are rapid. To monitor the nucleation, images should be captured with short interval. Since MT nucleation occurs in 3D, it is also necessary to take Z-stack images at each time point in order to quantify the nucleation events.

References

1. Salaycik, K. J., Fagerstrom, C. J., Murthy, K., Tulu, U. S., and Wadsworth, P. (2005) Quantification of microtubule nucleation, growth and dynamics in wound-edge cells, *J Cell Sci 118*, 4113–4122.
2. Bergen, L. G., Kuriyama, R., and Borisy, G. G. (1980) Polarity of microtubules nucleated by centrosomes and chromosomes of Chinese hamster ovary cells in vitro, *J. Cell Biol. 84*, 151–159.
3. Kirschner, M. and Mitchison, T. (1986) Beyond self-assembly: from microtubules to morphogenesis, *Cell 45*, 329–342.
4. Etienne-Manneville, S. and Hall, A. (2003) Cdc42 regulates GSK-3[beta] and adenomatous polyposis coli to control cell polarity, *Nature 421*, 753–756.
5. Keating, T. J., Peloquin, J. G., Rodionov, V. I., Momcilovic, D., and Borisy, G. G. (1997) Microtubule release from the centrosome, *Proceedings of the National Academy of Sciences of the United States of America 94*, 5078–5083.
6. Ahmad, F. J. and Baas, P. W. (1995) Microtubules released from the neuronal centrosome are transported into the axon, *J Cell Sci 108*, 2761–2769.
7. Khodjakov, A., Copenagle, L., Gordon, M. B., Compton, D. A., and Kapoor, T. M. (2003) Minus-end capture of preformed kinetochore fibers contributes to spindle morphogenesis, *The Journal of Cell Biology 160*, 671–683.
8. Bugnard, E., Zaal, K. J., and Ralston, E. (2005) Reorganization of microtubule nucleation during muscle differentiation, *Cell Motil Cytoskeleton 60*, 1–13.
9. Malikov, V., Kashina, A., and Rodionov, V. (2004) Cytoplasmic Dynein Nucleates Microtubules to Organize Them into Radial Arrays In Vivo, *Mol. Biol. Cell 15*, 2742–2749.
10. Komarova, Y. A., Vorobjev, I. A., and Borisy, G. G. (2002) Life cycle of MTs: persistent growth in the cell interior, asymmetric transition frequencies and effects of the cell boundary, *J Cell Sci 115*, 3527–3539.
11. Piehl, M., Tulu, U. S., Wadsworth, P., and Cassimeris, L. (2004) Centrosome maturation: Measurement of microtubule nucleation throughout the cell cycle by using GFP-tagged EB1, *PNAS 101*, 1584–1588.
12. Efimov, A., Kharitonov, A., Efimova, N., Loncarek, J., Miller, P. M., Andreyeva, N., Gleeson, P., Galjart, N., Maia, A. R., McLeod, I. X., Yates, J. R., Maiato, H., Khodjakov, A., Akhmanova, A., and Kaverina, I. (2007) Asymmetric CLASP-dependent nucleation of noncentrosomal microtubules at the trans-Golgi network, *Dev Cell 12*, 917–930.
13. Zhai, Y. and Borisy, G. G. (1994) Quantitative determination of the proportion of microtubule polymer present during the mitosis-interphase transition, *J Cell Sci 107*, 881–890.

Chapter 18

Investigating Microtubule Dynamic Instability Using Microtubule-Targeting Agents

Stéphane Honore and Diane Braguer

Abstract

This chapter presents protocols not only used to investigate the effect of microtubule-targeting agents on microtubule dynamic instability parameters, but also their impact on loading +TIPs at microtubule plus ends. These agents can be considered either as drugs to analyze their pharmacological effects on microtubule dynamics and their subsequent functions or as tools to improve basic knowledge on the regulation of microtubule dynamics. Deciphering the complexes of proteins that regulate microtubule dynamic instability may lead to the discovery of new potential targets for therapy.

Key words: Microtubule dynamics, Tubulin, +TIPs, Drugs, Live cell imaging, Fluorescence time-lapse microscopy

1. Introduction

1.1. Microtubule Functions in Mitosis and in Interphase

Microtubules are highly dynamic cylindrical structures in the cytoskeleton, consisting of α and β tubulin heterodimers. They are essential components that play critical roles in a variety of cell processes, including cell division, cell polarity, and migration or neuronal differentiation (1–3). The cellular function of microtubules results from their dynamic behavior termed "dynamic instability" which is highly regulated in response to extracellular and intracellular signals. Their role in cell division is especially important in cancer field since a misregulation of their dynamic behavior causes chromosome mis-segregation which is associated with tumor cell death (4). Microtubules play also a significant role in cell migration by regulating cell polarization, actin polymerization, and focal adhesion dynamics by an exquisite-regulated crosstalk involving microtubule plus-end-tracking proteins (+TIPs) (5, 6).

Anne Straube (ed.), *Microtubule Dynamics: Methods and Protocols*, Methods in Molecular Biology, vol. 777, DOI 10.1007/978-1-61779-252-6_18, © Springer Science+Business Media, LLC 2011

Moreover, microtubules are the main target of major anticancer drugs, i.e., microtubule-targeting agents, including the taxanes and the *vinca* alkaloids, which are largely used in the treatment of human cancers. It is now clear that microtubule-targeting agents act on signaling pathways leading to cell death and alteration of microtubules functions by disturbing microtubule dynamic instability parameters.

1.2. Microtubule Dynamic Instability

Dynamic instability is an intrinsic property of microtubules, characterized by an alternation between periods of polymerization and depolymerization. In living cells, such process is both spatially and temporally regulated by several factors, such as Microtubule-Associated Proteins (MAPs), +TIPs, and stathmin (1, 7). Many of these proteins are regulated by phosphorylation or other posttranslational modifications and are, thus, under the dependence of signaling pathways emanating from the microenvironment or during cell cycle progression (3). Microtubule dynamic instability can be described using at least six main parameters: the rates of microtubule growth and shortening, the frequency of transition from growth or pause to shortening (called catastrophe), and the frequency of transition from shortening to growth or pause (called rescue). Periods of pause are the condition in which changes in microtubule length are drastically attenuated or stopped (Fig. 1). The parameter called "dynamicity" is used to describe the overall rate of tubulin subunits exchange at microtubule ends (8, 9).

1.3. Microtubule-Targeting Agents

The microtubule-targeting agents are classified into two major groups, the microtubule-destabilizing agents and the microtubule-stabilizing agents, according to their effects at high concentrations on microtubule polymer mass (10). The so-called destabilizing agents inhibit microtubule polymerization. Most of these agents bind in one of two domains on tubulin, the "Vinca" domain and the "colchicine" domain. They include the vinca alkaloids (vinblastine, vincristine, vinorelbine, vindesine, and vinflunine), cryptophycins, halichondrins, dolastatins, estramustine, 2-methoxyestradiol, colchicine, nocodazole, noscapine, and combretastatins. The "microtubule-stabilizing" agents enhance microtubule polymerization and include paclitaxel (the first identified in this class), docetaxel, epothilones, discodermolide, eleutherobins, and laulimalide. The classification of drugs as either microtubule "stabilizers" or "destabilizers" is simple and useful, but is somewhat misleading. Indeed, both classes of drugs, those that increase and those that decrease microtubule polymerization at high concentrations ($>IC_{100}$), both potently suppress microtubule dynamics in a concentration-dependent manner at IC_{10} to IC_{100}. At these concentrations, they kinetically stabilize microtubules without changing the microtubule polymer mass by

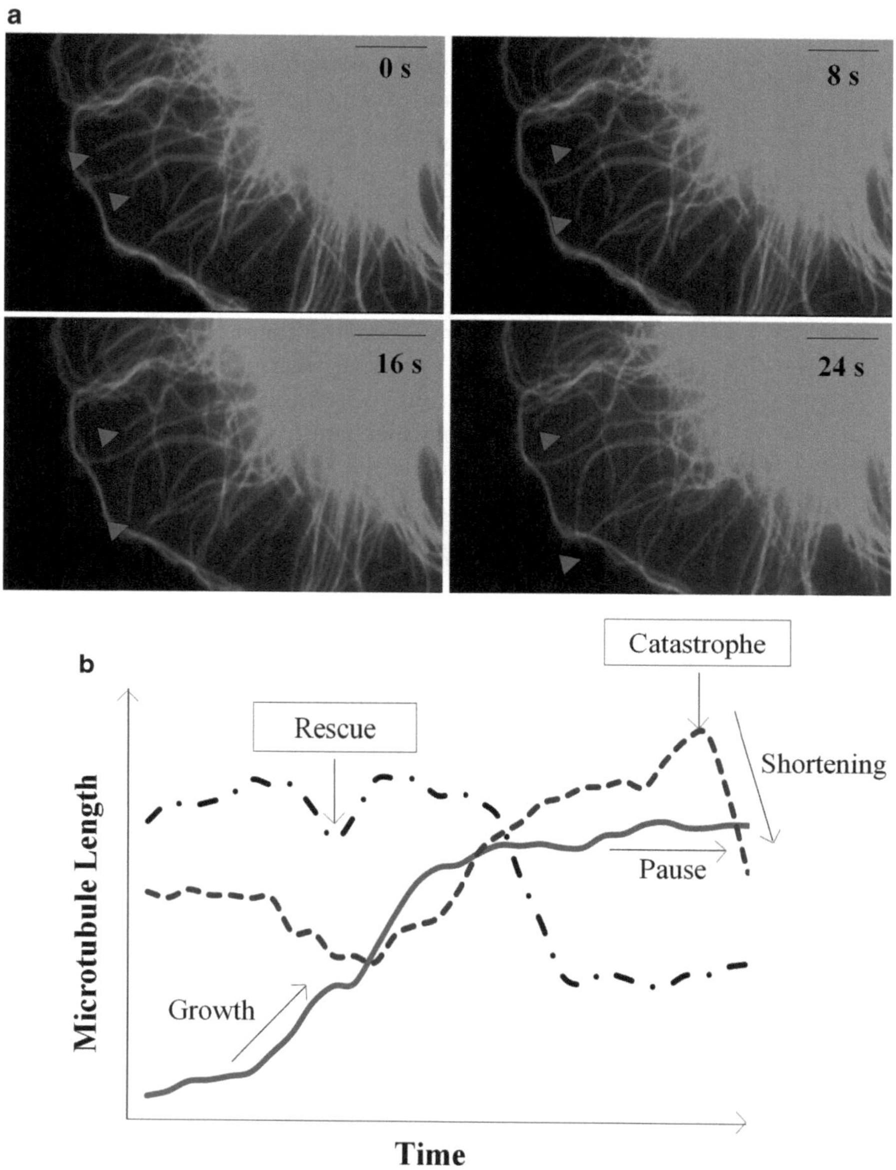

Fig. 1. Analysis of microtubule dynamic instability in living cells. (**a**) Time-lapse sequence of video frames of human dermal microvascular endothelial cell (HMEC-1) transiently transfected with GFP-tubulin, showing the plus-end position changes of several microtubules (*arrow heads*) over time. Bar: 5 μm. (**b**) Life history plot of the changes in microtubule length, representing the different phases and transitions of microtubule dynamic instability.

a direct action on the microtubule (4, 11–13). Moreover, at such concentrations the +TIPs network at microtubule plus ends (EB1; CLIP-170, p150 Glued) are completely disorganized ((14); S. Honore and D. Braguer personal data). However, in some cell types, at low antimigratory and noncytotoxic concentrations, *vinca* alkaloids and taxanes enhance microtubule dynamics mostly by disrupting the regulation and the influence of the cell microenvironment through disruption of +TIPs, such as EB1 (15–17).

As our knowledge of microtubule-targeting agents increases, we understand that the mechanism underlying their anticancer activity is very complex and mainly lies in alteration of microtubule and +TIPs dynamics, rather than in their effects on microtubule polymer mass changes.

Owing to the importance of microtubule dynamics for cell cycle and cell migration, future research on microtubules in terms of structure, localization, dynamic behavior, and regulation must be developed under different experimental conditions, such as the influence of the microenvironment (growth factors, extracellular matrix proteins) or drug treatment and/or overexpression and/or downregulation of endogenous regulatory proteins. Deciphering regulatory mechanisms that orchestrate microtubule dynamics and particularly the players of such regulation is essential for understanding cell biology and to identify more specific targets for anticancer therapy.

Measurement of microtubule dynamic instability in living cells by using well-known microtubule-targeting agents as tools to modify their dynamics is a common way to study microtubule behavior. This technique is also crucial to determine the mechanism of action of new potential microtubule-targeting agents as well as that of chemical or proteic compound able to modify signaling pathways involved in the regulation of microtubule behavior and functions (for example, tyrosine kinase inhibitors, anti-VEGF antibody, anti-EGFR antibodies).

To that end, we present methods to acquire microtubule and/or microtubule plus end proteins videos using time-lapse fluorescence microscopy and to quantify microtubule dynamic instability parameters using drugs.

2. Materials

2.1. Drugs

A large number of chemically diverse substances originating from natural sources or obtained by chemical synthesis bind to tubulin and affect microtubule dynamics by different ways. Among them, nocodazole and the anticancer drugs, such as the taxanes and the *vinca* alkaloids, have been extensively studied and their effects on microtubule dynamics are well-described. A nonexhaustive list of drugs and their effects on microtubule dynamic instability parameters are given in Tables 1 and 2. However, their effects on individual microtubule dynamic instability parameters may vary with their concentration and also according to the cell type and the cell microenvironment. Moreover, as previously mentioned, microtubule dynamic instability is a finely regulated process in cells which involves many regulators, such as the +TIPs and signaling molecules, including kinases or phosphatases. Thus, other drugs or

Table 1
Alteration of microtubule dynamic instability parameters by microtubule-targeting agents in living cells

Drug	Concentration (nM)	Cell type	GR (%)	GL (%)	SR (%)	SL (%)	CF min^{-1}	RF min^{-1}	CF μm^{-1}	RF μm^{-1}	Pause duration (%)	Dynamicity (%)	References
>IC_{50} cell proliferation (24 h)													
Vinblastine	32	BS-C-1	−20	−50	−67	−70	−60%	ns	−77%	+133%	+65	−75	[18]
IC_{50} mitotic arrest													
Paclitaxel	7.5	MCF-7	−41	−42	−57	−63	nd	ns	+28%	+382%	nd	−54	[19]
Epothilone B	3.5	MCF-7	−44	−59	−44	−67	ns	+62%	+78%	+165%	ns	−62	[19]
Vinblastine	1	MCF-7	−26	−57	−13	−28	ns	ns	nd	nd	nd	−48	[4]
Discodermolide	83	A549	−30	−34	−30	−54	−42%	+88%	ns	+172%	ns	−62	[12]
Halichondrin Analogue E7389	1	MCF-7	−27	−50	ns	ns	ns	ns	nd	nd	ns	−28	[20]
IC_{50} cell proliferation (72 h)													
Paclitaxel	2	A 549	ns	ns	−14	−30	ns	ns	ns	+86%	ns	−25	[11]
Discodermolide	7	A 549	ns	ns	−20	−42	ns	ns	ns	+93%	ns	−23	[11]
Vinflunine	50	SK-N-SH	−19	−23	−29	−15	ns	ns	nd	nd	ns	−35	[21]
Noncytotoxic, antiangiogenic, antimigratory													
Paclitaxel	1	HMEC-1	+42	+25	+113	+50	−45%	+50%	−65%	ns	−40	+100	[16]
Paclitaxel	0.1	HUVEC	+22	+33	+27	+32	ns	ns	−35%	−29%	−36	+54	[16]
Vinflunine “ECM:gelatin”	2	HMEC-1	+30	+11	+19	+18	+58%	ns	ns	ns	−25	+54	[15]
Vinflunine “ECM:fibronectin”	2	HMEC-1	ns	ns	ns	+87	ns	−50%	nd	nd	−33	+23	S. Honore and D. Braguer personal data
Antimigratory and cytotoxic													
Paclitaxel	50	NRK fibroblasts	−43	nd	−38	nd	−25%	ns	nd	nd	nd	−75	[22]
Nocodazole	50	NRK fibroblasts	−54	nd	−38	nd	−75%	ns	nd	nd	nd	−79	[22]

GR growing rate, *GL* growing length, *SR* shortening rate, *SL* shortening length, *CF* catastrophe frequency, *RF* rescue frequency, *nd* not determined, *ns* not statistically significant, *ECM* extracellular matrix

Table 2
Source and solubility of microtubule-targeting agents

Drug	Source	Solubility (maximal concentration)
Paclitaxel	Sigma Alexis	DMSO (50 mg/ml) Ethanol
Vinblastine	Sigma	DMSO (50 mg/ml) Water (10 mg/ml)
Nocodazole	Sigma	DMSO (10 mg/ml)
Discodermolide	Novartis Pharmaceuticals	DMSO
Vinflunine	Pierre Fabre Oncology	Water
Halichondrin analogue E7389	Eisai Research Institute	Water

molecules that do not directly bind to tubulin may also affect microtubule dynamic instability by acting through such regulators.

2.2. Extracellular Matrix Proteins (see Note 2)

1. 0.1% (w/v) poly-L-lysine in water (Sigma).
2. 1 mg/ml fibronectin, cell culture tested (Sigma), stored at −20°C.
3. 1 mg/ml laminin (EHS), stored at −20°C.

2.3. Labeling Microtubules or Microtubule Plus Ends in Living Cells

1. Cell culture (80% confluent).
2. To label the entire microtubule network in living cells, expression vector containing tubulin sequence is fused with a GFP fluorescent protein, such as pEGFP-α-tubulin or pAcGFP1-α-tubulin expression vector, commercially available from Clontech. Other commercially available expression vectors with various fluorescent proteins, such as mCherry- or DsRed-tubulin, can be used.
3. To label microtubule plus ends, expression vector containing EB1 or EB3 sequence with GFP or other fluorescent protein is fused on the C-terminal part of the protein. The fusion of GFP to the N-terminal part of EB1 alters its binding to microtubules as opposed to EB3 (23). Other fluorescently tagged proteins that localize to microtubule plus ends, such as GFP-CLIP-170, can also be used (see Note 4).
4. Cell transfection is performed using electroporation (AMAXA®, Cologne, Germany) or Lipofectamine 2000 transfection reagent (Invitrogen Corp) or other reagent suitable for transient transfection of mammalian cells. About 2–5 μg of purified plasmid DNA is used for each transfection reaction.

2.4. Fluorescence Time-Lapse Videomicroscopy of Microtubules and/or Microtubule Plus Ends

1. For live cell imaging, transiently or stably transfected cells are placed in a recording media, usually tissue culture media lacking phenol red supplemented with 15 mM HEPES, 1% FBS, 2% oxyrase, or 0.1 μg/ml ascorbic acid to reduce photodamage during acquisition of time series.
2. Live imaging of microtubules and/or microtubule plus ends is performed on an inverted fluorescence microscope, such as a Leica DM-IRBE or a Nikon TE2000 with 100× or 63× 1.4 NA objective lens, and equipped with appropriate filters according to the fluorescent protein used. Time-lapse acquisition is performed with a cooled CCD camera, such as COOLSNAP FX or COOLSNAP HQ (Roper Scientific), driven by a software which integrates individual images in time-lapse sequences ready for microtubule dynamics analysis, such as Metamorph software (Universal Imaging Corp.). Images acquisition must be performed at a temperature of 37 ± 1°C.

2.5. Dynamic Instability Parameters Analysis

Metamorph software (Universal Imaging Corp.) or any software allowing the tracking of microtubule plus end over time, such as image J is required. The methods are described with Metamorph 6.1 software.

3. Methods

Several techniques can be used to measure microtubule dynamic instability in living cells. Individual fluorescent microtubules can be readily visualized in thin peripheral regions of cells after microinjection of fluorescent tubulin (e.g., rhodamine-labeled tubulin or by expression of GFP-labeled tubulin). However, the use of GFP technology is more powerful in term of percentage of labeled cells and easier to perform. Microtubule growth and shortening dynamics are recorded by time-lapse video microscopy. To determine how microtubule length changes with time, individual microtubule ends are traced by a cursor on succeeding time-lapse frames (Fig. 1). Rates, lengths, and durations of growth and shortening events are calculated from the recorded *x*–*y* positions of the microtubule ends. While it is easy to measure dynamic instability of individual microtubules at the cell periphery, direct measurements in the cell interior are nearly impossible in most cells because of the cell shape and the high number of microtubules in this zone. The use of GFP-labeled proteins that specifically bind microtubule plus ends, such as EB1/EB3 or CLIP-170, is particularly useful for examining microtubule dynamics in the cell interior or during mitosis, but it is limited to quantification of microtubule growing events and catastrophe frequency. Elsewhere, +TIPs dynamics by itself represents an attractive new potential target in cancerology since they are altered by microtubule-targeting agents and play critical role in the regulation of microtubule dynamics and functions (17).

3.1. Expression of Fluorescently Labeled Proteins into the Cells

1. Transfection of cells with plasmid encoding fluorescently labeled proteins is performed by electroporation using the Amaxa electroporator (AMAXA®, Cologne, Germany) or using transfection reagents according to experimental protocol for each cell type provided by the manufacturer (see Note 1).
2. After transfection, cells are plated either directly onto the experimental glass coverslips or in 6-well plates for subsequent trypsinization and replating on experimental glass coverslips precoated with extracellular matrix proteins to allow cell adhesion and sufficient cell spreading. Coating of coverslips is performed with freshly prepared 10 μg/ml fibronectin solution or a mix of several extracellular matrix proteins for 2 h at 37°C or overnight at 4°C (see Note 2). Coverslips are then rinsed with sterile water and dried by aspiration. Fresh-coated coverslips are used for cell plating, and then cells are incubated in a CO_2 incubator at 37°C. Fluorescence can be detected within few hours after transfection. However, an optimal level of fluorescence is obtained after 24–72 h posttransfection. Either transiently transfected cells or stably transfected cells can be used.

3.2. Incubation of Cells with Drugs

1. It is thus crucial before performing microtubule dynamics analysis to determine the working concentrations of drugs by performing several functional assays in which microtubules play a role, such as mitotic arrest, inhibition of cell proliferation, alteration of cell migration, or cell differentiation (e.g., capillary-like tube formation on Matrigel® for endothelial cells, or neurite elongation for neuronal cells) (see Note 3).
2. Twenty-four hours after cell plating, cells were then incubated for 4 h with the desired concentration(s) of the microtubule-targeting agent(s) or with vehicle alone (control) into the cell culture media (see Tables 1 and 2).

3.3. Preparation of Cells for Time-Lapse Video Microscopy Recording

The experimental glass coverslip with cells expressing fluorescently tagged protein(s) and incubated with microtubule-targeting agent(s) are mounted into a close chamber containing the recording media at 37°C and placed on the stage of an inverted microscope with controlled temperature.

3.4. Time-Lapse Acquisition

Time-lapse series of images of labeled microtubules or microtubule plus ends are taken using a 60× or 100× objective lens. To accurately measure microtubule dynamic instability parameters, images are collected with short time interval between frames. The time interval may vary upon the cell lines tested, and their basal microtubule dynamics. Usually, 4 s intervals between frames are sufficient and allow reduced photobleaching. However, for cells with high basal level of microtubule dynamics a shorter time interval (1–3 s) is required. Exposure time depends on the brightness of the fluorescence signal and may vary from 200 to 1,500 ms.

3.5. Tracking Microtubule Dynamics of Labeled Microtubules with Metamorph

1. Make sure that you have the "track points" functionality installed on your Metamorph version.
2. Open Metamorph and the stack you wish to analyze. Go through all the planes and select some microtubules to analyze. It is important to track all the microtubules in one cell at single setting in order to avoid tracking a microtubule twice. It is also recommended to print the image of the first plane or the plane you begin the track. On this printed image, mark and number the microtubules you plan to track in order to avoid retracking the same microtubules.
3. Go to "measure" and select "calibrate distance" and calibrate according to your microscope settings, image binning, and objective lens characteristics. Then, go to "Apps", choose "Track Points". Within Track Points, go to configure log. Double click with the mouse to choose or excuse the different choices. You need only elapsed time, X, and Y.
4. Still on Track Points menu, select "set interval" to calibrate the elapsed time according to the time interval between two frames. Then, go to "set overlay". It is better to select a dot, size 3–5 and not a bigger symbol. Make sure that the symbol appears only on the last plane by selecting "display point on current plane".
5. Then, select "set origin". Choose a virtual origin point of the microtubule to be tracked. Make sure that you have a straight portion of the microtubule between this origin point and the microtubule end. At least, make sure that the microtubule has the maximal shortening episode within the straight part and that the virtual origin point is aligned with the line traced by the microtubule.
6. Click on "Add Track". Go to the first plane of the stack or the plane you begin your track. As you click on the end of the microtubule, "Track Points" records the data and also advances to the next plane. If you are unable to track the microtubule end until the last frame of the stack, simply click on "done" when you stop tracking the microtubule.
7. Because all data depend on the origin point selected, it is only possible to have one track at each time. It is, thus,

recommended to go to the next step (Subheading 3.6) before changing the origin point to track another microtubule.

3.6. Extraction of the Basic Parameters on the Life History Curves

1. On the "Track Points" menu, select "distance to origin", and then click on "Graph data". Metamorph displays a "microtubule life history curve" showing the changes in length (distance to the origin point to the microtubule plus end) over time.
2. Using the right button of the mouse, click on the graph and select "point statistics". Determine, on the graph, the episodes of microtubule growth, shortening and pause events. Usually, growth and shortening events are defined as changes in length >0.5 μm. Changes in length <0.5 μm in at least three planes are usually considered as microtubule dynamic attenuation or pause events. Depending upon the resolution of the images and the elapsed time, these limits can be reduced to 0.25 μm and two planes.
3. For each growth or shortening event identified, click on the first point, and then the last point. Metamorph displays the event duration, length, and rate. For pause events, you need only the event duration. These data can be copied manually or logged to a log file that will be processed later. In that case, growing lengths and rates appear as positive values, and shortening lengths and rates as negative values.
4. For each tracked microtubule, it is also necessary to visually determine the number of microtubule catastrophes (C) and rescues (R). A microtubule catastrophe is defined as the transition between a growth or pause event to a shortening event. This value can be determined visually or by counting the number of shortening events on the data log file. A microtubule rescue is defined as the transition between a shortening event to growth or pause. This value can be evaluated visually or determined on the log file. Indeed, $R = C$ if the last event of the tracked microtubule is a growth or pause event. However, $R = (C - 1)$ if the last event of the track is a shortening event.
5. Tracking 30 microtubules in four to six cells for each condition is usually sufficient to get reproducible results between different sets of experiments. At least three sets of individual experiments are needed for statistic analysis.

3.7. Calculation of the Microtubule Dynamic Instability Parameters

1. Results are usually expressed as the mean of all the individual events or mean of all individual microtubules. In the last case, a mean for each dynamic parameter must be calculated for each microtubule. However, depending on the microtubules tracked, all the parameters are not always accessible from one microtubule. For example, the catastrophe frequency cannot be calculated from a microtubule that is only growing during

the time of experiment. Similarly, the rescue frequency cannot be calculated from a microtubule that is only shortening. Thus, to avoid discarding such microtubules for the global analysis, such parameters can be calculated for the whole population of microtubules and not per microtubule. Before pooling all the events and/or microtubules in a single analysis, it is important to ensure that the data for each similar condition are not statistically different. If they are different, it may be necessary to increase the number of microtubules analyzed per set of experiments. Alternatively, the results can be expressed as a mean of three independent experiments.

2. Whatever the method used, the basic parameters of dynamic instability are directly accessible by calculating the mean. These parameters are the mean growth length, duration, and rate; the mean shortening length, duration, and rate; and the mean duration of pauses. Other parameters must be calculated according to the following formulas per microtubule if possible or for the entire population.
3. To determine the percentage of time spent in each phase, apply the following formulas: % of time spent growing = $100 \times \Sigma_{\text{growth duration}} / (\Sigma_{\text{shortening duration}} + \Sigma_{\text{growth duration}} + \Sigma_{\text{pause duration}})$. % of time spent shortening = $100 \times \Sigma_{\text{shortening duration}} / (\Sigma_{\text{shortening duration}} + \Sigma_{\text{growth duration}} + \Sigma_{\text{pause duration}})$ and the % of time spent in pause = $100 \times \Sigma_{\text{pause duration}} / (\Sigma_{\text{shortening duration}} + \Sigma_{\text{growth duration}} + \Sigma_{\text{pause duration}})$.
4. The transition frequencies can be determined based on both time and length.

   Catastrophe frequency based on time ($C^{\text{time}-1}$) and rescue frequency based on time $R^{\text{time}-1}$ are calculated according to the formulas: $C^{\text{time}-1} = \Sigma_{\text{Catastrophes}} / (\Sigma_{\text{growth duration}} + \Sigma_{\text{pause}} \text{duration})$. $R^{\text{time}-1} = \Sigma_{\text{Rescues}} / \Sigma_{\text{shortening duration}}$.

   Catastrophe frequency based on length ($C^{\text{length}-1}$) and rescue frequency based on length ($R^{\text{length}-1}$) are calculated according to the formulas: $C^{\text{length}-1} = \Sigma_{\text{Catastrophes}} / \Sigma_{\text{growth length}}$. $R^{\text{length}-1} = \Sigma_{\text{Catastrophes}} / \Sigma_{\text{shortening length}}$.
5. Finally, determine the overall dynamicity (OD) using the following formula:

   $\text{OD} = \Sigma_{\text{shortening length}} + \Sigma_{\text{growth length}} / (\Sigma_{\text{shortening duration}} + \Sigma_{\text{growth duration}} + \Sigma_{\text{pause duration}})$.

3.8. Tracking Microtubule Dynamics of Plus-End-Labeled Microtubules with Metamorph

The principle of the technique is that labeled +TIPs, such as EB1-GFP, specifically form comet-like structure on growing microtubule plus ends. Moreover, EB1-GFP is also able to mark microtubules in pause or attenuated state (which may occur during the growing phase). However, in that case, the size of the comet is strongly reduced and more likely appears as a dot. Finally,

EB1-GFP completely disappears from microtubule plus end when the microtubule begins to shrink, which indicates a catastrophe (see Note 4).

The manual tracking of microtubule plus-end dynamics using labeled +TIPs, such as EB1-GFP is performed as described in Subheading 3.5. However, when selecting the origin of the track (see Subheading 3.5, step 5), choose "first point in track".

3.9. Calculation of the Dynamic Instability Parameters After Tracking +TIPs

The tracking of the displacement of EB1-GFP comets over time allows the calculation of dynamic instability parameters, such as the mean growth rate, the mean growth length, and the mean growth duration (24). Moreover, the determination of duration or length of growing events until catastrophe allows the calculation of the catastrophe frequency based on time or on distance, respectively. The following formulas are used:

$$C^{\text{time}-1} = \sum_{\text{Catastrophes}} \Big/ \sum_{\text{growth + pause duration until catastrophe}}$$

$$C^{\text{length}-1} = \sum_{\text{Catastrophes}} \Big/ \sum_{\text{growth length until catastrophe}}$$

Illustration of the results can be made by presenting kymographs (Fig. 2). Dynamic instability parameters can also be determined directly on kymographs (25).

3.10. EB1 Comet Length and Fluorescence Intensity Measurements

As mentioned above, microtubule-targeting agents affect the localization of EB1 and other +TIP proteins at microtubule plus ends. The concentration-dependent reduction of the length and/or fluorescence intensity of +TIPs comets at microtubule plus ends by microtubule-targeting agents may also be useful to analyze their effect on the microtubule cytoskeleton. Such effect can be analyzed on double-transfected cells with mCherry-tubulin and EB1-GFP plasmids or using indirect immunofluorescence on endogenous proteins. Information on EB1 comet length or EB1 comet fluorescence intensity can be directly measured on acquired images (Fig. 3).

4. Notes

1. Expression levels of fluorescently tagged proteins usually vary from cell to cell. Good images are obtained with intermediate (optimal) level of expression.
2. There are no real guidelines for the use of extracellular matrix proteins, growth factors, and culture medium during the experiment. The better is to use the same experimental conditions as those chosen for other experiments (i.e., cell proliferation, cell migration, biochemistry, etc.). Most experiments on microtubule dynamics are performed using extracellular

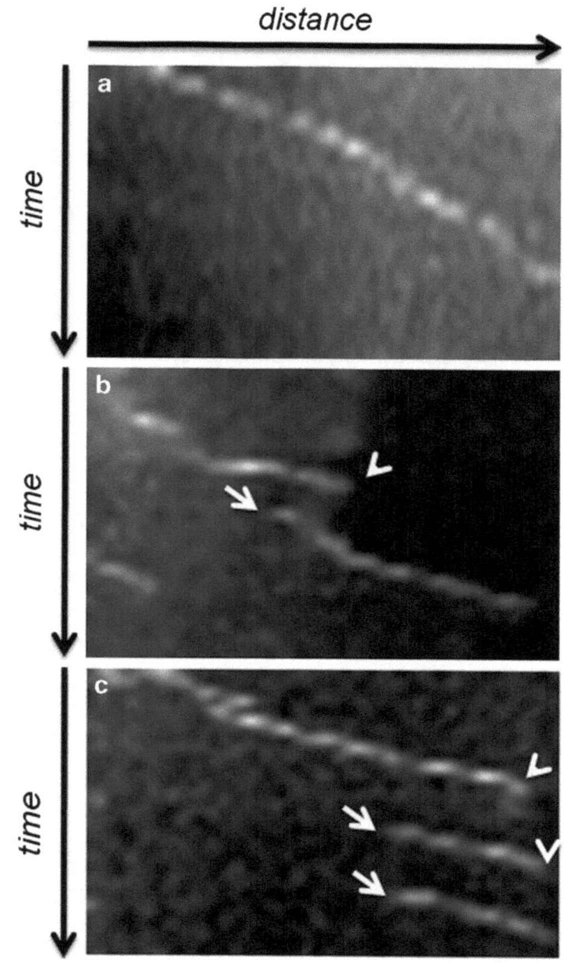

Fig. 2. Kymographs of EB1-GFP comet displacement over time. (**a**) The displacement of the EB1 comet over time indicates that the microtubule was growing continuously during the time of observation. (**b**, **c**) The displacement of the EB1 comet over time Indicates that the microtubule underwent one or several catastrophes (*arrow heads*) and a Rescue (*arrows*) during the time course of experiment.

matrix proteins, e.g., fibronectin and/or growth factors to enhance cell spreading. If not enough, the mix of several extracellular matrix proteins may be useful. For example, the mix of 50 μg/ml poly-L-lysin, 10 μg/ml laminin, and 10–20 μg/ml fibronectin is often used. However, changing the substrate or adding growth factors may differently change the basal microtubule dynamic instability parameters and the effects of microtubule-targeting agents, particularly when drugs are used at low noncytotoxic concentrations. The concentration and the volume of the coating solution and the incubation time must be kept constant in the various sets of experiments.

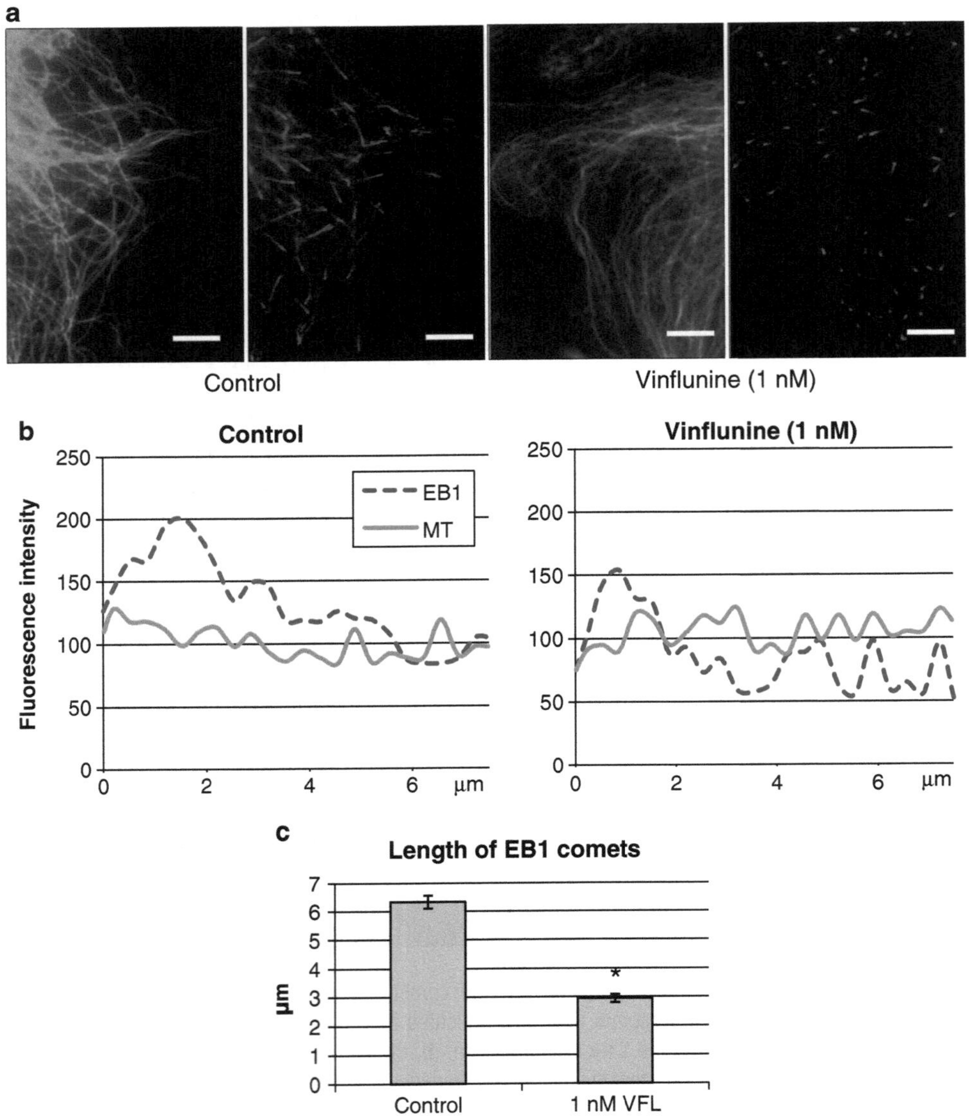

Fig. 3. Vinflunine altered EB1 localization at microtubule (+) ends. HMEC-1 cells were incubated with 1 nM vinflunine for 4 h and then immunostained for α-tubulin (*left panel*) and EB1 (*right panel*). (**a**) In control cells, EB1 appeared as long comets at the peripheral microtubule (+) ends. In contrast, in cells incubated with vinflunine, the length of EB1 comets was drastically reduced and more likely appeared as dots at the (+) end of microtubules. Scale bars: 10 μm. (**b**) Fluorescence intensity of EB1 and tubulin along a virtual line of 8 μm starting from the microtubule plus end in control (*left panel*) or in HMEC-1 cells incubated with 1 nM vinflunine (*right panel*). Such graph allows the determination of EB1 comet length and as well as the fluorescence intensity of each comet. EB1 comet length was 6 μm long in the control with a maximal fluorescence intensity of 200. In contrast, EB1 comet was 1.9 μm long with a maximal intensity of 150. (**c**) Effect of 1 nM vinflunine on EB1 comet length determined by measuring more than 300 EB1 comets per condition (mean ± SEM). $^*p < 0.05$.

3. The effect of drugs on microtubule dynamic instability parameters is dependent on the cell type. Indeed, some peculiar effects occur with low drug concentrations in some cell types, such as endothelial and neuronal cells and some tumor cells,

in which we can observe biphasic effects, i.e., increased microtubule dynamics at low noncytotoxic concentrations and suppressed microtubule dynamics at cytotoxic concentrations (15, 16). For each phase, each parameter of dynamic instability is affected in a concentration-dependent manner but not in the same concentration range.

4. Most microtubule-targeting agents decrease the binding of many microtubule plus end proteins at microtubule ends in a concentration-dependent manner leading to reduced length of the typical "comets" formed by the +TIPs (14, 17). However, in a large range of concentrations of drugs, the plus-end localization of these labeled +TIPs remains detectable and can thus be used to follow microtubule plus end dynamics ((17); S. Honore and D. Braguer personal data).

Acknowledgments

The authors would like to thank Mary Ann Jordan and Leslie Wilson to introduce us in the "microtubule dynamics world" and Kathy Kamath for technical collaboration.

References

1. Honore S, Pasquier E, Braguer D (2005) Understanding microtubule dynamics for improved cancer therapy. Cell Mol Life Sci 62(24):3039–56.
2. Poulain FE, Sobel A (2010) The microtubule network and neuronal morphogenesis: Dynamic and coordinated orchestration through multiple players. Mol Cell Neurosci 43(1):15–32.
3. Etienne-Manneville S (2010) From signaling pathways to microtubule dynamics: the key players. Curr Opin Cell Biol 22(1):104–11.
4. Jordan MA, Kamath K (2007) How do microtubule-targeted drugs work? An overview. Curr Cancer Drug Targets 7(8):730–42.
5. Akhmanova A, Stehbens SJ, Yap AS (2009) Touch, grasp, deliver and control: functional cross-talk between microtubules and cell adhesions. Traffic 10(3):268–74.
6. Zaoui K, Honoré S, Isnardon D, et al (2008) Memo-RhoA-mDia1 signaling controls microtubules, the actin network, and adhesion site formation in migrating cells. J Cell Biol 183(3):401–8.
7. Akhmanova A, Steinmetz MO (2008) Tracking the ends: a dynamic protein network controls the fate of microtubule tips. Nat Rev Mol Cell Biol 9(4):309–22.
8. Toso RJ, Jordan M.A, Farrell KW et al (1993) Kinetic stabilization of microtubule dynamic instability in vitro by vinblastine. Biochemistry 32(5):1285–93.
9. Honore S, Pasquier E, Braguer D (2005) Understanding microtubule dynamics for improved cancer therapy. Cell Mol Life Sci 62(24):3039–56.
10. Calligaris D, Verdier-Pinard P, Devred F et al (2010) Microtubule targeting agents: from biophysics to proteomics. Cell Mol Life Sci 67(7):1089–104.
11. Honore S, Kamath K, Braguer D et al (2004) Synergistic suppression of microtubule dynamics by discodermolide and paclitaxel in non-small cell lung carcinoma cells. Cancer Res 64(14):4957–64.
12. Honore S, Kamath K, Braguer D et al (2003) Suppression of microtubule dynamics by discodermolide by a novel mechanism is associated with mitotic arrest and inhibition of tumor cell proliferation. Mol Cancer Ther 2(12):1303–11.
13. Smith JA, Wilson L, Azarenko O et al (2010) Eribulin binds at microtubule ends to a single site on tubulin to suppress dynamic instability. Biochemistry 49(6):1331–7.

14. Rovini A, Carré M, Bordet T et al (2010) Olesoxime prevents microtubule-targeting drug neurotoxicity: selective preservation of EB comets in differentiated neuronal cells. Biochem Pharmacol in press.
15. Pourroy B, Honoré S, Pasquier E et al (2006) Antiangiogenic concentrations of vinflunine increase the interphase microtubule dynamics and decrease the motility of endothelial cells. Cancer Res 66(6):3256–63.
16. Pasquier E, Honore S, Pourroy B et al (2005) Antiangiogenic concentrations of paclitaxel induce an increase in microtubule dynamics in endothelial cells but not in cancer cells. Cancer Res 65(6):2433–40.
17. Honoré S, Pagano A, Gauthier G et al (2008) Antiangiogenic vinflunine affects EB1 localization and microtubule targeting to adhesion sites. Mol Cancer Ther 7(7):2080–9.
18. Dhamodharan R, Jordan MA, Thrower D, Wilson L, Wadsworth P (1995) Vinblastine suppresses dynamics of individual microtubules in living interphase cells. Mol Biol Cell 6(9):1215–29.
19. Kamath K, Jordan MA (2003) Suppression of microtubule dynamics by epothilone B is associated with mitotic arrest. Cancer Res 63(18):6026–31.
20. Jordan MA, Kamath K, Manna T, Okouneva T, Miller HP, Davis C, Littlefield BA, Wilson L (2005). The primary antimitotic mechanism of action of the synthetic halichondrin E7389 is suppression of microtubule growth. Mol Cancer Ther 4(7):1086–95.
21. Pourroy B, Carré M, Honoré S, Bourgarel-Rey V, Kruczynski A, Briand C, Braguer D (2004). Low concentrations of vinflunine induce apoptosis in human SK-N-SH neuroblastoma cells through a postmitotic G1 arrest and a mitochondrial pathway. Mol Pharmacol 66(3):580–91.
22. Mikhailov A, Gundersen GG. Relationship between microtubule dynamics and lamellipodium formation revealed by direct imaging of microtubules in cells treated with nocodazole or taxol. Cell Motil Cytoskeleton 41(4):325–40.
23. Skube SB, Chaverri JM, Goodson HV (2010) Effect of GFP tags on the localization of EB1 and EB1 fragments in vivo. Cytoskeleton 67(1):1–12.
24. Komarova Y, De Groot CO, Grigoriev I et al (2009) Mammalian end binding proteins control persistent microtubule growth. J Cell Biol 184(5):691–706.
25. Smal I, Grigoriev I, Akhmanova A et al (2009) Accurate estimation of microtubule dynamics using kymographs and variable-rate particle filters. Conf Proc IEEE Eng Med Biol Soc 721–1078:1012–5.

Chapter 19

Laser Ablation of the Microtubule Cytoskeleton: Setting Up and Working with an Ablation System

Nicola Maghelli and Iva M. Tolić-Nørrelykke

Abstract

Laser ablation is a powerful tool that can be used to study a variety of biological mechanisms. Microscopes with high optical performances are nowadays available, and lasers that could be used to perform ablations have become accessible to every laboratory. Setting up a laser ablation system is a relatively straightforward task; however, it requires some basic knowledge of optics. We illustrate the fundamental components of the experimental setup and describe the most common pitfalls and difficulties encountered when designing, setting up, and working with a laser ablation system.

Key words: Laser ablation, Optical manipulation, Microdissection, Nanosurgery

1. Introduction

Optical manipulation of living specimen has been successfully employed to investigate a variety of phenomena. Laser ablation can be used to either remove an organelle or a part of it (1, 2), or to perturb the force balance of the cellular cytoskeleton to study the mechanical equilibrium and the force generators inside a cell (3–12). In the first case, ablation allows to investigate the function of a subcellular element using an approach that is complementary to the well-established genetic tools. In the second case, the short-term modifications of the sample structure after the ablation reflect the forces acting in the sample while long-term reactions carry information about compensation or adaptation mechanisms that might be triggered by the modification.

Anne Straube (ed.), *Microtubule Dynamics: Methods and Protocols*, Methods in Molecular Biology, vol. 777,
DOI 10.1007/978-1-61779-252-6_19, © Springer Science+Business Media, LLC 2011

2. Materials

2.1. Optics

Custom-built two-photon setup (13), lenses, mirrors and optics holders (Thorlabs, Inc.), dichroic mirrors, filters (Chroma Technology Corp.), microscope objectives (Carl Zeiss, AG) precision mechanical hardware (OWIS GmbH).

2.2. Yeast Media (14)

1. Vitamins solution (1,000×): 1 g/l pantothenic acid (Sigma-Aldrich), 10 g/l nicotinic acid (Sigma-Aldrich), 10 g/l inositol (Sigma-Aldrich), and 10 mg/l biotin (Sigma-Aldrich).
2. Minerals solution (10,000×): 5 g/l boric acid (Sigma-Aldrich), 4 g/l manganese sulfate (Sigma-Aldrich), 4 g/l zinc sulfate hydrate ($ZnSO_4 \cdot 7H_2O$) (Sigma-Aldrich), 2 g/l iron chloride hydrate ($FeCl_2 \cdot 6H_2O$) (Sigma-Aldrich), 0.4 g/l molybdic acid (Sigma-Aldrich), 1 g/l potassium iodine (Sigma-Aldrich), 0.4 /l copper sulfate hydrate ($CuSO_4 \cdot 5H_2O$) (Sigma-Aldrich), and 10 g/l citric acid (Sigma-Aldrich).
3. Salts solution (50×): 52.5 g/l magnesium chloride hydrate ($MgCl_2 \cdot 6H_2O$) (Sigma-Aldrich), 0.735 g/l calcium chloride hydrate ($CaCl_2 \cdot 2H_2O$) (Sigma-Aldrich), 50 g/l potassium chloride (Sigma-Aldrich), and 2 g/l sodium sulfate (Sigma-Aldrich).
4. Yeast extract with supplement (YES): 5 g/l of yeast extract (Bacto), 30 g/l glucose monohydrate (Merck), 225 mg/l of Adenine, Leucine, and Uracil (Sigma-Aldrich), 150 mg/l G418 (Sigma-Aldrich), 0.67 mg/l thiamine (Sigma-Aldrich), and 1 ml/l vitamin solution.
5. Edinburgh minimal medium (EMM): 3 g/l potassium hydrogen phthalate (Sigma-Aldrich), 2.2 g/l sodium hydrogen phosphate (Sigma-Aldrich), 5 g/l ammonium chloride (Sigma-Aldrich), 20 g/l glucose monohydrate (Merck), 20 ml/l salts solution, 1 ml/l minerals solution, and 1 ml/l vitamin solution.

3. Methods

3.1. Optical Setup

To perform optical manipulations, it is necessary to visualize the structure of interest by using a microscope. Many contrast techniques, such as dark-field, DIC, or phase contrast, can be employed. However, if the sample can be labeled with fluorescent markers, then epifluorescence or confocal microscopy are of great help to clearly distinguish the feature to be ablated. In this case, the emission, excitation, and dichroic filters used to separate the excitation light from the emitted fluorescence should be taken

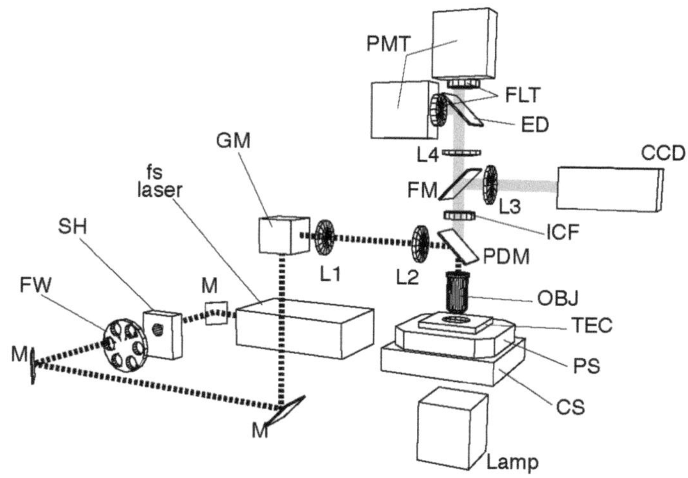

Fig. 1. Schematic of a custom-built laser ablation setup based on a near-infrared fs-pulsed laser (13). Legend: *M* mirror, *SH* shutter, *FW* filter wheel, *GM* galvanometer mirrors, *L1–L2* Keplerian telescope, *PDM* primary dichroic mirror, *OBJ* objective, *TEC* thermoelectric element, *PS* piezoelectric stage, *CS* coarse-motorized stage, *ICF* infrared cutoff filter, *FM* flip mirror, *L3–L4* collecting lenses, *ED* emission dichroic, *FLT* emission filters, *PMT* photomultiplier.

into account when designing the optical path of the ablation beam (see Note 1; Fig. 1).

3.2. Mechanical Setup

To take full advantage of the high spatial resolution that laser microdissection can achieve, the optical setup must rest on a stable and vibration-free platform. It is advisable to use a damped optical table and to avoid placing any source of mechanical noise (e.g., cooling fans, pumps, etc.) in close proximity of the ablation setup. Any high-end commercial microscope has a good intrinsic mechanical stability, whereas when developing a custom-built setup the mechanical components must also possess a good mechanical stability.

3.3. Laser Sources

Different laser sources have been successfully employed to ablate microtubules in a broad variety of organisms. The physical processes underlying the ablation process differ according to the power density of the focused light, the pulse duration, and the wavelength (15–19).

Roughly, it is possible to group the lasers into three main categories, according to their wavelength:

1. UV and near-UV lasers: The wavelength typically ranges from 325 nm emitted by an He-Cd laser (CW) to 405 nm of deep-violet laser diodes (CW or pulsed). The power ranges from 10 mW to 100 mW. Owing to the short wavelength, it is generally possible to focus the laser to a smaller area, therefore increasing the ablation accuracy compared to ablation

setups relying on visible or near-infrared laser sources (5, 6, 10, 20–22). However, when using deep-UV lasers, the optical performances of the microscope might become an issue:

- The transmission of commercially available microscope objectives starts to decrease below 400 nm, and can be as low as 50% or less for wavelength below 350 nm. Therefore, the laser power delivered to the sample that can be effectively used for ablation might be only a fraction of the total power emitted by the laser.
- The objectives are usually well-corrected against spherical and chromatic aberrations in the visible range (400–700 nm). In the UV range, the performances are worse, therefore minimizing the resolution gain given by the shorter wavelength.

Since many biologically relevant molecules have a pronounced absorption in the ultraviolet, the penetration depth of UV lasers is worse in comparison to visible or near-infrared lasers. If the structure to be ablated is deep within the sample, the ablation beam is progressively attenuated while travelling through the sample. As a consequence, it might be difficult to deliver enough power to efficiently perform ablations in thick sample using UV lasers. Under these circumstances, increasing the intensity of the ablation beam to achieve sufficient power at the ablation spot might result in a heavy collateral damage due to the sample absorption (see Notes 2 and 3). When working in vivo, it is often crucial to minimize the sample stress; therefore, UV-based setups might not be the optimal choice to ablate structures in the interior of a thick sample.

2. Visible lasers: The lasers emitting in the visible range most frequently used in ablation setups are the He–Cd laser (441 nm), the Ar–Kr ion lasers (488 and 514.5 nm), and the frequency-doubled Nd:YAG laser (532 nm). These lasers can deliver on the average more power than UV and near-UV laser sources (~100 mW to several watts). Moreover, the optical components used to deliver and focus the ablation beam on the sample have high transmissions and present low residual aberrations in the visible range. As a result, the ablation beam can be focused almost to diffraction limit (21, 23–25). The main drawback of using visible lasers for ablation is that often they emit in a range that is also employed to observe the sample. Therefore, it becomes difficult to prevent cross talks with the imaging channel. As an example, to observe eGFP-labeled samples requires a bandpass filter centered on the eGFP emission peak. Usually, the emission filters used for imaging eGFP transmit light from around 500 nm to around 550 nm. If the wavelength of the laser beam used to perform ablations falls within the transmission window of the emission

filter, and no special precautions are taken, back reflections from the optical surfaces or from the sample itself will reach the detector used for taking images, potentially damaging light-sensitive devices, such as CCDs, APD, or PMT (or in the worst case the eye of the user). These issues can be addressed by using shutters that protect the detector while the sample is exposed for ablation; however, this case requires to accurately synchronize the image acquisition with the ablation.

3. Near-infrared lasers: Recently, near-infrared-pulsed lasers have been successfully employed to perform ablations (1–4, 13, 15, 17, 26). The physical processes taking place in the sample when exposed to a short burst of highly intense light are complex, and not yet fully understood. However, multiphoton absorption, as well as optical breakdown and plasma formation might play a role (3, 15). Multiphoton absorption is achieved only at extremely high radiances (~10^{10}–10^{12} W/cm^2); to achieve such high photon densities, the laser emits short (~100 fs) pulses at high (~100 MHz) repetition rate, reaching a peak power of many kW. The most widespread femtosecond (fs)-pulsed, near-infrared laser source is the Titanium-Sapphire (Ti:Sa) laser. Its emission can be continuously tuned over a broad range of wavelengths, typically ranging from around 700 nm to around 1,000 nm. Ablation setups relying on near-infrared laser sources benefit of several advantages:
   - The scattering decreases with the inverse fourth power of the wavelength; therefore, longer wavelengths are less affected by the optical properties of the sample, and can penetrate deeper into the tissues. When performing ablations in thick samples, such as embryos, the penetration depth is a critical factor, and only near-infrared lasers are capable of efficiently ablating structures well below the sample surface.
   - Multiphoton absorption only takes places around the focal plane, and not above or below it. The energy density required for an efficient multiphoton absorption is only achieved at the focal plane, where the laser beam is focused to the smallest possible area. Above and below the focal plane, the beam diameter is larger than at the focal plane, and the energy density is not high enough to trigger multiphoton absorption. Therefore, the ablation is performed in a spatially *confined* region.
   - Many biological relevant molecules have very low absorption coefficients in the near-infrared range. Therefore, the risk of inducing unspecific damage to the sample is smaller when using near-infrared lasers than when using UV or visible lasers. Keeping the damage low is particularly

important when performing ablations of sensitive samples, and critical when working with single cells.

However, several issues must be considered when using femtosecond-pulsed, near-infrared lasers:

- Using wavelengths in the near-infrared range decreases the attainable resolution.
- The transmission and correction of microscope objectives in the near-infrared range are worse than that in the visible range, although the performance degradation is not as severe as in the case of UV and near-UV lasers.
- A femtosecond pulse undergoes distortion when travelling through materials that have a wavelength-dependent refractive index, such as the glasses commonly used for manufacturing optical components. The result is a broadening of the pulse that becomes more pronounced as the original pulse length decreases. The pulse broadening decreases the peak power and consequently lowers the multiphoton absorption efficiency (see Note 4). The net result is a less efficient ablation. For this reason, it is problematic to deliver the short pulses of the laser to the ablation setup using optical fibers. Working with free-spaced setups, and using high-power lasers might be hazardous if proper precautions concerning laser safety are not taken.
- Typically, an fs-pulsed laser is more expensive than a gas- or solid-state laser.

3.4. Positioning Control

To accurately ablate a specific target, it is necessary to precisely control the position the ablation spots on the sample. The axial position of the ablation beam with respect to the sample is usually controlled by moving the sample with respect to the objective. Simultaneous imaging while performing ablations requires the focus of the ablation beam to coincide with the focal plane of the imaging system. A relative offset, typically present when using UV or near-infrared ablation lasers as a result of chromatic aberration, can be corrected by adjusting the divergence of the ablation beam. The lateral positioning can instead be achieved either by keeping the laser beam fixed and moving the sample, or by steering the ablation beam while keeping the sample at rest with respect to the microscope. The first method is the most straightforward and simple to implement. It is usually employed when the ablation precision is not critical (see Note 5), or to ablate the sample along an extended path that would normally not fit in a single field of view of the setup. The second method is technically more complex, but offers more flexibility and permits to better control the ablation area. Usually, a couple of galvanometer scanners are employed to steer the ablation beam (see Note 6).

3.5. Coupling of the Ablation Beam

To guarantee a high spatial accuracy and coupling efficiency, the $1/e$ diameter of the ablation beam should ideally match the size of the back aperture of the microscope objective (see Note 7). The laser for ablation is typically available either as a free beam, with diameters of around 1 mm, or as output of an optical fiber. In both cases, it is necessary to insert some rely optics that will collimate and expand the beam to its correct diameter (see Notes 8 and 9). In addition, if the ablation setup implements a beam steering unit, care must be taken that the beam pivoting point is on a conjugate telecentric plane. Otherwise, scanning the beam induces a movement of the beam at the objective back aperture, changing the coupling efficiency, and therefore the total power of the laser delivered to the sample (see Note 9). As a result, the ablation efficiency depends on the position of the ablation beam in the microscope field of view. A typical solution employed to magnify and couple a collimated beam to the microscope objective is to use either a Galilean or a Keplerian telescope. The first one uses a convex and a concave lens, has a compact design, but cannot be used with a beam steering unit. The second one uses two convex lenses and can be used with a beam steering unit; however, this design needs more space.

3.6. Performing Ablations

Performing ablation in a living specimen requires a delicate balance: it is necessary to deliver enough energy to reliably disrupt the ablation target, yet the sample must not suffer too much "collateral damage." Otherwise, it becomes difficult to discriminate between the consequences of the ablation, and the artifacts due to more unspecific damage. To choose the optimal parameter set for ablation (exposure time, power, wavelength, scanned area), it is advisable to perform several test runs changing a single parameter at a time while monitoring the ablation efficiency and the insurgence of any possible unspecific damage (4, 20). The actual experiment is then performed choosing the set of parameters that maximized the ablation efficiency while still keeping the sample damage low. Once the best parameter set is found, a typical ablation experiment performed on fission yeast cells (Fig. 2) follows this protocol:

1. Grow fission yeast cells in liquid YES (50 ml) at 25°C to logarithmic phase. To ensure the homogeneity, the culture is placed on a temperature-regulated shaker.
2. Place a small aliquot (~50–100 μl) of the culture on a 35-mm glass-bottom Petri dish (Matek), which was previously coated with ~2 μl of 2 mg/ml lectin BS-1 (Sigma-Aldrich).
3. Allowed to adhere for 10 min.
4. Rinse the sample twice with 0.5 ml EMM.
5. Fill the Petri dish with 3 ml EMM with appropriate supplements for imaging.

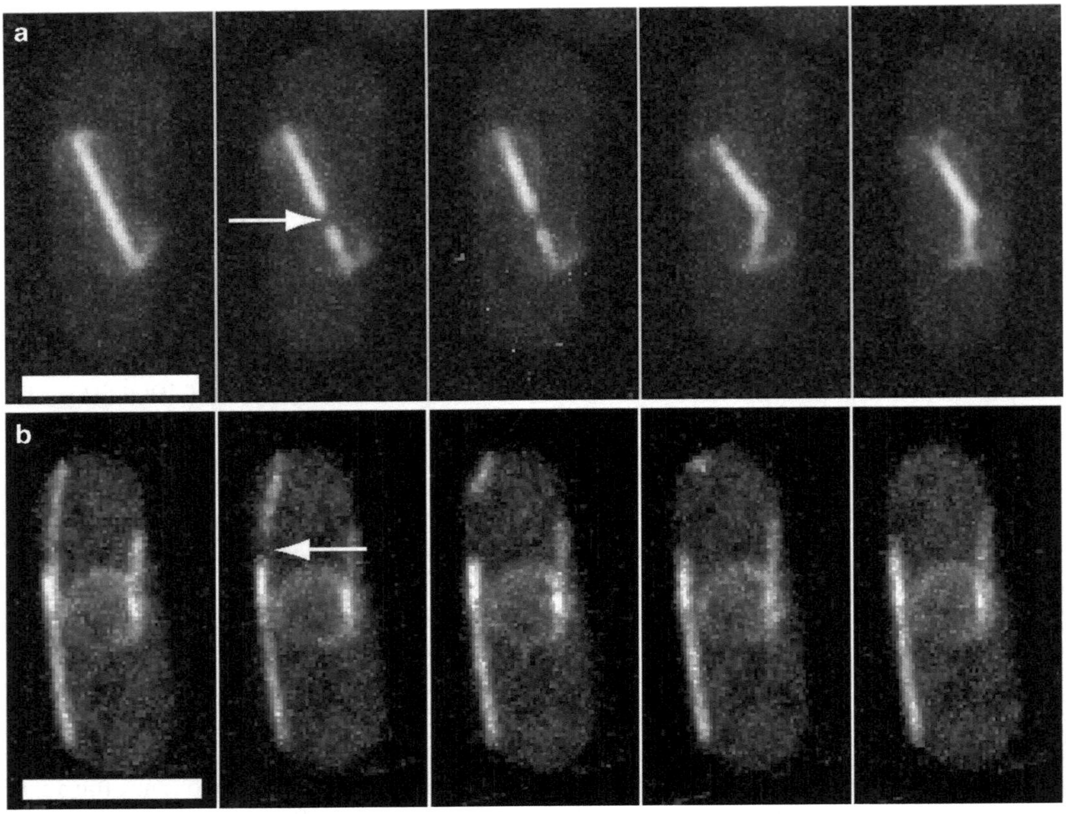

Fig. 2. Laser ablations of microtubules in *Schizosaccharomyces pombe*. The images sequences illustrate the effect of targeted irradiations (marked by the *arrows*) on microtubules. In (**a**), the mitotic spindle is destabilized. In (**b**), a microtubule is severed, and it is possible to follow the depolymerization of the detached fragment. Scale bar 5 μm, time between frames is 15 s for (**a**) and 2.5 s for (**b**). Laser ablation performed using the custom-built system described in (13). Cells expressing tubulin-GFP and cut11-GFP grown in YES medium, imaging and ablation performed in EMM medium.

6. Mount the sample on the ablation setup. Before proceeding, the thermal state of the sample is allowed to settle for approximately 10 min.
7. A suitable cell is selected. During the search, low-damage imaging (brightfield) is used.
8. Pre-ablation imaging. The sample is imaged to make sure that the structure to ablate presents no abnormalities and to provide an unperturbed reference to compare with the ablation outcome. Moreover, the user decides at this stage which region of the sample to ablate.
   In Fig. 2, pre-ablation imaging was achieved by scanning a stack of ten planes, 0.5 μm apart, at a pixel size of 100 nm/pixel using the custom-built two-photon setup described in (13). A Ti:Sa femtosecond-pulsed laser (Chameleon XR by Coherent, 150 fs nominal pulse width, 90 MHz rep. rate) tuned to 895 nm was used. The average power at the sample

plane was 3 mW. The ablation region was selected in the plane were the structure was most in focus.

9. Performing the actual ablation. During this phase the sample is exposed with the ablation beam over the area previously designated, using the optimal power and exposure time. In our case, to ablate the microtubules in Fig. 2, the average power at the sample was 100 mW while the exposure time was 50 ms for ablating the mitotic spindle (Fig. 2a) and 20 ms for interphase microtubules (Fig. 2b).
10. Post-ablation imaging. After ablating, the sample is imaged to assess the ablation outcome and to monitor the changes induced by the ablation. The same settings used for pre-ablation imaging were used.

4. Notes

1. If using polarization-sensitive components, such as Nomarski or Wollaston prisms, care must be taken to properly design the optical path of the ablation beam because the laser used for ablation might emit highly polarized light.
2. Absorption by fluorescent markers might give an important contribution to the ablation efficiency; the ablation requires less power if the wavelength is efficiently absorbed by the fluorescent marker (15, 16).
3. Depleting oxygen, by example when using oxygen scavengers to prevent fast bleaching, might decrease the ablation efficiency (15).
4. An acusto-optical modulator can provide quick control of the power of the ablation beam. However, the beam quality might be negatively influenced and, in case of a femtosecond laser, the pulses might be excessively broadened.
5. When ablating small structures, simply parking the beam over it might not be efficient. Scanning the ablation beam over a slight bigger area might give better results.
6. If the beam steering unit is mechanical (galvanometer or piezoelectric mirrors), the inertia of the moving parts may introduce residual errors in the beam positioning. To rule out these errors, it is possible to record the beam position at various positioning speed and compensate the steering system consequently.
7. Slightly overfilling the objective with the ablation beam (i.e., magnifying the ablation beam to a diameter larger than the diameter of the back aperture of the microscope objective)

might help compensating for any residual misalignment introduced by a beam steering system.

8. To facilitate the coupling of the ablation beam, it is possible to start using a low magnification objective (5× or 10×) to check the coarse alignment and then switch to higher magnifications for the fine adjustments.
9. Aligning a near-infrared laser might be troublesome. If using a Ti:Sa laser, it is sometimes possible to tune the laser to a visible wavelength, perform the alignment, and then switch back to near-infrared for ablation. Alternatively, a visible laser might be aligned to the infrared beam and then used as a reference.
10. If a steering unit controls the ablation beam, it is critical to calibrate the beam position with respect to the imaging field of view. A quick method consists in bleaching spots over a uniformly fluorescent polymer.

References

1. Stiess, M., Maghelli, N., Kapitein, L. C., Gomis-Ruth, S., Wilsch-Brauninger, M., Hoogenraad, C. C., Tolic-Norrelykke, I. M., and Bradke, F. (2010) Axon extension occurs independently of centrosomal microtubule nucleation, *Science 327*, 704–707.
2. Sacconi, L., O'Connor, R. P., Jasaitis, A., Masi, A., Buffelli, M., and Pavone, F. S. (2007) In vivo multiphoton nanosurgery on cortical neurons, *J Biomed Opt 12*.
3. Vogel, S. K., Pavin, N., Maghelli, N., Julicher, F., and Tolic-Norrelykke, I. M. (2009) Self-organization of dynein motors generates meiotic nuclear oscillations, *PLoS Biol 7*, e1000087.
4. Sacconi, L., Tolic-Norrelykke, I. M., Antolini, R., and Pavone, F. S. (2005) Combined intracellular three-dimensional imaging and selective nanosurgery by a nonlinear microscope, *J Biomed Opt 10*, 14002.
5. Colombelli, J., Reynaud, E. G., and Stelzer, E. H. (2005) Subcellular nanosurgery with a pulsed subnanosecond UV-A laser, *Med Laser Appl 20*, 217–222.
6. Colombelli, J., Reynaud, E. G., Rietdorf, J., Pepperkok, R., and Stelzer, E. H. (2005) In vivo Selective Cytoskeleton Dynamics Quantification in Interphase Cells Induced by Pulsed Ultraviolet Laser Nanosurgery, *Traffic 6*, 1093–1102.
7. Tolic-Norrelykke, I. M., Sacconi, L., Thon, G., and Pavone, F. S. (2004) Positioning and elongation of the fission yeast spindle by microtubule-based pushing, *Curr Biol 14*, 1181–1186.
8. Khodjakov, A., La Terra, S., and Chang, F. (2004) Laser microsurgery in fission yeast; role of the mitotic spindle midzone in anaphase B, *Curr Biol 14*, 1330–1340.
9. Botvinick, E. L., Venugopalan, V., Shah, J. V., Liang, L. H., and Berns, M. W. (2004) Controlled ablation of microtubules using a picosecond laser, *Biophys J 87*, 4203–4212.
10. Spurck, T., Stonington, O., Snyder, J., Pickett-Heaps, J., Bajer, A., and Mole-Bajer, J. (1990) UV microbeam irradiations of the mitotic spindle. II. Spindle fiber dynamics and force production, *J Cell Biol 111*, 1505.
11. Grill, S. W., Howard, J., Schaffer, E., Stelzer, E. H., and Hyman, A. A. (2003) The distribution of active force generators controls mitotic spindle position, *Science 301*, 518–521.
12. Tolic-Norrelykke, I. M. (2008) Push-me-pull-you: how microtubules organize the cell interior, *Eur Biophys J 37*, 1271–1278.
13. Maghelli, N., and Tolic-Norrelykke, I. M. (2008) Versatile laser-based cell manipulator, *J Biophotonics 1*, 299–309.
14. Moreno, S., Klar, A., and Nurse, P. (1991) Molecular genetic analysis of fission yeast Schizosaccharomyces pombe, *Methods Enzymol 194*, 795–823.
15. Vogel, A., Noack, J., Hüttman, G., and Paltauf, G. (2005) Mechanisms of femtosecond laser nanosurgery of cells and tissues, *Appl Phys B 81*, 1015–1047.

16. Vogel, A., and Venugopalan, V. (2003) Mechanisms of pulsed laser ablation of biological tissues., *Chem Rev 103*, 577–644.
17. König, K., Riemann, I., Fischer, P., and Halbhuber, K. J. (1999) Intracellular nanosurgery with near infrared femtosecond laser pulses, *Cellular and molecular biology (Noisy-le-Grand, France) 45*, 195–201.
18. Berns, M. W., Wright, W., and Wiegand Steubing, R. (1991) Laser microbeam as a tool in cell biology, *Int Rev Cytol 129*, 1–44.
19. Heisterkamp, A., Maxwell, I. Z., Mazur, E., Underwood, J. M., Nickerson, J. A., Kumar, S., and Ingber, D. E. (2005) Pulse energy dependence of subcellular dissection by femtosecond laser pulses, *Opt Express 13*, 3690–3696.
20. Raabe, I., Vogel, S. K., Peychl, J., and Tolic-Norrelykke, I. M. (2009) Intracellular nanosurgery and cell enucleation using a picosecond laser, *J Microsc 234*, 1–8.
21. Aist, J. R., Liang, L. H., and Berns, M. W. (1993) Astral and spindle forces in PtK2 cells during anaphase B: a laser microbeam study, *J Cell Sci 104*, 1207–1216.
22. Leslie, R., and Pickett-Heaps, J. (1983) Ultraviolet microbeam irradiations of mitotic diatoms: investigation of spindle elongation, *J Cell Biol 96*, 548–561.
23. Moore, J. K., Magidson, V., Khodjakov, A., and Cooper, J. A. (2009) The spindle position checkpoint requires positional feedback from cytoplasmic microtubules, *Curr Biol 19*, 2026–2030.
24. Aist, J. R., and Berns, M. W. (1981) Mechanics of chromosome separation during mitosis in Fusarium (Fungi imperfecti): new evidence from ultrastructural and laser microbeam experiments., *J Cell Biol 91*, 446–458.
25. Berns, M. W., Rattner, J., Brenner, S., and Meredith, S. (1977) The role of the centriolar region in animal cell mitosis. A laser microbeam study, *J Cell Biol 72*, 351–367.
26. König, K., Liang, L. H., Berns, M. W., and Tromberg, B. J. (1995) Cell damage by near-IR microbeams, *Nature 377*, 20–21.

Chapter 20

Isolation, Identification, and Validation of Microtubule-Associated Proteins from *Drosophila* Embryos

Robin Antrobus and James G. Wakefield

Abstract

The microtubule (MT) cytoskeleton is required for many aspects of cell function, including the transport of intracellular materials, maintenance of cell polarity, and the regulation of mitosis. These functions are coordinated by MT-associated proteins (MAPs), which work in concert with each other, binding MTs and altering their properties. We have used an MT co-sedimentation assay, combined with 1D and 2D PAGE and mass spectrometry, to isolate MAPs from early *Drosophila* embryos. This technique has identified many novel proteins and an association with MTs for many known proteins, previously not described as associating with MTs.

Key words: Microtubule, MAP, Drosophila, Mass spectrometry

1. Introduction

Microtubules (MTs) play diverse and important roles in many cellular processes. They are dynamic protein polymers of two related proteins, α- and β- tubulins, that exist in a soluble, cytosolic pool and that incorporate into existing MTs with predetermined polarity (1). In most animal cells, MTs are nucleated from and anchored at an organising centre, the centrosome, which resides juxtaposed to the nuclear envelope. The intrinsic polarity of the MT can, therefore, be translated into sub-cellular positional information, allowing the efficient transport of materials from one location to another (2, 3). In addition, the ability of MTs to grow and shrink dynamically provides a single cell with the capacity to form multiple populations of MTs possessing different functional properties at a single point in time (4, 5). The elaborate temporal and

Anne Straube (ed.), *Microtubule Dynamics: Methods and Protocols*, Methods in Molecular Biology, vol. 777,
DOI 10.1007/978-1-61779-252-6_20, © Springer Science+Business Media, LLC 2011

spatial organisation of MT function within the cell is primarily regulated by proteins that interact with MTs, the so called Microtubule-Associated Proteins or MAPs. *Drosophila* syncytial-stage embryos are ideal tissues from which to isolate MTs and associated MAPs of this nature as they contain sufficient quantities of individual proteins to undergo multiple cell cycles prior to zygotic transcription (6–8). In addition, large quantities can be obtained with relative ease (9). This chapter describes the co-isolation of MAPs with MTs, via an MT co-sedimentation assay, followed by their identification using mass spectrometry and verification by western blotting and/or immunofluorescence.

2. Materials

2.1. Fly Handling and Embryo Collection

1. Large population of *Drosophila* (e.g., 12–24 standard bottles of strain w1118 flies) maintained in 2–4 10-cm diameter collection chambers within LMS 1200 temperature- and humidity-controlled incubator (25°C, 70% relative humidity) with fixed day (12 h) and night (12 h) cycles.
2. Standard apple juice/molasses agar plates.
3. Yeast paste: Ordinary baker's yeast dissolved in water to form a paste.
4. PBS-T: PBS Tablets (Sigma-Aldrich, Cat. No. P4417) dissolved in water, 0.05% v/v Triton X-100 (Sigma-Aldrich, Cat. No. T8787) (see Notes 1 and 2).
5. Standard thin bleach.
6. Liquid N_2.
7. Filter unit fitted onto vacuum flask.
8. Motor-driven Dounce homogeniser with tight-fitting Teflon pestle (20-mL volume).

2.2. MT Co-sedimentation Assay Reagents

1. C Buffer: 50 mM HEPES (pH 7.4), 50 mM KCl, 1 mM $MgCl_2$, 1 mM EGTA, 0.1% NP-40, complete protease inhibitor cocktail (Roche, Cat. No. 11836153001). C Buffer can be prepared in advance and stored at 4°C until use. Protease inhibitors are added just before use.
2. C Buffer + 40% sucrose: 50 mM HEPES (pH 7.4), 50 mM KCl, 1 mM $MgCl_2$, 1 mM EGTA, 0.1% NP-40, 40% w/v sucrose. C Buffer + 40% sucrose can be prepared in advance and stored at 4°C until use.
3. Taxol (Sigma-Aldrich, Cat. No. T7402) is dissolved in DMSO (Sigma-Aldrich, Cat. No. 41648) at 10 mM and stored at –20°C (see Note 3).

4. GTP (Sigma-Aldrich, Cat. No. G8877) is prepared as 500 mM in water and stored at −20°C (see Note 3).
5. Dithiothreitol (DTT) (Sigma-Aldrich, Cat. No. D9779) is prepared as 1 M in water and stored at −20°C.
6. Protein sample buffer: 5 M urea, 2 M thiourea, 2 mM tributylphosphine, 65 mM DTT, 4% CHAPS, 150 mM NDSB-256, 1 mM sodium vanadate, 0.1 mM sodium fluoride, and 1 mM benzamidine.

2.3. 1D and 2D Gel Electrophoresis for MS

1. Carrier ampholytes: SERVALYT® 3-10 (SERVA, Cat. No. 42951.01); SERVALYT® 2-4 (SERVA, Cat. No. 42902.01); SERVALYT® 9-11 (SERVA, Cat. No. 42909.01).
2. IPG Non-linear 18 cm pH 3–10 IPG DryStrips (GE Healthcare, Cat. No. 17-1235-01).
3. Immobiline DryStrip Reswelling Tray for 7–18 cm IPG strips (GE Healthcare, Cat. No. 80-6371-84).
4. Immobiline Dry Strip Cover Fluid (GE Healthcare, Cat. No. 17-1335-01).
5. Multiphor II system (GE Healthcare, Cat. No. 18-1018-06).
6. 2D PAGE equilibration solution: 4 M urea, 2 M thiourea, 50 mM Tris–HCl (pH 6.8), 30% (v/v) glycerol, 2% (w/v) SDS, 130 mM DTT, 0.002% (w/v) bromophenol blue.
7. Gel fix: 40% (v/v) ethanol, 10% (v/v) acetic acid.
8. Gel prime: 7.5% (v/v) acetic acid, 0.05% (w/v) SDS.
9. Gel stain: 7.5% (v/v) acetic acid, 1.2 mg/l OGTMP17 (see Note 4).
10. Destain 1: dH_2O.
11. Destain 2: 50% (v/v) EtOH in dH_2O.
12. NuPAGE® Novex 4–12% Bis-Tris–HCl gel (Invitrogen, Cat. No. NP0322BOX).
13. NuPAGE® MOPS SDS Running buffer 20× (Invitrogen, Cat. No. NP0050).
14. NuPAGE® LDS Sample Buffer 4× (Invitrogen, Cat. No. NP0007).

2.4. Tryptic Digestion for MS

1. Digestion reaction plate (Intavis, Cat. No. 40.010).
2. Digestion collection plate (Abgene, Cat. No. TUL-962-011N).
3. In-gel destain solution: 50% (v/v) acetonitrile in dH_2O.
4. In-gel reduction solution: 10 mM DTT, 100 mM NH_4HCO_3. Can be stored at −20°C.

5. In-gel alkylation solution: 55 mM iodoacetamide, 100 mM NH_4HCO_3. Make fresh.
6. In-gel digestion solution: 12.5 ng/μL trypsin (Roche, Cat. No. 03 708 969 001), 50 mM NH_4HCO_3.
7. Peptide extraction solution 1: 50% (v/v) acetonitrile in 25 mM NH_4HCO_3.
8. Peptide extraction solution 2: 50% (v/v) acetonitrile in 2.5% (v/v) formic acid.

2.5. LC-MSMS

1. Acetonitrile CHROMASOLV® HPLC gradient grade (Sigma-Aldrich, Cat. No. 34851).
2. Water HPLC grade (Rathburn Chemicals Ltd., Cat. No. RH1020).
3. Formic acid Super Purity Solvent (Romil, Cat. No. H353).
4. Mass spectrometry (MS) injection solution: 0.1% (v/v) formic acid.
5. MS solvents A and C: 5% (v/v) acetonitrile, 0.1% (v/v) formic acid.
6. MS solvent B: 95% (v/v) acetonitrile, 0.1% (v/v) formic acid.
7. Pre-column: Acclaim PepMap 300, 300 μm × 5 mm C18 cartridge (LC Packings, Cat. No. 163589).
8. Analytical column: Acclaim PepMap 100, 75 μm × 25 cm C18 nanocolumn (LC Packings, Cat. No. 160326).

2.6. Analysis Software

1. ProteinLynx Global Server 2.1.5 (Waters, Milford, MA, USA).
2. Mascot Daemon 2.1.0 (Matrix Science, London, UK).

2.7. Western Blotting for Verification

1. BioRad mini-PROTEAN gel system.
2. Gel running buffer (10×): 250 mM Tris–HCl, 1.92 M glycine, 1% (w/v) SDS. Store at room temperature.
3. Bio Rad precast 10% Any kD Tris–HCl Ready TGX gel (Bio Rad, Cat. No. 456-9033).
4. Transfer buffer: 25 mM Tris–HCl (do not adjust pH), 190 mM glycine, 20% (v/v) methanol, 0.05% (w/v) SDS.
5. Bio Rad mini trans-blot cell for wet blotting.
6. Amersham Hybond ECL nitrocellulose membrane (GE Healthcare, Cat. No. RPN203D).
7. Bio Rad blot-absorbent thick filter paper (Bio Rad, Cat. No. 170-3932).
8. TBS-T: Prepare 10× stock: 1.37 M NaCl, 27 mM KCl, 250 mM Tris–HCl, pH 7.4, 1% Tween-20 (see Note 2). Dilute to 1× with H_2O before use.
9. Blocking buffer: 5% (w/v) non-fat dry milk in TBS-T.

10. Secondary antibody: Goat anti-rabbit IgG (H + L) peroxidase-conjugated (Pierce, Cat. No. 31460) or goat anti-mouse IgG (H = L) peroxidase-conjugated (Pierce, Cat. No. 31430).
11. Supersignal West Pico Chemiluminescent Substrate (Pierce, Cat No. 34087).
12. Amersham Hyperfilm ECL (GE Healthcare, Cat. No. 28-9068-35).

2.8. Immunofluorescence for Verification

1. *Drosophila* S2 tissue culture cells (available from the DGRC: see https://dgrc.cgb.indiana.edu/).
2. Schneider's *Drosophila* medium, modified (Lonza, Cat. No. 04-351Q).
3. FBS (Invitrogen, Cat. No. 10108-165).
4. Microscope coverslips: 22 × 22 mm (Scientific Laboratory Supplies, Cat. No. MIC3110).
5. Microscope slides (Thermo Scientific, Cat. No. BS7011/2).
6. Concanavalin A (Sigma Aldrich, Cat. No. C5275-5MG).
7. Phosphate-buffered saline (PBS): See Subheading 2.1, item 4.
8. Methanol (Fisher, Cat. No. M/4000/PC17).
9. PBS-T: PBS, 0.5% (v/v) Triton X-100.
10. Antibody dilution buffer: PBS, 3% (w/v) BSA.
11. Primary antibody: DM1A mouse anti-tubulin antibody (Sigma Aldrich, Cat. No. T6199).
12. Secondary antibodies: Alexa Fluor 488 goat anti-mouse IgG antibody (Molecular Probes, Cat. No. A1101) and Alexa Fluor 555 goat anti-rabbit IgG antibody (Molecular Probes, Cat. No. A21428).
13. Mounting medium: Vectashield with DAPI (Vector Laboratories, Cat. No. H1200).

3. Methods

The MT co-sedimentation assay relies on the ability of tubulin present in a soluble, cytosolic supernatant to polymerise into dense MTs under the right conditions. Re-centrifugation of the supernatant following polymerisation then results in the specific sedimentation of MAPs (10). Initial studies using brain tissue and HeLa cells identified core MAPs with roles in stabilising MTs, such as MAP1 and MAP2 (11–13). However, this procedure was greatly enhanced by the use of Taxol, a drug derived from the Yew tree, which both reduces the critical concentration of free tubulin subunits in equilibrium with polymerised tubulin and selectively stabilises MTs (14). Over the next decade, the sedimentation of

MTs in this way led to the identification of a number of MAPs in a number of biological systems. This approach culminated with the identification of over 50 proteins from embryos with the ability to bind MTs from the *Drosophila* syncytial embryo (6). These embryos undergo many rounds of synchronous, rapid cell cycles in the absence of zygotic transcription; thus, the microtubules and MAPs required for the formation of thousands of mitotic spindles are already present in the laid egg (15). However, it was only with the advent of mass spectrometry as a method for identifying isolated proteins that this technique became high-throughput. Over the last few years, a number of studies have used the Taxol-induced MT stabilisation and co-sedimentation assay to purify MAPs, combined with mass spectrometry for their identification, e.g. (16, 17). We recently applied this technique to revisit the *Drosophila* embryo with the intention of identifying novel proteins functioning in mitotic spindle organisation (7). We identified 270 MAPs, many of which we have validated as having a role in this process.

3.1. Embryo Collection

1. 0- to 3-h-old *Drosophila* embryos are collected by placing fresh apple juice/molasses agar plate with a lump of yeast paste on the surface into the population cages every 3 h (see Note 5).
2. Embryos are washed off the plate with PBS-T using a paint brush and poured into the filter unit.
3. The filter unit is switched on to remove PBS-T and then switched off.
4. ~20 mL of thin bleach is added to the embryos and swirled gently for 3 min to remove the chorion. The filter unit is switched on to remove bleach and then switched off.
5. ~20 mL of PBS-T is added to the embryos to wash off bleach and swirled. The filter unit is switched on to remove wash and then switched off.
6. Step 5 is repeated, ensuring that all PBS-T is removed through suction.
7. Embryos are scraped off the base of the filter unit using a razor blade, and the dry weight determined (see Note 6).
8. Embryos are frozen in liquid nitrogen and stored at −80°C until enough tissue has been collected. For a large-scale co-sedimentation assay, 2 g of tissue is sufficient.

3.2. Large-Scale MT Co-sedimentation Assay

1. The glass homogeniser is placed in an ice bucket with ice in advance. 3 mL ice-cold C Buffer + one mini protease inhibitor cocktail tablet are added.
2. The ice bucket is taken to the −80°C freezer. Sufficient tubes of embryos to make up to 2 g are quickly emptied into the homogeniser.

3. The embryos are homogenised using six strokes of the motor-driven pestle.
4. The homogenate is equally distributed between 1.5-mL microfuge tubes and centrifuged at top speed for 10 min in a benchtop microfuge (approximately 15,000 × *g*).
5. Using a needle and a syringe, the clarified low-speed supernatant is carefully removed from each tube and combined (see Note 7).
6. The sample is redistributed into tubes, suitable for the ultracentrifuge of use, and centrifuged at 100,000 × *g* for 45 min (see Note 8).
7. The clarified extract is transferred, as in step 5, into a fresh tube and centrifuged at 100,000 × *g* for a further 15 min.
8. The clarified supernatant is transferred to a 15-mL Falcon tube. DTT and GTP are added to 1 mM final concentration each and mixed by inversion.
9. The sample is incubated in a water bath at 25°C for 5 min to allow MTs to polymerise from the endogenous tubulin.
10. The sample is then split into two equal volumes. One is placed on ice. This constitutes your –Taxol (control) samples. To the other (+Taxol), Paclitaxel is added to a final concentration of 10 μM and mixed immediately by inversion.
11. The samples are placed back at 25°C for 15 min and then transferred to ice.
12. 3 mL of ice-cold C Buffer + 40% sucrose are added to two fresh centrifuge tubes.
13. With a syringe and a needle, the –Taxol and + Taxol samples are carefully layered on top of their respective sucrose cushions. The level of the interface is marked with a pen (see Note 9).
14. The samples are centrifuged for 30 min at 100,000 × *g*.
15. After centrifugation, almost all of the top layer (i.e. almost to the interface mark) is removed, taking care not to disturb the interface itself (see Note 10). Transfer 200 μL of this "supernatant" to a tube containing 200 μL of 2× protein sample buffer. This can be used for testing the depletion of MAPs via western blotting.
16. The interface of the sucrose cushion is carefully washed with 1 mL of C Buffer (see Note 11).
17. Half of the remaining sucrose cushion is removed. The wash is repeated as above.
18. The final half of the sucrose cushion is removed. A further 1 mL of C Buffer is added to the side of the tube, allowing it to run to the bottom, and immediately removed. Take care not to disturb any pellet at the bottom of the tube (see Note 12).

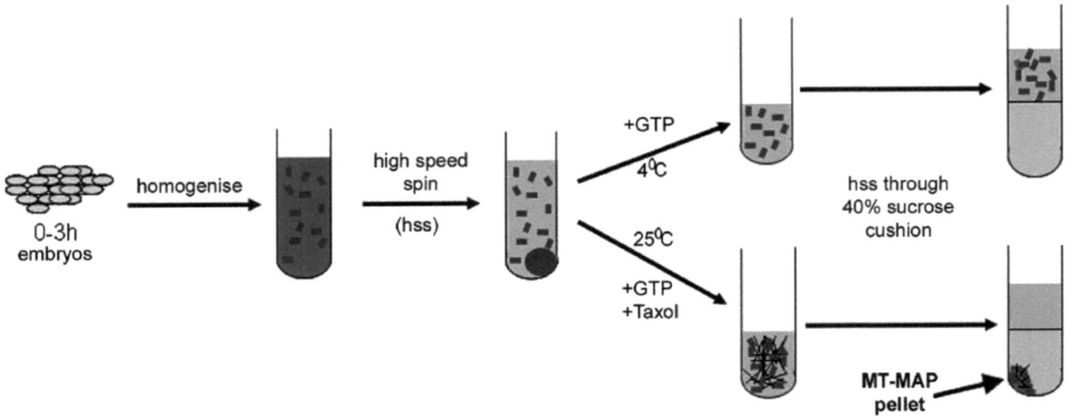

Fig. 1. A representation of the methodology behind the MT co-sedimentation assay (see Subheading 3.2 for details). Only MTs and their associated proteins pellet through the sucrose cushion – all other proteins remain in the supernatant (taken from (7)).

19. The "pellet" is then re-suspended in 200 μL of 1× protein sample buffer. There should, therefore, be four samples: –Taxol supernatant; –Taxol pellet; +Taxol supernatant; and +Taxol pellet.

Figure 1 provides a diagrammatical representation of the methodology described in this section.

3.3. Protein Analysis by 2D SDS-PAGE

1. Prior to 2-dimensional electrophoresis (2DE), the MTSD sample is made up to 375 μL and a maximum protein concentration of 1.33 mg/mL (see Note 13).
2. Carrier ampholytes are added to the prepared MTSD sample to 0.9% (v/v) pH 3–10, 0.45% (v/v) pH 2–4, and 9–11.
3. 18-cm DryStrips are rehydrated overnight in 375 μL prepared MTSD sample. Rehydration is performed in Reswelling Tray with the DryStrip gel side down and overlaid with DryStrip Cover Fluid.
4. Rehydrated DryStrips are carefully drained of excess Cover Fluid and the strips arranged gel side up on the Multiphor II IEF system using DryStrip aligner (see Note 14).
5. 2-cm paper wicks, moistened with 100 μL dH_2O, are positioned at both ends of the DryStrip in partial contact with the gel surface and electrodes aligned over the wicks (see Note 15).
6. Strips are overlaid with Cover Fluid, and isoelectric focusing is performed using a stepped run of 300 V for 2 h followed by 3,500 V to a total of 70 kVh at 17°C.
7. Focused strips are incubated in equilibration solution for 15 min.

8. The equilibrated strip is carefully blotted to remove equilibration solution, placed on top of the 9–16% (w/v) gradient SDS-PAGE gel, and embedded within 0.5% (w/v) melted agarose (see Note 16).
9. Proteins are resolved in the second dimension using 20 mA per gel for 1 h, followed by 40 mA per gel for 4 h at 10°C (see Note 17).
10. Following electrophoresis, gels are fixed overnight, primed for 30 min, and stained with fluorescent dye for 4–5 h.
11. Gels are destained for 5 min using destain solution 1 followed by a further 5 min with destain solution 2. Once destained, gels are scanned using an Apollo II linear fluorescence scanner at a wavelength of 488 nm to obtain 16-bit monochrome gel images at 200 μm resolution.
12. Scanned gel images are analysed using a custom version of Melanie II (BioRad, Hemel Hempstead, UK).
13. Protein spots selected for LC-MSMS are excised using a software-driven robotic cutter (see Note 18).
14. Excised gel plugs are deposited into a digestion reaction plate for automated in-gel digestion using a DigestPro workstation (Intavis, Koeln, Germany).

Figure 2 shows a 2D gel of an MT co-sedimentation assay, as described in this section.

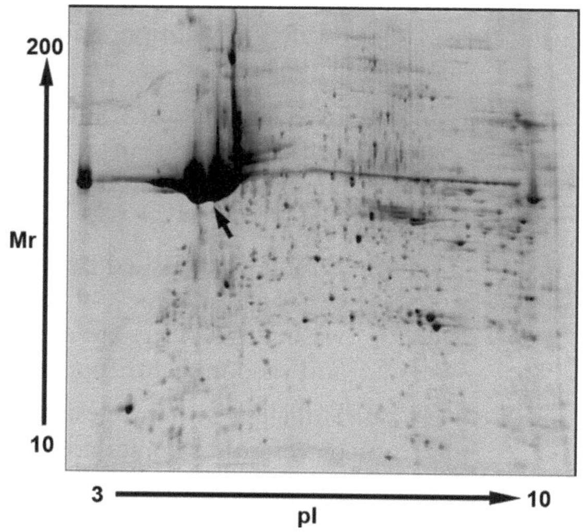

Fig. 2. 2D gel electrophoresis of an MT–MAP pellet isolated using co-sedimentation. Approximately 900 features are visible in the 2D gel. Tubulin migrates as distinct features (e.g. *arrow*), which should be avoided when choosing spots for mass spectrometry (taken from (7)).

3.4. 1D Gel Electrophoresis for LC-MSMS

1. Prepared MTSD in 2D sample buffer are mixed with 4× 1D PAGE sample buffer.
2. Samples are electrophoresed on precast mini gels at 150 V (constant voltage) at room temperature until the dye front reaches the bottom of the gel.
3. Following electrophoresis, gels are fixed overnight, primed for 30 min, and stained with fluorescent dye for 4–5 h.
4. Gels are destained for 5 min using destain solution 1 followed by a further 5 min with destain solution 2. Once destained, 8-bit monochrome images of stained gels are captured using a Fuji CCD CameraLAS-1000 Pro (Tokyo, Japan).
5. Protein bands are visualised for manual excision using a Dark Reader Transilluminator (Clare Chemical Research, CO, USA).
6. The entire lane is cut into 31 approximately equal slices using a clean scalpel and each slice cut into 1 × 1-mm cubes to be processed for LC-MSMS.

3.5. In-Gel Digestion

1. Following the method described by Shevchenko et al. (18), gel pieces are destained at room temperature for 15 min with shaking using In-gel destain solution. Multiple washes may be necessary to fully destain samples (see Note 19).
2. Destained samples are equilibrated in 100 mM NH_4HCO_3; after 5-min equilibration, an equal volume of acetonitrile is added and incubated at room temperature for 15 min with shaking.
3. Equilibration solution is removed and replaced with in-gel reduction solution. Samples are reduced at 56°C for 45 min with shaking.
4. In-gel reduction solution is removed and replaced with an equal volume of in-gel alkylation solution. Samples are alkylated for 30 min at room temperature in the dark with shaking.
5. Samples are washed twice with 100 mM NH_4HCO_3 followed by equal volumes of acetonitrile as in step 2. Following washes, gel pieces are dehydrated and dried using acetonitrile and vacuum centrifuge (see Note 20).
6. Sufficient in-gel digestion solution is added to cover the gel pieces, and samples are incubated for 30 min on ice to allow gel pieces to rehydrate. Excess in-gel digestion solution is removed from the rehydrated gel pieces and replaced with fresh 50 mM NH_4HCO_3. Samples are digested overnight at 37°C (see Note 21).
7. Digested peptides are extracted once with peptide extraction solution 1 and twice with peptide extraction solution 2.

All extraction steps are performed at room temperature with shaking. The three washes are pooled in a 96-well digestion collection plate and dried down using a vacuum centrifuge (see Note 22).

3.6. LC-MSMS

1. Dried samples are redissolved in 7.5 μL MS injection solution.
2. 6 μL is injected onto a CapLC (Micromass, UK) and trapped on a pre-column for concentration and desalting by 20 μL/min MS solvent C for 5 min.
3. Peptides are resolved on an analytical column at a flow rate of 200 nl/min. Peptides are eluted using a gradient of MS solvent B rising from 5 to 40% (v/v) by 35 min and to 95% solvent B by 36 min with a total run time of 45 min.
4. Eluted tryptic peptides are sprayed into a QTof 1 (Micromass, UK) at a capillary voltage of 3.3 kV and a cone voltage of 45 V. The mass spectrometer is operated using MassLynx 4.0, and spectra are acquired in positive mode with MS to MSMS switching controlled in an automatic data-dependent fashion (see Note 23).
5. Raw data files are processed using ProteinLynx Global Server 2.1.5 (Micromass, UK) to produce .pkl files (see Note 24).
6. Processed files are searched against the SwissProt, MSDB, and NCBI databases using an in-house search engine. All searches were restricted to the *Drosophila* taxonomy with carbamidomethyl cysteine defined as a fixed modification and oxidised methionine as a potential variable modification. Data are searched allowing 0.5 Da error for both peptide and fragment masses and a maximum of two missed tryptic cleavage sites.
7. Confident protein identification requires two peptides with ion scores indicating identity or extensive homology. Low-scoring peptides or single-peptide identifications are manually validated (see Note 25).

3.7. Western Blotting for Verification

Once you have identified potential MAPs, their identities in the MT–MAP pellet can be verified using SDS-PAGE and western blotting, assuming that a suitable antibody is available. Novel MAPs require antibodies to be produced or verification using alternative methods (i.e. expression of GFP-fusion proteins in S2 tissue culture cells). These instructions assume the use of a Bio Rad PROTEAN mini gel and Bio Rad mini transblot transfer systems, but are easily adaptable to other formats. Figure 3 shows a number of western blots carried out using samples isolated using the methodology described in this chapter.

1. Prepare the running buffer by diluting 100 mL of the 10× running buffer with 900 mL of water in a measuring cylinder.

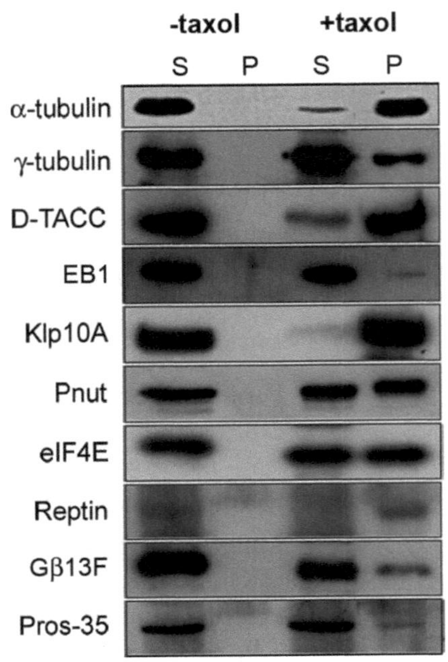

Fig. 3. Western blot analysis of control and MT co-sedimentation, "supernatant" and "pellet", samples using a selection of antibodies generated against candidate MAPs (taken from (7)).

2. Prepare the gel by cutting the protective plastic coating along the line and pulling the tab to remove the bottom strip of plastic.
3. Place the gel in the cassette such that the smaller plate faces inwards, place the cassette into the holder swinging the arms inwards to lock the gel in place, and transfer the apparatus into the gel tank.
4. Add the running buffer to the upper and lower chambers of the gel unit and load the 10 μL of each sample in a well. Include one well for pre-stained molecular weight markers (see Note 26).
5. Complete the assembly of the gel unit and connect to a power supply. The gel can be run at 100 V at room temperature or 150 V with cooling, until the dye front runs off (approximately 2.5 h at 100 V, 1.5 h at 150 V).
6. Make your transfer buffer, placing most in the transfer tank. Add the foam supports and the two sheets of thick blotting paper.
7. Submerge a sheet of nitrocellulose, cut just larger than the size of the separating gel, in a tray containing the rest of blotting buffer.
8. Disconnect the gel unit from the power supply and disassemble it. Remove and discard the stacking gel. Lay the separating gel on top of the nitrocellulose membrane (see Note 27).

9. Open a blotting cassette, clear side down and black side up. Place a wet foam support on the clear side, followed by a sheet of wet blotting paper.
10. Carefully lift the nitrocellulose/gel sheet from the tray and place on top of the wet blotting paper.
11. Lay the remaining piece of wet blotting paper on top, smoothing down with fingers to ensure that no air bubbles get trapped.
12. Lay the second wet foam support on top and close the transfer cassette.
13. Place the cassette in the transfer tank such that the nitrocellulose membrane is between the gel and the anode. It is vitally important to ensure that this orientation or the proteins will be lost from the gel into the buffer rather than transferred to the nitrocellulose.
14. Put the lid on the tank and turn the power supply on. Transfer the protein for an equivalent of 250 mA (see Note 28).
15. Once the transfer is complete, take the cassette out of the tank and carefully disassemble. The gel will have left an indentation on the nitrocellulose, and the coloured molecular weight markers should be clearly visible. Cut around the area where the gel was with a razor blade. The gel, excess nitrocellulose, and blotting paper can be discarded. Wash the foam supports and cassette in water for subsequent use.
16. Transfer the nitrocellulose to a tray of suitable size and briefly wash with water (see Note 29).
17. Incubate the membrane with a suitable volume (usually ~10 mL) of blocking buffer for 30 min at room temperature on a rocking platform.
18. Discard the blocking buffer and add primary antibody solution (blocking buffer + suitably diluted antibody) (see Note 30).
19. Incubate for 2 h at room temperature on a rocking platform or overnight at 4°C.
20. The primary antibody is then removed and the membrane washed three times for 10 min each with 20 mL TBS-T with rocking (see Note 31).
21. Prepare the secondary antibody solution as a 1:10,000-fold dilution in blocking buffer and add to the membrane for 2 h at room temperature on a rocking platform.
22. Discard the secondary antibody solution and wash the membrane three times for 15 min each with TBS-T.
23. During the final wash, 2 mL aliquots of each portion of the ECL reagent are warmed separately to room temperature, and the remaining steps are done in a dark room under safe light conditions. Once the final wash is removed from the blot, the

ECL reagents are mixed together and then immediately added to the blot, which is then rotated by hand for 1 min to ensure even coverage.

24. The blot is removed from the ECL reagents, blotted with tissue, and then placed between the leaves of an acetate sheet protector that has been cut to the size of an X-ray film cassette.
25. The acetate containing the membrane is then placed in an X-ray film cassette with film for a suitable exposure time, typically a few minutes. An example of the results produced is shown in Fig. 3.

3.8. Immuno fluorescence for Verification

Drosophila S2 cells are cultured in Schneider *Drosophila* media with 10% FBS. Cell density is kept between 5×10^5 and 5×10^6 mL, measured using a haemacytometer. Cells are passaged approximately 1 in 3 when the number of cells exceeds these limits. S2 cells are only slightly adherent and therefore, prior to visualisation by immunofluorescence, they must be adhered to concanavulin-coated coverslips. Figure 4 shows a number of immunofluorescence images of S2 cells treated using the methodology described in this section.

1. Prepare concanavalin-coated coverslips: Dip coverslips in 70% ethanol, allow to dry in a suitable rack, immerse in 1 mg/mL concanavalin A, and allow to dry. Wash 5×5 min with PBS and allow to dry before use (see Note 32).
2. Approximately 1×10^6 cells are allowed to settle for 3 h on concanavalin-coated coverslips on the base of six-well plates.
3. The media removed and gently washed by addition and immediate removal of 2 mL PBS.
4. Cells are fixed by pipetting 3 mL methanol (−20°C) directly on top of the washed cells, after which the plate is left on dry ice for 20 min.
5. The plate is then removed from dry ice, equilibrated to room temperature for a further 20 min, the methanol is removed, and each well is briefly washed with PBS as before.
6. Cells are blocked for 30 min in PBS-T/3% BSA with gentle agitation.
7. Primary antibodies in blocking solution are added to the coverslip either for 2 h at room temperature or overnight at 4°C, in a humid box (see Note 33).
8. The cells are then washed 3×10 min in PBS-T before addition of the appropriate secondary antibody in blocking solution for 2 h at room temperature.
9. The cells are then washed 3×15 min in PBS-T, prior to mounting in 4 μL VectaShield Mounting Media with DAPI (see Note 34).

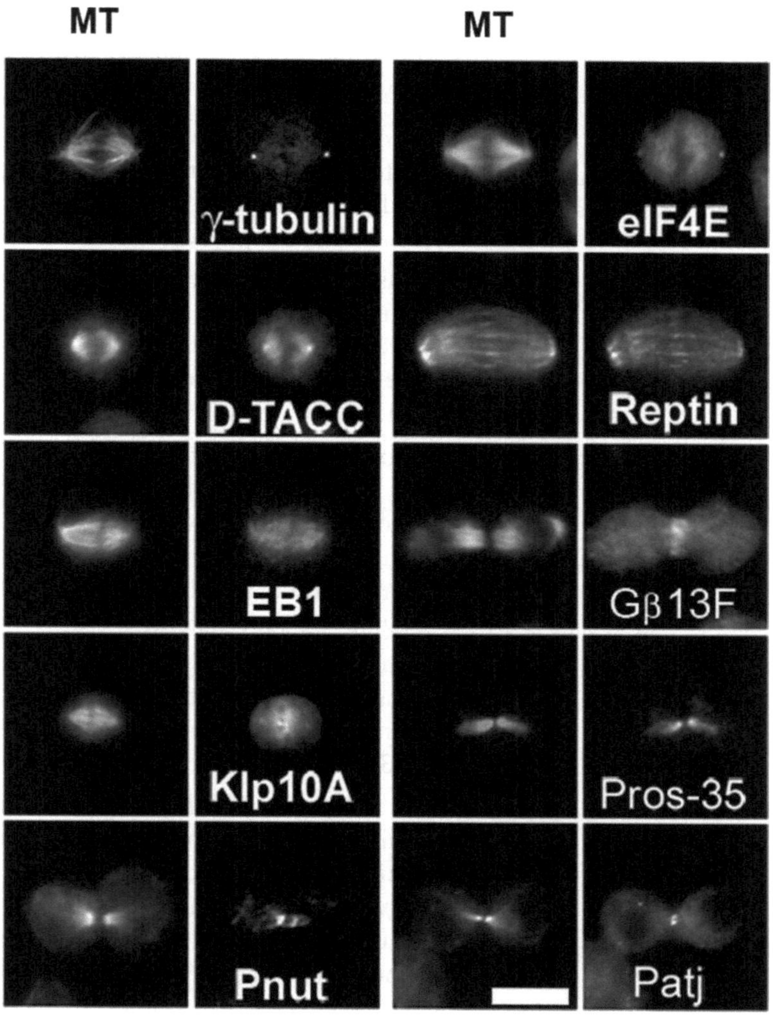

Fig. 4. Immunofluorescence analysis of candidate MAPs. S2 cells were co-stained with antibodies recognising tubulin and candidate MAPs (taken from (7)).

4. Notes

1. Unless stated otherwise, all solutions should be prepared in water that has a resistivity of 18.2 MΩ cm and total organic content of less than five parts per billion. This standard is referred to as "water" in this chapter.
2. Both Triton X-100 and Tween-20 can take a while to dissolve in water. We make 20% stock of each and dilute when needed, as appropriate.
3. Both Taxol and GTP have a finite shelf life. To ensure reproducibility of the assay, we routinely use aliquots once, rather than refreezing, and make fresh stocks from powder every 6 months.

4. The fluorescent dye used, OGTMP17 (Oxford Glycoscience, Abingdon, UK), is a proprietary stain with similar sensitivity to SYPRO stains.

5. The quantity of embryos collected varies greatly from day to day and from batch-to-batch. On average, we find that 4–9-day-old flies from six standard bottles in a single 10-cm diameter collection chamber produce 0.1–0.2 g of embryos per collection.

6. To assess the quantity of embryos, we weigh an empty 1.5-mL microfuge tube, scrape the embryos into the tube, and re-weigh. In our initial studies, we used 2 g of embryos for a large-scale MT co-sedimentation assay. Recently, we have reduced this quantity to 1 g.

7. Following the spin, three distinct layers should be present; a thin layer of fat at the top, a dense pellet of material at the bottom, and the low-speed supernatant in between. When taking the supernatant, move the needle to approximately ¾ down the tube, avoiding disturbing the fatty layer, and slowly draw up the supernatant.

8. We use the Beckman 50.4 Ti fixed angle rotor and Quick-seal Ultraclear tubes (Beckman Coulter: 344320), which require sealing via a cordless tube topper kit (Beckman Coulter: 358314).

9. The best way to add the sample without disturbing the sucrose cushion is to place the edge of the needle against the side of the tube, just above the cushion and, in one movement, gently expel the whole sample, moving upwards ahead of the liquid as you do so. To minimise the mixing of the layers prior to the spin, bring the cooled centrifuge rotor to the bench on which you layer the samples, so the tubes can be placed into the rotor immediately after layering and sealing. The rotor can then carefully be taken to the centrifuge.

10. Reverse the previous process; take the rotor to the bench where the samples will be taken, prior to removing the centrifuge tubes. Cut the tubes open at the top using a razor blade to get access to the supernatant. While removing most of the top layer, take 200 μL and add to 200 μL of 2× protein sample buffer. This can be used as the "supernatant" sample for testing the depletion of MAPs via western blotting.

11. Carefully add 1 mL extract buffer in the same way as the sample was added to minimise disruption of the sucrose cushion, and then immediately remove the same top 1 mL.

12. You should see a large, glassy pellet at the bottom of the MTSD tube, constituting the MT–MAP pellet. There may also be a small pellet at the bottom of the control tube.

13. To minimise keratin contamination, all sample preparation, processing, and gel manipulation are performed in a clean room and dedicated, proteomics grade reagents should be used.

14. Care must be taken to orientate DryStrips correctly with acidic end (marked +) towards the anode (red).
15. Wicks should be extensively blotted to remove excess water and reduce protein streaking during focusing.
16. Care must be taken to avoid trapping bubbles between DryStrip and gel and if the agarose is too hot, protein carbamylation may occur.
17. Gels were maintained at 10°C using a recirculating chiller. Different types of direct cooling, such as a water bath at 4°C, can also be used. If unavailable, second-dimension gels should be run in a cold room.
18. Spots are selected for excision which are absent from the control gel and distinct from the abundant tubulin feature. A single representative spot is excised, where a cluster of related spots are observed.
19. For automated digestion, the gel plug size is consistent and the volume of each solution added remains constant. For manual digestion, the volume of each solution added should be sufficient to just cover the gel cubes.
20. If a vacuum centrifuge is not available, it is sufficient to dehydrate the gel pieces with acetonitrile.
21. Care must be taken to avoid drying out the gel pieces during digestion, and heat blocks without a heated lid should be avoided.
22. Dried samples can be stored at −20°C for 1 month with no appreciable loss of peptides.
23. A 1-s survey scan of the range m/z 400–1,800 is followed by three 1-s MSMS scans of the range m/z 50–2,000 of the most intense ions fulfilling the peptide ion criteria. Ions selected for MSMS are excluded from further fragmentation for 2 min, and the most common trypsin autolysis products are excluded from selection for MSMS.
24. Spectra are smoothed (2× smooths, Savitzky-Golay, ±3 channels) and centred (peak width at half height 4 channels, centroid at 80%).
25. For manually validated peptide IDs, MASCOT output files are examined for consistent *y*-ions (or *b*-ions for C-terminal peptides), fragment ion isotopes, signal/noise ratio, and sequence-specific fragmentation markers, such as proline residues.
26. You should have four samples from the MTSD: supernatant (−Taxol); pellet (−Taxol); supernatant (+Taxol); and pellet (+Taxol), see Subheading 3.2, step 19.
27. It is crucial that no air bubbles exist between gel and membrane, indeed between any layers of the setup. Bubbles result

in uneven transfer of proteins onto the nitrocellulose and complicate analysis. Setting up the sandwich under buffer makes it relatively easy to spot and rectify air bubbles.

28. This can be 250 mA for 1 h, 20 mA overnight or anywhere in between.

29. Try to choose a tray just larger than the nitrocellulose membrane to avoid using more primary and secondary antibody solutions than necessary.

30. Most antibodies are diluted in the region of 1:500–2,000 depending on their affinity and avidity. Each antibody is different.

31. In general, primary antibodies can be saved for subsequent experiments by storing at −20°C. Recurrent use of diluted antibody solutions leads to increased lowering of the antibody titre in the solution and associated longer exposure times at the ECL step.

32. Coverslips should be prepared in the tissue culture hood to ensure sterility.

33. For economy, only 100 μL of diluted antibody per sample needs to be used. An aspirator can be used to dry the dish around the coverslip thoroughly. Careful addition of the antibody solution establishes a good meniscus on top of the coverslip, without it dissipating into the rest of the well. Placing the plate in a humid box stops evaporation of the solution.

34. Careful inversion of the coverslip onto a slide onto which a drop of Vectashield has been placed helps to avoid trapping air bubbles in the sample.

Acknowledgements

We would like to thank Bevin Gangadharan and Tommy Duncan for their input into the above methods and for critical reading of the manuscript.

References

1. Mitchison T and Kirschner M (1984) Dynamic instability of microtubule growth. Nature 312:232–237
2. Siegrist S and Doe CQ (2007) Microtubule-induced cortical cell polarity. Genes Dev 21:483–496
3. Hehnley H and Stamnes M (2007) Regulating cytoskeleton-based vesicle motility. FEBS Lett 581:2112–2118
4. Avila J (1990) Microtubule dynamics. FASEB J 4:3284–3290
5. Maiato H, Sampaio P, and Sunkel CE (2004) Microtubule-associated proteins and their essential roles during mitosis. Int Rev Cytol. 241:53–153
6. Kellogg D, Field C, and Alberts BM (1989) Identification of microtubule-associated proteins in the centrosome, spindle and kinetochore

of the early Drosophila embryo. J Cell Biol 109:2977–2991

7. Hughes JR, Meireles AM, Fisher KH, Garcia A, Antrobus PR, Wainman A, Zitzmann N, Deane C, Ohkura H, and Wakefield JG (2008) A microtubule interactome: complexes with roles in cell cycle and mitosis. PLoS Biol 6:e98
8. Sullivan W, Ashburner M, and Hawley RS (eds.) 2000. Drosophila Protocols. Cold Spring Harbor Laboratory Press, Cold Spring Harbor, N.Y
9. Greenspan RJ (1997) Fly Pushing: The Theory and Practice of Drosophila Genetics. Cold Spring Harbor Laboratory Press, Cold Spring Harbor, N.Y
10. Weisenberg RC (1972) Microtubule formation in vitro in solutions containing low calcium concentrations. Science 177:1104–5
11. Murphy DB and Borisy GG (1975) Association of high-molecular-weight proteins with microtubules and their role in microtubule assembly in vitro. Proc Natl Acad Sci USA 72: 2696–700
12. Murphy DB, Vallee RB and Borisy GG (1977) Identity and polymerization-stimulatory activity of the nontubulin proteins associated with microtubules. Biochemistry 16:2598–605
13. Bulinski and Borisy GG (1979) Self-assembly of microtubules in extracts of cultured HeLa cells and the identification of HeLa microtubule-associated proteins. Proc Natl Acad Sci USA 76:293–7
14. Vallee RB (1982) A taxol-dependent procedure for the isolation of microtubules and microtubule-associated proteins (MAPs). J Cell Biol 92:435–42
15. Foe VE and Alberts BM (1983) Studies of nuclear and cytoplasmic behaviour during the five mitotic cycles that preced gastrulation in Drosophila embryogenesis. J Cell Sci 61:31–70
16. Chuong SD, Good AG, Taylor GJ, Freeman MC, Moorhead GB, and Muench DG (2004) Large-scale identification of tubulin-binding proteins provides insight on subcellular trafficking, metabolic channeling, and signaling in plant cells. Mol Cell Proteomics 3:970–983
17. Sakamoto T, Uezu A, Kawauchi S, Kuramoto T, Makino K, Umeda K, Araki N, Baba H, and Nakanishi H (2008) Mass spectrometric analysis of microtubule co-sedimented proteins from rat brain. Genes Cells 13:295–312
18. Shevchenko A, Tomas H, Havliš H, Olsen JV, and Mann M (2007) In-gel digestion protocol for mass spectrometric characterization of proteins and proteomes. Nature Protocols 6:2856–2860

Chapter 21

Isolation of Novel +TIPs and Their Binding Partners Using Affinity Purification Techniques

Ka Lou Yu, Nanda Keijzer, Casper C. Hoogenraad, and Anna Akhmanova

Abstract

Microtubule organization and dynamics are controlled by a large set of cellular factors. An important group of microtubule regulators is microtubule plus-end-tracking proteins (+TIPs), which accumulate specifically at the growing microtubule ends, affect different phases of dynamic instability, and link microtubules to various cellular structures. +TIPs include a very diverse set of proteins with widely different structural properties. One of the most conserved and ubiquitous +TIP families are end-binding (EB) proteins, which can track growing microtubule ends autonomously in the absence of any other factors. In contrast, the majority of other known +TIPs cannot recognize the growing microtubule plus ends on their own; instead, they "hitchhike" to the plus ends by interacting with one of the members of the EB family. Therefore, the association with EBs and the ability to track growing microtubule ends are tightly linked, and binding to the EBs can be used to identify new +TIPs. In this chapter, we describe two affinity purification techniques, glutathione S-transferase and biotinylation tag-based pull-down assays that proved to be very useful for the identification of new EB-interacting +TIPs and their binding partners by mass spectrometry. We also discuss cytological techniques that can be applied to confirm plus-end localization of newly identified proteins.

Key words: Microtubule dynamics, EB1, GST, Biotinylation, Mass spectrometry

1. Introduction

Microtubules (MTs) grow and disassemble from their ends, and numerous microtubule end-binding (EB) proteins regulate different aspects of microtubule function (1, 2). Microtubule plus-end-tracking proteins (+TIPs) are defined by their specific comet-like accumulation at the ends of growing MTs (3, 4). +TIPs include a wide variety of proteins that form a complex interaction network. Recent studies showed that the core of this network is

Anne Straube (ed.), *Microtubule Dynamics: Methods and Protocols*, Methods in Molecular Biology, vol. 777, DOI 10.1007/978-1-61779-252-6_21, © Springer Science+Business Media, LLC 2011

formed by small (~30 kDa) and highly conserved EB proteins, which are represented in mammals by three homologues, EB1, EB2, and EB3 (4–6). The N-terminal part of the EBs consists of a Calponin Homology domain that binds to MTs (7) while the C-terminal part includes a coiled coil domain that ends with a four-helix bundle and an acidic tail that resembles the tail of α-tubulin (8–11). EB C-terminus binds to two major types of partners: CAP-Gly domain proteins, such as the CLIPs (12), and various structurally unrelated proteins with basic and serine-rich regions harboring a short linear motif SxIP (13). Both types of interactions show affinities in a low micromolar range and are relatively unstable, allowing +TIP complexes to remodel and exchange rapidly at the MT tips (4, 13–15). Because +TIP interactions are relatively weak and transient, the EBs are usually difficult to co-precipitate together with their partners from cell extracts unless they are overexpressed. On the other hand, complexes of EBs with other +TIPs easily form in vitro. Furthermore, unlike many other +TIPs, EB proteins with different tags can be readily purified from bacteria in their native form. Therefore, glutathione S-transferase (GST) pull-down assays are ideally suited for the identification of new EB partners: recombinant GST–EB fusions can be easily prepared and used to capture different EB partners in cell extracts. Numerous studies employed this approach to confirm the interactions between EBs with other proteins (13, 16–21). In combination with mass spectrometry, this technique was successfully used to identify new +TIPs, such as RhoGEF2, transmembrane ER-resident protein STIM1, and a neuronal adaptor p140Cap (22–24).

While GST pull-down assays proved very successful for identifying EB interaction partners, the success of their application to other proteins is variable because it strongly depends on the possibility to purify intact and properly folded proteins from *Escherichia coli*, and also because many protein complexes do not assemble easily in cell extracts. Overexpression of tagged proteins and their immunoprecipitation is a common alternative approach. In this chapter, we describe the use of the biotinylation tag, ~20 amino acid polypeptide, which contains a lysine residue that can be very specifically biotinylated by the bacterial biotin ligase BirA (25, 26). Coexpression of BirA together with the tagged protein of interest results in its efficient biotinylation in eukaryotic cells. Biotinylated proteins can be purified by a one-step procedure through their binding to streptavidin (27–29). Since the affinity of biotin to streptavidin is very high ($K_d = 10^{-15}$ M, which is several orders of magnitude higher than a typical antibody), purification is efficient even when the protein of interest is expressed at low levels. The procedure permits use of different cell lysis buffers, and can be applied to cross-linked materials (30). Since the binding is fast, unstable proteins and transient protein complexes can be readily purified. Furthermore, since no antibodies are used, their presence does not complicate mass spectrometry analysis. It is also important to note that biotin is a covalently attached cofactor of some metabolic enzymes, but their number is small (31) and their presence among the isolated proteins does not pose significant problems with identification of the relevant partners. However, streptavidin beads generally show higher levels of unspecific protein binding compared to glutathione sepharose. Therefore, proper controls, such as pull downs from cells expressing BirA alone or in combination with irrelevant biotinylation-tagged protein, like GFP, should be carried out. Because of high reproducibility of biotinylation-tag affinity purification assays, it is possible to assemble lists of common background proteins in specific cell types and use them as a reference.

Here, we describe the preparation of samples for GST-EB and biotinylation tag-based pull downs that can be used for mass spectrometry analysis. We provide lists for the most common background proteins frequently observed in these assays (Tables 1–3). Finally, we also briefly describe the most important cytological validation techniques, which are based on making GFP fusions of the newly identified proteins and the dual-color imaging in fixed and live cells.

Table 1
Common background proteins in GST pull-down assays from HeLa cells

Description	Entrez gene	Acc./id
Carbonyl reductase 1 [Homo sapiens]	CBR1	gi|75061940
Eukaryotic translation elongation factor 1 [Homo sapiens]	EEF1A1	gi|48734733
Glutathione S-transferase [Homo sapiens]	GSTM3	gi|14250650
Heat shock 70 kDa protein [Homo sapiens]	HSPA	gi|5123454
Keratin [Homo sapiens]	KRT1 KRT2 KRT6A KRT9 KRT14	gi|11935049 gi|47132620 gi|5031839 gi|55956899 gi|119581149

A list of proteins observed in several GST pull downs using GST alone and Triton-X100 extracts of HeLa cells. Only proteins that showed a Mascot Score higher than 300 were included

Table 2
Common background proteins in pull-down assays with strepatividin beads from HeLa cells

Description	Entrezgene			Acc./id		
Acetyl-CoA carboxylase [Homo sapiens]	ACACB	ACACA		gi|94963132	gi|38679960	
Actin, alpha [Homo sapiens]	ACTA1			gi|4501881		
Actin, beta [Homo sapiens]	ACTB	ACTBL2		gi|4501885	gi|63055057	
Actin, gamma [Homo sapiens]	ACTG1			gi|4501887		
Actinin, alpha [Homo sapiens]	ACTN1	ACTN4		gi|4501891	gi|2804273	
AHNAK nucleoprotein isoform 1 [Homo sapiens]	AHNAK			gi|535177		
Serum albumin [Homo sapiens]	ALB			gi|28592		
Annexin A2 [Homo sapiens]	ANXA2			gi|18645167		
ATP synthase beta subunit [Homo sapiens]	ATP5B			gi|32189394		
Hypothetical protein LOC126353 [Homo sapiens]	C19orf21			gi|27735067		
Carbamoyl-phosphate synthetase AD protein [Homo sapiens]	CAD			gi|18105007		
F-actin capping protein [Homo sapiens]	CAPZA1	CAPZB		gi|5453597	gi|4826659	
Chaperonin containing TCP1 [Homo sapiens]	CCT2 CCT5	CCT3 CCT6A	CCT4 CCT8	gi|5453603 gi|24307939	gi|58761484 gi|62089036	gi|38455427 gi|48762932
CD55 antigen [Homo sapiens]	CD55			gi|119613891	gi|1857867	
Clathrin heavy chain 1 [Homo sapiens]	CLTC			gi|4758012	gi|119614803	
Coronin, actin-binding protein 1C [Homo sapiens]	CORO1C			gi|7656991		
Amplaxin [Homo sapiens]	CTTN			gi|182087		
Drebrin 1 [Homo sapiens]	DBN1			gi|18426913		
DEAD (Asp-Glu-Ala-Asp) box polypeptide 5 [Homo sapiens]	DDX5			gi|4758138		
Desmoglein 2 [Homo sapiens]	DSG2			gi|416178		

Eukaryotic translation elongation factor [Homo sapiens]	EEF1A1	EEF1AL3	EEF2	gi|4503471	gi|74746925	gi|4503483
Eukaryotic translation initiation factor 4A [Homo sapiens]	EIF4A1			gi|4503529		
Enolase 1 [Homo sapiens]	ENO1			gi|4503571		
Epiplakin [Homo sapiens]	EPPK1			gi|37196760		
Glutamyl-prolyl tRNA synthetase [Homo sapiens]	EPRS			gi|62241042		
Hypothetical protein LOC222584 [Homo sapiens]	FAM83B			gi|61676089		
Family with sequence similarity FAM83H [Homo sapiens]	FAM83H			gi|157311635		
Fatty acid synthase [Homo sapiens]	FASN			gi|41872631		
Flightless-I homolog [Homo sapiens]	FLII			gi|440177		
Filamin [Homo sapiens]	FLNA	FLNB		gi|116063573	gi|3282771	
Flotillin 1 [Homo sapiens]	FLOT1			gi|5031699		
Fascin 1 [Homo sapiens]	FSCN1			gi|4507115		
Fragile X mental retardation syndrome-related protein 1	FXR1			gi|114590562		
Guanine nucleotide-binding protein alpha polypeptide 3 [Homo sapiens]	GNAI3			gi|5729850		
Heat shock 90 kDa protein [Homo sapiens]	HSP90AA1	HSP90AB1	HSP90B1	gi|62914009	gi|306891	gi|62088648
Heat shock 70 kDa protein [Homo sapiens]	HSPA1A HSPA8	HSPA1B HSPA9	HSPA5	gi|4529893 gi|5729877	gi|4529892 gi|292059	gi|16507237
Heat shock protein 27 [Homo sapiens]	HSPB1	HSPD1		gi|662841	gi|31542947	
Mitofilin [Homo sapiens]	IMMT			gi|8131894		
Plakoglobin [Homo sapiens]	JUP			gi|762885		
Phostensin [Homo sapiens]	KIAA1949			gi|145386517		

(continued)

Table 2 (continued)

<table>
<tr><th>Description</th><th>Entrezgene</th><th></th><th></th><th>Acc./id</th><th></th><th></th></tr>
<tr><td>Keratin [Homo sapiens]</td><td>KRT1
KRT9
KRT7
KRT6A
KRT14</td><td>KRT2
KRT10
KRT8
KRT6B
KRT16</td><td>KRT5
KRT13
KRT17
KRT18</td><td>gi|11935049
gi|55956899
gi|12803727
gi|46812692
gi|12803709</td><td>gi|386849
gi|21961605
gi|4504919
gi|386849
gi|24430192</td><td>gi|18999435
gi|158261511
gi|4557701
gi|4557888</td></tr>
<tr><td>LIMA1 protein [Homo sapiens]</td><td>LIMA1</td><td></td><td></td><td>gi|7705373</td><td></td><td></td></tr>
<tr><td>LIM and calponin homology domains-containing protein 1 [Homo sapiens]</td><td>LIMCH1</td><td></td><td></td><td>gi|163310741</td><td></td><td></td></tr>
<tr><td>Lamin A protein [Homo sapiens]</td><td>LMNA</td><td></td><td></td><td>gi|386856</td><td></td><td></td></tr>
<tr><td>LIM domain only 7 [Homo sapiens]</td><td>LMO7</td><td></td><td></td><td>gi|119600955</td><td></td><td></td></tr>
<tr><td>LUZP1 protein [Homo sapiens]</td><td>LUZP1</td><td></td><td></td><td>gi|30354561</td><td></td><td></td></tr>
<tr><td>3-methylcrotonyl-CoA carboxylase [Homo sapiens]</td><td>MCCC1</td><td>MCCC2</td><td></td><td>gi|12276066</td><td>gi|11545863</td><td></td></tr>
<tr><td>Myosin phosphatase-Rho interacting protein isoform 1 [Homo sapiens]</td><td>MPRIP</td><td></td><td></td><td>gi|50980307</td><td></td><td></td></tr>
<tr><td>Myosin heavy chain nonmuscle [Homo sapiens]</td><td>MYH10</td><td>MYH9</td><td></td><td>gi|641958</td><td>gi|12667788</td><td></td></tr>
<tr><td>Myosin [Homo sapiens]</td><td>MYO10
MYO1B
MYO1G</td><td>MYO6
MYO1C</td><td>MYO18A
MYO1E</td><td>gi|9910111
gi|56204015
gi|54873627</td><td>gi|27529702
gi|68533085</td><td>gi|24660442
gi|55956916</td></tr>
<tr><td>Polyadenylate-binding protein [Homo sapiens]</td><td>PABPC1</td><td>PABPC4</td><td></td><td>gi|693937</td><td>gi|4504715</td><td></td></tr>
<tr><td>Poly (ADP-ribose) polymerase [Homo sapiens]</td><td>PARP1</td><td></td><td></td><td>gi|22902366</td><td></td><td></td></tr>
<tr><td>Pyruvate carboxylase precursor [Homo sapiens]</td><td>PC</td><td></td><td></td><td>gi|106049292</td><td></td><td></td></tr>
<tr><td>Propionyl-Coenzyme A carboxylase [Homo sapiens]</td><td>PCCA</td><td>PCCB</td><td></td><td>gi|65506442</td><td>gi|62087598</td><td></td></tr>
<tr><td>Plectin [Homo sapiens]</td><td>PLEC1</td><td></td><td></td><td>gi|41322916</td><td></td><td></td></tr>
</table>

Plastin 3 [Homo sapiens]	PLS3			gi|7549809		
Protein phosphatase 1, regulatory (inhibitor) subunit 12A [Homo sapiens]	PPP1R12A			gi|4505317		
Tumor necrosis factor type 1 receptor-associated protein [Homo sapiens]	PSMD2			gi|687239		
Novel retinal pigment epithelial cell protein [Homo sapiens]	RAI14			gi|7274242		
Raichu404X [Homo sapiens]	RAP1A			gi|14595132		
RNA-binding motif protein 14 [Homo sapiens]	RBM14			gi|5454064		
Replication protein A1, 70 kDa [Homo sapiens]	RPA1			gi|4506583		
S3 ribosomal protein [Homo sapiens]	RPS3			gi|7765076		
Structure protein NSP5a3b [Homo sapiens]	SPECC1			gi|56789962		
Spectrin, alpha (alpha-fodrin) [Homo sapiens]	SPTAN1			gi|31565122		
Spectrin, beta [Homo sapiens]	SPTBN1			gi|62089082		
Sperm-specific antigen 2 [Homo sapiens]	SSFA2			gi|134047924		
Stomatin peptide [Homo sapiens]	STOM			gi|181184		
Tropomodulin 3 [Homo sapiens]	TMOD3			gi|6934244		
Tropomyosin [Homo sapiens]	TPM1	TPM4	TPM4ALK	gi|29792232	gi|4507651	gi|13274400
Alpha-tubulin [Homo sapiens]	TUBA1A	TUBA1B		gi|37492	gi|340021	
Beta-tubulin [Homo sapiens]	TUBB	TUBB2C	TUBB3	gi|338695	gi|5174735	gi|62897639
Vimentin [Homo sapiens]	VIM			gi|62414289		
WDR1 protein [Homo sapiens]	WDR1			gi|9257257		
Kaiso [Homo sapiens]	ZBTB33			gi|5803229		

A list of proteins observed in several streptavidin bead pull downs using HeLa cells expressing BirA in combination with biotinylation-tagged GFP. Only proteins that showed a Mascot Score higher than 300 were included

Table 3
Common background proteins in pull-down assays with strepatividin beads from HEK293T cells

Description	Entrezgene			Acc./id		
Acetyl-Coenzyme A carboxylase alpha isoform 2 [Homo sapiens]	ACACA			gi|38679967		
Actin, gamma 1 [Homo sapiens]	ACTG1			gi|4501887		
Serum albumin [Homo sapiens]	ALB			gi|4502027		
Ataxin-2-related domain protein [Homo sapiens]	ATXN2L			gi|18071115		
GPI-anchored membrane protein 1 [Homo sapiens]	CAPRIN1			gi|42558250		
Chaperonin containing TCP1 [Homo sapiens]	CCT2 CCT6A	CCT3 CCT7	CCT4 CCT8	gi|5453603 gi|4502643	gi|14124984 gi|5453607	gi|38455427 gi|1136741
Desmoplakin I [Homo sapiens]	DSP			gi|58530840	gi|1147813	
Eukaryotic translation elongation factor [Homo sapiens]	EEF1A1	EEF1AL3	EEF2	gi|4503471	gi|55665593	gi|4503483
Eukaryotic translation initiation factor 4A isoform 1 [Homo sapiens]	EIF4A1			gi|4503529		
Enolase 1 [Homo sapiens]	ENO1			gi|693933		
Fatty acid synthase [Homo sapiens]	FASN			gi|41872631		
Filamin A [Homo sapiens]	FLNA			gi|53791219		
Ras-GAP SH3-binding protein [Homo sapiens]	G3BP1	G3BP2		gi|5031703	gi|3098601	
Heterogeneous nuclear ribonucleoprotein [Homo sapiens]	HNRNPF HNRNPM	HNRNPH1 HNRNPU	HNRNPK HNRNPUL1	gi|4826760 gi|187281	gi|5031753 gi|32358	gi|14165437 gi|12803479

Heat shock 90 kDa protein [Homo sapiens]	HSP90AA1			gi|62914009	gi|17865490	gi|83699649
	HSP90AB1			gi|306891		
Heat shock 70 kDa protein [Homo sapiens]	HSPA1A HSPA8	HSPA1B	HSPA5	gi|386785 gi|5729877	gi|4885431	gi|16507237
Heat shock protein beta-1 [Homo sapiens]	HSPB1	HSPD1		gi|4504517	gi|31542947	
Keratin [Homo sapiens]	KRT1 KRT7 KRT10 KRT15 KRT18 KRT73 KRT77	KRT2 KRT8 KRT13 KRT16 KRT19 KRT75	KRT5 KRT9 KRT14 KRT17 KRT6A KRT77	gi|11935049 gi|12803727 gi|21961605 gi|24430190 gi|4557888 gi|28173564 gi|113414871	gi|47132620 gi|181400 gi|131412225 gi|1195531 gi|7594734 gi|3901030	gi|18999435 gi|55956899 gi|12803709 gi|4557701 gi|908801 gi|119617049
Methylcrotonoyl-Coenzyme A carboxylase [Homo sapiens]	MCCC1	MCCC2		gi|12276066	gi|11545863	
Non-POU domain containing [Homo sapiens]	NONO			gi|34932414		
Pyruvate carboxylase [Homo sapiens]	PC			gi|106049292		
Plectin [Homo sapiens]	PLEC1			gi|1477646		
Peroxiredoxin [Homo sapiens]	PRDX1	PRDX4		gi|4505591	gi|5453549	
Siah-binding protein 1 [Homo sapiens]	PUF60			gi|1809248		
Raichu404X [Homo sapiens]	RAP1A			gi|14595132		
S3 ribosomal protein [Homo sapiens]	RPS3			gi|7765076		
Splicing factor proline/glutamine rich [Homo sapiens]	SFPQ			gi|4826998		
Solute carrier family 25 [Homo sapiens]	SLC25A3	SLC25A5	SLC25A6	gi|4505775	gi|45829841	gi|15928608

(continued)

Table 3 (continued)

<table>
<tr><th>Description</th><th>Entrezgene</th><th>Acc./id</th></tr>
<tr><td>Sjogren syndrome/scleroderma autoantigen 1 [Homo sapiens]</td><td>SSSCA1</td><td>gi|5453838</td></tr>
<tr><td>Stomatin peptide [Homo sapiens]</td><td>STOM</td><td>gi|181184</td></tr>
<tr><td>Alpha-tubulin [Homo sapiens]</td><td>TUBA1A TUBA1B TUBA1C</td><td>gi|37492 gi|158259731 gi|193786502</td></tr>
<tr><td>Tubulin, beta [Homo sapiens]</td><td>TUBB TUBB2C TUBB3
TUBB4 TUBB6</td><td>gi|18088719 gi|5174735 gi|50592996
gi|21361322 gi|194388028</td></tr>
<tr><td>Vimentin [Homo sapiens]</td><td>VIM</td><td>gi|340219 gi|62414289</td></tr>
</table>

A list of proteins observed in several streptavidin bead pull downs using HeLa and HEK293 cells expressing BirA in combination with biotinylation-tagged GFP. Only proteins that showed a Mascot Score higher than 300 were included

2. Materials

2.1. Equipment

1. Incubator shaker for growing bacteria (Innova 4300, New Brunswick Scientific).
2. Spectrophotometer (Ultrospec III, Pharmacia).
3. Centrifuges:

   Eppendorf microcentrifuge 5415R for 1.5/2-ml Eppendorf tubes

   Eppendorf centrifuge 5810R for 15- and 50-ml Falcon tubes

   High-speed centrifuge RC5C, equipped with SLA-1500 and SS-34 rotors (Sorvall Instruments)
4. Tissue culture incubator (37°C, 5% CO_2) (Sanyo).
5. Laminar flow cabinet for tissue culture.
6. Ultrasonic disintegrator Soniprep 150 (MSE).
7. Rotating wheel.
8. Dynal MPC-S magnet.
9. Microscope suitable for imaging fixed and/or live cells, equipped with a high NA 100× objective. Description of microscopy equipment goes beyond the scope of this chapter.
10. The description of mass spectrometry methods and equipment is beyond the scope of this chapter. Most of our analyses are currently performed with NanoLC-MS/MS on a 1100 series capillary LC system (Agilent Technologies) coupled to an LTQ ion trap mass spectrometer (Thermo) operating in positive mode and equipped with a nanospray source.

2.2. Preparation of GST Proteins from E. coli

1. GST fusions of EB1, EB2, and EB3: EB1, EB2, and EB3 open reading frames were generated by PCR and cloned into *Bam*HI and *Eco*RI sites of the pGEX-3× vector (18, 32).
2. LB Broth EZMix Powder, dust-free, fast-dissolving fermentation medium (Sigma).
3. Ampicillin (working concentration: 100–200 μg/ml).
4. Isopropyl β-d-1-thiogalactopyranoside (IPTG; working concentration: 0.8–1.6 mM; stock solution prepared in water).
5. *E. coli* strain BL21 competent cells: An overnight culture of BL21 *E. coli* is used to inoculate a fresh culture (dilution 1:40) and grown for 1.5–3 h at 37°C with agitation. Cells are collected by centrifugation, washed in ice-cold 0.1 M $CaCl_2$ (10 ml per 250 ml culture), resuspended in ice-cold 0.1 M $CaCl_2$ solution containing 15% glycerol (5 ml per 250 ml culture), aliquoted, and stored at −80°C.

6. Glutathione Sepharose 4B (GE Healthcare).
7. Bacteria lysis buffer: 1% Triton X-100 in phosphate-buffered saline (PBS), supplemented with the complete protease inhibitor cocktail (Roche). Prepare fresh and keep ice cold.
8. Washing buffer: 0.1% Triton in PBS. Prepare fresh and keep ice cold.
9. High-salt washing buffer: 0.1% Triton in PBS, 500 mM NaCl. Prepare fresh and keep ice cold.

2.3. Pull-Down Assays with GST-EB Fusion Proteins from Cell Extracts

1. Dulbecco's Phosphate-Buffered Saline (DPBS) without calcium and magnesium (BioWhittaker, Lonza).
2. Cell culture medium: 45% Dulbecco's modified Eagle's medium (DMEM) with ultraglutamine and 4.5 g/l glucose (BioWhittaker, Lonza), 45% Ham's F10 without thymidine and with ultraglutamine (BioWhittaker, Lonza), 10% fetal calf serum (FCS; Lonza), supplemented with penicillin and streptomycin (PS, 100× stock; Lonza).
3. Cell lysis buffer: 1% Triton X-100 in PBS supplemented with the complete protease inhibitor cocktail (Roche). Prepare fresh and keep ice cold.
4. Washing buffer: 20 mM Tris–HCl, pH 8.0, 150 mM NaCl, 0.05% Triton, 1 mM DTT. Prepare fresh and keep ice cold.
5. NuPAGE Novex precast gels and buffers (Invitrogen).
6. Colloidal Blue staining kit (Invitrogen).

2.4. Biotinylation Tag-Based Pull-Down Experiments from Transfected Cells

1. Biotinylation tag fusion of the protein of interest: We most frequently use an N-terminal biotinylation tag with the sequence MASGLNDIFEAQKIEWHEGGG; the lysine residue (underlined) can be biotinylated by BirA and the three glycines serve as flexible linkers. The sequence can be inserted in the plasmid of interest as a synthetic oligonucleotide linker.
2. The expression construct for biotin ligase BirA under the CMV promoter (33).
3. Cell culture medium: 45% DMEM with ultraglutamine and 4.5 g/l glucose (BioWhittaker, Lonza), 45% Ham's F10 without thymidine and with ultraglutamine (BioWhittaker, Lonza), 10% FCS (Lonza), supplemented with penicillin and streptomycin (PS, 100× stock; Lonza).
4. DPBS without calcium and magnesium (BioWhittaker, Lonza).
5. Polyethylenimine (PEI) "Max" (nominally Mw 40,000) – High-Potency Linear PEI (Equivalent to Mw 25,000 in Free Base Form; Polysciences).
6. Dynabeads M-280 Streptavidin (Invitrogen).
7. Dynabead blocking buffer: 20 mM HEPES, pH 7.5, 150 mM KCl, 0.1% Triton X-100, 20% glycerol, 0.2 μg/μl albumin

from chicken egg white (Grade V, minimum 98% agarose gel electrophoresis; Sigma). Albumin from chicken egg white should be added freshly to the blocking buffer. Prepare a 20 mg/ml stock in PBS and store at −20°C.

8. Cell lysis buffer: 20 mM HEPES, pH 7.5, 150 mM KCl, 1% Triton X-100 supplemented with complete protease inhibitor cocktail (Roche). Prepare fresh and keep ice cold.
9. Washing buffer: 20 mM HEPES, pH 7.5, 150 mM KCl, 0.1% Triton X-100. Prepare fresh and keep on ice.

2.5. Immunofluorescent Staining for +TIPs and Adaptor Proteins

1. Cell culture medium: 45% DMEM with ultraglutamine and 4.5 g/l glucose (BioWhittaker, Lonza), 45% Ham's F10 without thymidine and with ultraglutamine (BioWhittaker, Lonza), 10% FCS (Lonza), supplemented with penicillin and streptomycin (PS, 100× stock; Lonza).
2. Trypsin–EDTA (0.25% Trypsin in EDTA; Lonza).
3. Labtek Chamber Slides (Thermo Scientific Nunc; 4-well glass slides).
4. Ham's F10 without thymidine and with ultraglutamine (BioWhittaker, Lonza).
5. Polyfect Transfection Reagent (Qiagen).
6. Paraformaldehyde (PFA; Sigma Aldrich). Prepare 4% PFA solution in PBS prior to the experiment. Dissolve 4 g of PFA in 100 ml PBS at 65°C and cool to room temperature.
7. Blocking buffer: 1% albumin from bovine serum (Fraction V; Sigma)/0.1% Tween 20/PBS. Prepare fresh and keep on ice.
8. Washing buffer: 0.1% Tween-20 in PBS.
9. Secondary Alexa 350- and 594-conjugated goat antibodies against mouse IgG and rabbit IgG antibodies (Molecular Probes, Invitrogen). Store at −20°C.
10. Vectashield (with DAPI, H-1200, or without DAPI, H-1000, Vector laboratories).
11. Coverslips 24 × 60 mm (Menzel-Gläser).

2.6. +TIP Imaging in Cultured Cells

1. 25-mm diameter round coverslips (N1, Menzel-Glaser).
2. Cell culture medium: 45% DMEM with ultraglutamine and 4.5 g/l glucose (BioWhittaker, Lonza), 45% Ham's F10 without thymidine and with ultraglutamine (BioWhittaker, Lonza), 10% FCS (Lonza), supplemented with penicillin and streptomycin (PS, 100× stock; Lonza).
3. Ham's F10 without thymidine and with ultraglutamine (BioWhittaker, Lonza).
4. Polyfect Transfection Reagent (Qiagen).
5. Attofluor cell chamber (A-7816, Molecular Probes).

3. Methods

3.1. Preparation of GST Proteins from E. coli

1. Transform GST, GST-EB1, GST-EB2, and GST-EB3 plasmids into competent BL21 *E. coli* cells by heat shock at 42°C, plate on LB agar plates with ampicillin, and grow overnight at 37°C.
2. Inoculate individual colonies in 10-ml LB liquid medium with ampicillin and grow overnight at 37°C (see Note 1).
3. Transfer 10 ml of the overnight bacterial culture to a 1-l Erlenmeyer flask with 250-ml LB medium with ampicillin.
4. Grow the cultures at 37°C for 3 h with agitation. Take 1 ml of culture and measure the optical density (OD) at 600 nm. Use 1 ml of LB medium as a blank.
5. When OD 600 is 0.6–0.8, induce protein expression by adding IPTG to a final concentration of 0.8–1.6 mM. Continue growing for 4 h at 37°C with agitation.
6. Transfer bacterial cultures to centrifuge buckets (Sorvall SLA-1500 rotor) and centrifuge at 2,000 × *g* for 30 min at 4°C. Discard the supernatant and store the pellets at –20°C or proceed with purification.
7. Resuspend bacterial pellets in 4 ml of freshly prepared ice-cold lysis buffer.
8. Sonicate the lysates on ice eight times for 10 s with 30 s intervals while keeping the tubes on ice. At this stage, the lysates should become semitransparent; additional cycles of sonication can be performed if necessary.
9. Preclear the lysates by centrifuging for 10 min at 20,000 × *g* at 4°C (using the Sorvall SS-34 rotor).
10. Take a 1.5 ml aliquot of glutathione sepharose beads (50% slurry); add 1 ml ice-cold lysis buffer, mix, centrifuge briefly at 800 × *g* in a 15-ml Falcon tube at 4°C, and discard the supernatant; repeat the procedure 2–3 times.
11. Transfer the precleared bacterial lysate to the tube with the washed glutathione sepharose beads (bead volume ~0.75 ml) and incubate for 2 h at 4°C on a rotating wheel (see Note 2).
12. After a 2-h incubation, centrifuge the beads for 3 min at 800 × *g* in a 15-ml Falcon tube at 4°C and remove the supernatants. Wash the beads three times with the ice-cold washing buffer (5 ml per wash), three times with ice-cold high-salt washing buffer (5 ml per wash), and then again three times with the ice-cold washing buffer (5 ml per wash). For each washing step, mix the beads with the buffer, centrifuge for 3 min at 800 × *g* at 4°C, and aspirate the supernatant (see Note 3).

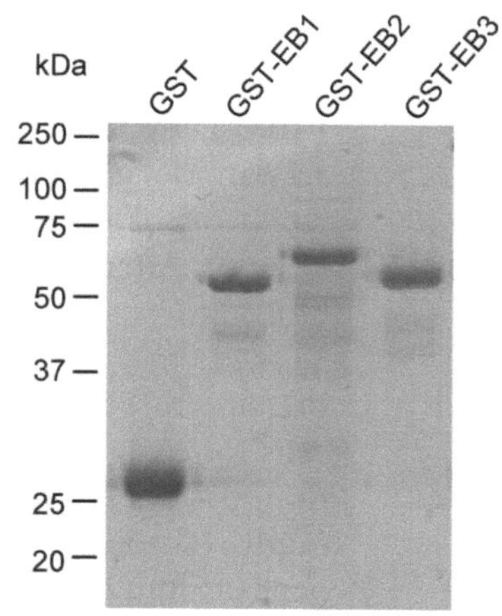

Fig. 1. Characterization of the GST fusion proteins. GST alone, GST-EB1, GST-EB2, and GST-EB3 were purified from *E. coli* and analyzed on a Coomassie-stained gel. Note that due to the presence of ~40 amino acid N-terminal extension, EB2 is longer than EB1 and EB3. Image courtesy of S. Montenegro Gouveia (Erasmus MC, Rotterdam).

13. Resuspend the beads as 2× slurry in ice-cold PBS, 0.1% Triton, and 10% glycerol. Keep 20 μl of the beads apart for running SDS PAGE. The rest of the beads (~1.5 ml of 2× slurry) can be aliquoted, snap-frozen in liquid nitrogen, and stored at −80°C.
14. Run the proteins on 12% SDS PAGE, fix in 25% methanol, 10% acetic acid, and stain with Coomassie blue staining solution. Destain the gel in 25% methanol (Fig. 1) (see Note 4).

3.2. Pull-Down Assays with GST-EB Fusion Proteins from Cell Extracts

1. Prepare two 100-mm tissue culture dishes with cells (>80% confluence) for each GST pull-down assay. Scrape the cells in DPBS (10 ml/dish) with a plastic disposable scrapper and transfer into a 50-ml Falcon tube. Centrifuge at 200 × *g* for 5 min at 4°C, aspirate the supernatant, and lyse the cell pellet in 150–200 μl ice-cold cell lysis buffer per dish. Incubate the lysate on ice for 10 min and preclear by centrifuging in a microcentrifuge for 10 min at 16,100 × *g* at 4°C.
2. Thaw on ice purified GST fusion proteins bound to glutathione sepharose beads that were stored at −80°C.
3. Prepare 200 μl aliquots of each purified GST protein (2× slurry) in 1.5-ml Eppendorf tubes and wash once with 1 ml fresh ice-cold lysis buffer; centrifuge at 400 × *g* in a microcentrifuge at 4°C, discard the supernatant, add the precleared cell lysate (~300 μl per tube), and incubate for 2 h at 4°C on a rotating wheel.
4. Centrifuge the beads for 3 min at 400 × *g* at 4°C and aspirate the supernatants. Wash the beads four times with 1 ml ice-cold washing buffer (20 mM Tris–HCl, pH 8.0, 150 mM NaCl, 0.05% Triton, 1 mM DTT) (see Note 5).

5. Resuspend the beads together with the bound proteins in 100 μl 1× NuPAGE LDS sample buffer supplemented with 1 mM DTT, boil and run on NuPAGE Novex T precast polyacrylamide gel, stain with the Colloidal Blue staining kit (Invitrogen), and send for mass spectrometry analysis (see Note 6). Alternatively, analyze the bound proteins by Western blotting.

3.3. Biotinylation Tag-Based Pull-Down Experiments from Transfected Cells

1. Prepare three 100-mm tissue culture dishes of HEK293T cells for each individual pull down. One day before transfection, passage 100% confluent (1×10^7 cells per dish) by thorough resuspension with a glass pipette. Seed 2.5×10^6 cells per 100-mm dish for transfecting them the next day or 1×10^6 cells if you intend to perform transfection after 2 days.
2. On the day of transfection, HEK293T cells should reach 70–85% confluence. Transfect the biotinylated-tagged fusion construct (i.e., Bio-GFP-CLASP2α) together with a plasmid-expressing BirA ligase using PEI transfection reagent. For each 100-mm dish, incubate a mixture of the two plasmids (2.5 μg each) and 15 μl PEI transfection reagent in two separate tubes containing 0.5 ml DPBS for 5 min at room temperature. Next, mix gently the DNA/DPBS and PEI/DPBS solutions and incubate for another 20 min at room temperature. Add 5-ml cell culture medium to the transfection mixture and transfer to the dish of HEK293T cells. Incubate the cells together with the transfection medium at 37°C, 5% CO_2 for 24–48 h (see Notes 7–9).
3. One hour before starting cell lysis and pull-down experiment, resuspend the Dynabeads in the original vial by vortexing. For each pull down, make a 50 μl aliquot of Dynabead suspension in a 1.5-ml Eppendorf tube (see Note 10).
4. Place the tube into the Dynal MPC-S magnet and remove the supernatant. Incubate the Dynabeads with 250 μl ice-cold blocking buffer for 1 h at 4°C on a rotating wheel.
5. Remove the medium from the dishes of transfected HEK293T cells, resuspend the cells by vigorous pipetting in 5 ml of DPBS per dish, transfer to 50-ml Falcon tubes, centrifuge at $200\times g$ for 5 min at 4°C, and aspirate the supernatant. For strongly adherent cell lines, scrape the cells in DPBS (10 ml/dish) with a plastic disposable scrapper, transfer into a 50-ml Falcon tube, centrifuge, and aspirate as described above.
6. Resuspend the cell pellet thoroughly in ice-cold cell lysis buffer (300 μl per 100-mm dish) (see Note 11). Centrifuge the lysate in a microcentrifuge at $16{,}100\times g$ for 10 min at 4°C. Collect 50 μl lysate and mix with 2× SDS sample buffer as the input fraction for analysis on SDS PAGE and Western blotting.

Use the rest of the cell lysate as input for the pull-down experiment.

7. Place the tubes with Dynabeads in the blocking buffer in the magnet for 1 min and aspirate the supernatant.
8. Add the cell lysate to the Dynabeads and incubate the mixture for 1 h at 4°C on a rotating wheel.
9. Place the tubes in the magnet for 1 min and remove the supernatant. Wash the Dynabeads 4–6 times with ice-cold washing buffer (1 ml per wash). For each washing step, remove the tubes from the magnet and mix by inverting them several times, place back onto the magnet, and aspirate the supernatant (see Note 5).
10. After the last washing step, remove the supernatant, resuspend the bound Dynabeads in 50 μl 1× NuPAGE sample buffer, and boil for 10 min at 100°C.
11. Use 5 μl of the bound Dynabeads samples for running SDS PAGE and stain with Coomassie or perform Western blotting (Fig. 2). The remaining Dynabeads can be used to run an NuPAGE Novex precast polyacrylamide gel. Stain the gel with the Colloidal Blue staining kit and send for mass spectrometry analysis (see Note 6).

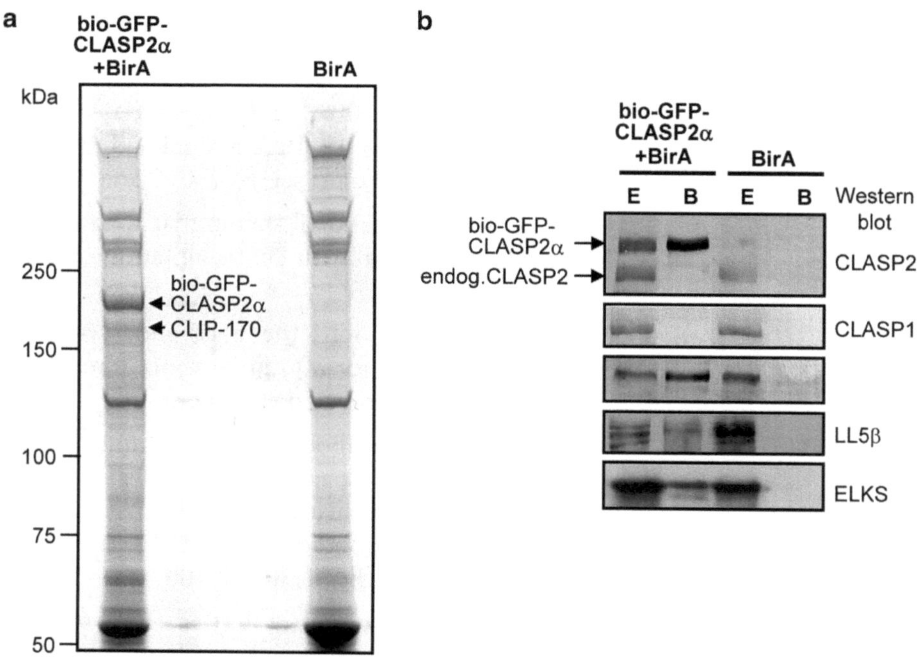

Fig. 2. Identification of CLASP binding partners by pull-down assays using biotinylation tag. Pull-down assays with streptavidin beads were performed with lysates of HeLa cells, coexpressing bio-GFP-CLASP2α together with BirA or BirA alone. Proteins bound to streptavidin beads were analyzed on a Coomassie-stained gel (**a**) or by Western blotting with the indicated antibodies (**b**). In panel (**a**), only the fraction bound to the beads was loaded. In panel (**b**), the extract before the pull down (lanes marked "E") and proteins bound to the beads (lanes marked "B") are shown. Lanes marked "E" correspond to 25% of the extract, used to obtain lanes marked "B" reproduced from (34) with permission from Elsevier.

3.4. Immunofluorescent Staining for +TIPs and Adaptor Proteins

1. One day before transfection, split a 100-mm dish of confluent HeLa cells (1×10^7 cells per dish). Trypsinize cells with 1 ml of trypsin–EDTA and resuspend in 9 ml of cell culture medium. Prepare HeLa cells in a dilution of 1:40. Seed 1 ml of cell suspension into one well of a four-well Labtek chamber slide; incubate the cells overnight with normal medium at 37°C in a humidified incubator with 5% CO_2.
2. Prewarm the cell culture media at 37°C and centrifuge all plasmid DNA for 3 min at $16,100 \times g$ in a microcentrifuge. Prepare transfection mixture: for one well of a four-well Labtek Chamber slide, add 30 μl of Ham's F10 medium, 0.5 μl of plasmid DNA (1 μg/μl), and 3 μl Polyfect transfection reagent. Incubate for 15 min at room temperature (see Note 12).
3. Add 300 μl cell culture medium (50% DMEM/50% Ham's F10/10% FCS/PS) to the transfection mixture, mix, and add to the chamber with cells after removing the culture medium. Place the cells back into incubator and incubate for 16–24 h at 37°C with 5% CO_2.
4. Aspirate the transfection medium and add fresh medium (1 ml/chamber); incubate for 1.5 h at 37°C with 5% CO_2.
5. Aspirate the medium from Labtek chamber slides, wash the cells with DPBS (1 ml/chamber), and detach the plastic disposable chamber.
6. Fix the cells for 10 min by inserting the slide into a Hellendahl staining jar containing 100% methanol precooled to −20°C. Transfer immediately to 4% PFA maintained at room temperature for another 10 min (see Notes 13 and 14).
7. Wash the cells in a Hellendahl staining jar with 1.5% Trition X-100 in PBS for 10 min with gentle agitation on a shaker at room temperature.
8. Pipet blocking buffer onto the glass slides (~100 μl per chamber) and incubate for 30 min at room temperature in a humidified and light-protected chamber.
9. Thaw on ice primary antibodies and centrifuge for 3 min at $16,100 \times g$ at 4°C. Prepare antibody dilutions in the blocking buffer (50 μl antibody mixture/chamber). Typically, commercial monoclonal antibodies are diluted 1:100–1:300, and crude polyclonal antibodies 1:300–1:400.
10. Carefully remove the blocking buffer from the slides by absorbing it with a tissue paper (avoid touching the cells). Apply 50 μl of diluted primary antibodies per well and incubate for 1 h at room temperature in a humidified and light-protected chamber.
11. Carefully remove the primary antibody solution from the slides with a tissue paper (avoid touching the cells). Wash the

slides three times with 0.1% Tween-20 in PBS in a Hellendahl staining jar.

12. Thaw on ice secondary antibodies and centrifuge for 3 min at 16,100×*g* at 4°C. Prepare antibody dilutions in the blocking buffer (50 μl antibody mixture/chamber). Typically, commercial secondary antibodies are diluted 1:300–1:500.
13. Apply 50 μl of diluted secondary antibodies per well and incubate for 45 min at room temperature in a humidified and light-protected chamber.
14. Carefully remove the secondary antibody solution from the slides with a tissue paper (avoid touching the cells). Wash the slides three times with 0.1% Tween-20 in PBS in a Hellendahl staining jar.
15. Remove the gelatin frames from the slides using sharp forceps or a knife. Rinse the slides in 70% ethanol and then 100% ethanol, air-dry, mount in Vectashield (with or without DAPI), and cover with glass coverslips. Slides can be stored indefinitely in a slide holder at −20°C in dark.
16. Use an appropriate fluorescent microscope for taking images (Fig. 3).

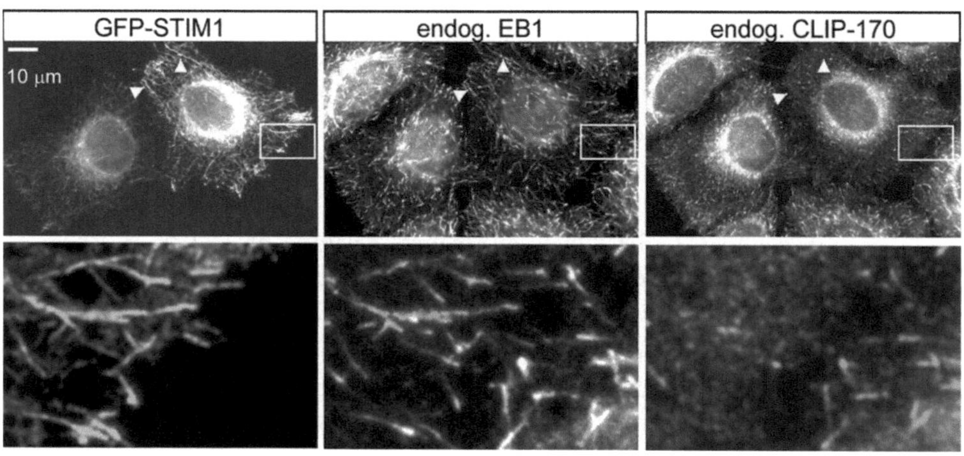

Fig. 3. Detection of GFP fusion protein at the MT plus ends by immunofluorescent cell staining. STIM1, an ER-resident protein essential for store-operated calcium entry, was identified as binding partner of EB1 by GST pull-down assays from extracts of B16F1 melanoma cells (24). To test if STIM1 associates with MT ends, HeLa cells were transfected with GFP-STIM1, fixed, and stained for the endogenous +TIPs. Antibodies against EB1 (monoclonal mouse EB1 antibody, dilution 1:200, BD Biosciences) and CLIP-170 (polyclonal rabbit CLIP-170 antibodies #2221 and #2360 (37), dilution 1:400) were used. Images were obtained with a Leica DMRBE microscope with a PL Fluotar 100× 1.3 NA objective, equipped with a Hamamatsu CCD camera (C4880) and processed in Adobe Photoshop. Insets show enlargements of the boxed areas. Note that in the cell with a higher GFP-STIM1 expression, CLIP-170 signal on GFP-STIM1-positive MT tips is reduced (*arrowheads*) because of competition between GFP-STIM1 and CLIP-170 for the binding sites on EB1 C-terminus. When performing these experiments, it is thus important to stain for endogenous EB proteins and not for any other +TIPs, which might be affected even by mild expression of exogenous plus-end-binding proteins due to competition for the binding to the EBs.

3.5. +TIP Imaging in Live Cultured Cells

1. Plate cells onto coverslips at the appropriate density (see Note 15). Different coverslips can be used depending on the observation chamber, including glass bottom Petri dishes. 25-mm diameter round coverslips are placed into six-well cell culture plates. Typically, cells should be at 25–30% confluence on the day of transfection and 50–75% confluence during imaging.
2. Prewarm the cell culture media at 37°C and centrifuge all plasmid DNA for 3 min at 16,100 × *g* in a microcentrifuge. Prepare transfection mixture: for one well of a six-well cell culture plate, add 100 μl of Ham's F10 medium, 1.5–2 μl of plasmid DNA (1 μg/μl), and 10 μl Polyfect transfection reagent. Incubate for 15 min at room temperature (see Note 12).
3. Add 1.5-ml cell culture medium (50% DMEM/50% Ham's F10/10% FCS/PS) to the transfection mixture, mix, and add to the well with cells after removing the culture medium. Place the cells back into incubator and incubate for 16–24 h at 37°C with 5% CO_2.
4. Aspirate the transfection mixture and add fresh culture medium (50% DMEM/50% Ham's F10/10% FCS/PS) before imaging. Mount the coverslip with cells into the observation chamber. Never allow the cells to dry during any preparation steps.
5. For a typical inverted microscope setup, which is most convenient for imaging live mammalian cells, add immersion oil on top of the lens. Use a 63× or a 100× objective.
6. Insert the chamber with cells into the heating incubator or onto the stage (this depends on the microscope configuration). Cells should be maintained at 37°C during imaging.
7. Collect live cell images. Imaging at 1–2 frames per second gives optimal results, but one frame per 2–3 s is sufficient (see Note 16). Collect 50–100 frames.
8. Watch the movie or generate a maximum intensity projection of consecutive frames; this helps to detect +TIP comets if the signal-to-noise ratio is low (Fig. 4).

4. Notes

1. Frozen stocks of BL21 cells transformed with GST constructs can be stored at −80°C and used for inoculating overnight cultures.
2. 1 ml of glutathione sepharose bead volume binds approximately 5 mg of GST protein.

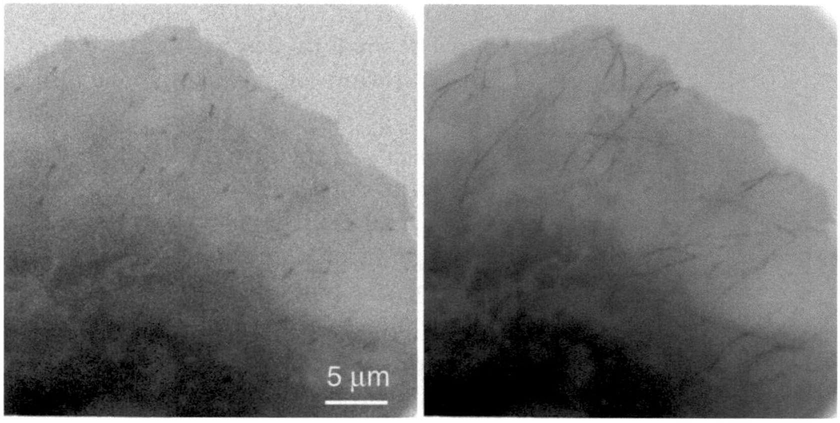

Fig. 4. Detection of MT plus-end tracking of a GFP fusion protein by live cell imaging. Images of a COS-7 cell transiently expressing a GFP fusion of the C-terminal fragment of MACF2 (13). Wide-field time-lapse imaging was performed on an inverted research Nikon Eclipse TE2000E (Nikon) microscope with a CFI Apo TIRF 100× 1.49 NA oil objective (Nikon). The microscope was equipped with a QuantEM EMCCD camera (Roper Scientific) and controlled by the MetaMorph 7.1 software package (Molecular Devices). For excitation, we used HBO 103 W/2 Mercury Short Arc Lamp (Osram) and Chroma ET-GFP filter cube. 16-bit images were projected onto the CCD chip at a magnification of 0.067 μm/pixel, with intermediate magnification of 2.5×. Images were collected with 500 ms exposure without delay between frames. A single frame is shown on the left, and maximum intensity projection of 100 consecutive frames is shown on the right. Note that although the GFP-labeled microtubule plus ends might be difficult to distinguish in individual frames due to high cytoplasmic background, they become apparent as characteristic linear tracks in maximum intensity projections of several frames. Image courtesy of I. Grigoriev (Erasmus MC, Rotterdam).

3. High-salt washing buffer reduces unspecific protein binding to the glutathione sepharose beads during purification.
4. The yield of GST-EB2 is lower than GST-EB1 or EB3, and its purity is typically less good.
5. Increasing salt concentration in the washing buffer can help to reduce unspecific binding during pull-down assays with cell lysates.
6. During preparation of samples and gels for mass spectrometry analysis, keratin contaminations should be avoided as much as possible: use clean tubes and glassware and wear gloves at all times.
7. For cell transfection, column-purified plasmid DNA always works better. We use QIAGEN QIAfilter DNA purification kits.
8. Transfection with Lipofectamine 2000 (Invitrogen) also works well for HEK293T cells. Although more expensive, it is preferable when large proteins (>150 kDa) need to be expressed.
9. Keep in mind that while some tissue culture media contain biotin (for example, Ham's F10), other media do not (for example, DMEM). If the culture medium contains no

biotin, exogenous biotin should be added to achieve efficient biotinylation. The medium described in Subheading 2.3 can be used for culturing all cell lines mentioned in this chapter.

10. The binding capacity of the Dynabeads (6.7×10^8 beads/ml, 10 mg/ml) is approximately 200 pmol biotinylated peptides per 1 mg. The actual protein yield observed in these experiments is ~2–10 μg protein per 50 μl beads. Protein binding can depend on protein size because of the sterical hindrance.
11. Biotinylation tag-based pull-down assays work in different buffers, including those with ionic detergents (for example, radioimmunoprecipitation assay (RIPA) buffer with 1% NP-40, 1% Deoxycholate, and 0.05% SDS).
12. When several plasmid DNAs are cotransfected, add DNA to the transfection mixture first, and only then add the transfection reagent to insure efficient cotransfection. Numerous other transfection reagents are commercially available; for example, Fugene 6 (Roche) is a good alternative to Polyfect.
13. Make sure that PFA solution has neutral pH and is at room temperature; otherwise, certain protein epitopes and direct GFP fluorescence may be negatively affected.
14. Combined cold methanol/PFA fixation is optimal for labeling EB and CLIP comets. Cold methanol insures very quick fixation and optimal preservation of microtubule cytoskeleton, but destroys membrane structures, such as vesicles. Postfixation with PFA prevents protein extraction after fixation; when it is omitted, staining for EBs and CLIPs still works but the signal is weaker. However, in some cases, like, for example, when staining for CLASPs or LL5β (34), PFA postfixation blocks epitope accessibility and inhibits labeling. Finally, the classic fixation with 4% PFA alone at room temperature is good for preserving membrane compartments, but cannot be used for visualizing microtubule plus ends because +TIPs rapidly dissociate from microtubule ends during fixation. Alternative protocols suitable for +TIP fixation are available (35, 36). For new proteins and epitopes, different fixation conditions should be tested and optimized.
15. Different cell types can be used, but cells that are easy to transfect, are flat and have relatively sparse microtubule cytoskeleton, such as COS-7 monkey kidney cells or MRC5 human lung fibroblasts, are preferable.
16. Keep in mind that in mammalian cells cultured at 37°C, microtubules grow with a velocity of 0.1–0.5 μm/s.

Acknowledgments

This work was supported by the Netherlands Organisation for Scientific Research grants ZonMW-VIDI and CW-ECHO to C.C.H and ALW-VICI and ZonMW-TOP to A.A. We thank Susana Montenegro Gouveia and Daniël Splinter for their contribution to establishing affinity purification protocols, Ilya Grigoriev for establishing live cell imaging protocols, and Jeroen Demmers for mass spectrometry analysis.

References

1. Desai, A., and Mitchison, T. J. (1997) Microtubule polymerization dynamics. *Annu Rev Cell Dev Biol 13*, 83–117.
2. Howard, J., and Hyman, A. A. (2003) Dynamics and mechanics of the microtubule plus end. *Nature 422*, 753–758.
3. Schuyler, S. C., and Pellman, D. (2001) Microtubule "plus-end-tracking proteins": The end is just the beginning. *Cell 105*, 421–424.
4. Akhmanova, A., and Steinmetz, M. O. (2008) Tracking the ends: a dynamic protein network controls the fate of microtubule tips. *Nat Rev Mol Cell Biol 9*, 309–322.
5. Su, L. K., and Qi, Y. (2001) Characterization of human MAPRE genes and their proteins. *Genomics 71*, 142–149.
6. Juwana, J. P., Henderikx, P., Mischo, A., Wadle, A., Fadle, N., Gerlach, K., Arends, J. W., Hoogenboom, H., Pfreundschuh, M., and Renner, C. (1999) EB/RP gene family encodes tubulin binding proteins. *Int J Cancer 81*, 275–284.
7. Hayashi, I., and Ikura, M. (2003) Crystal structure of the amino-terminal microtubule-binding domain of end-binding protein 1 (EB1). *J Biol Chem 278*, 36430–36434.
8. Slep, K. C., Rogers, S. L., Elliott, S. L., Ohkura, H., Kolodziej, P. A., and Vale, R. D. (2005) Structural determinants for EB1-mediated recruitment of APC and spectraplakins to the microtubule plus end. *J Cell Biol 168*, 587–598.
9. Honnappa, S., John, C. M., Kostrewa, D., Winkler, F. K., and Steinmetz, M. O. (2005) Structural insights into the EB1-APC interaction. *Embo J 24*, 261–269.
10. Honnappa, S., Okhrimenko, O., Jaussi, R., Jawhari, H., Jelesarov, I., Winkler, F. K., and Steinmetz, M. O. (2006) Key interaction modes of dynamic +TIP networks. *Mol Cell 23*, 663–671.
11. Hayashi, I., Wilde, A., Mal, T. K., and Ikura, M. (2005) Structural basis for the activation of microtubule assembly by the EB1 and p150Glued complex. *Mol Cell 19*, 449–460.
12. Steinmetz, M. O., and Akhmanova, A. (2008) Capturing protein tails by CAP-Gly domains. *Trends Biochem Sci 33*, 535–545.
13. Honnappa, S., Gouveia, S. M., Weisbrich, A., Damberger, F. F., Bhavesh, N. S., Jawhari, H., Grigoriev, I., van Rijssel, F. J., Buey, R. M., Lawera, A., Jelesarov, I., Winkler, F. K., Wuthrich, K., Akhmanova, A., and Steinmetz, M. O. (2009) An EB1-binding motif acts as a microtubule tip localization signal. *Cell 138*, 366–376.
14. Bieling, P., Kandels-Lewis, S., Telley, I. A., van Dijk, J., Janke, C., and Surrey, T. (2008) CLIP-170 tracks growing microtubule ends by dynamically recognizing composite EB1/tubulin-binding sites. *J Cell Biol 183*, 1223–1233.
15. Dragestein, K. A., van Cappellen, W. A., van Haren, J., Tsibidis, G. D., Akhmanova, A., Knoch, T. A., Grosveld, F., and Galjart, N. (2008) Dynamic behavior of GFP-CLIP-170 reveals fast protein turnover on microtubule plus ends. *J Cell Biol 180*, 729–737.
16. Berrueta, L., Tirnauer, J. S., Schuyler, S. C., Pellman, D., and Bierer, B. E. (1999) The APC-associated protein EB1 associates with components of the dynactin complex and cytoplasmic dynein intermediate chain. *Curr Biol 9*, 425–428.
17. Bu, W., and Su, L. K. (2003) Characterization of functional domains of human EB1 family proteins. *J Biol Chem 278*, 49721–49731.
18. Komarova, Y., Lansbergen, G., Galjart, N., Grosveld, F., Borisy, G. G., and Akhmanova, A. (2005) EB1 and EB3 control CLIP dissociation from the ends of growing microtubules. *Mol Biol Cell 16*, 5334–5345.

19. Mimori-Kiyosue, Y., Grigoriev, I., Lansbergen, G., Sasaki, H., Matsui, C., Severin, F., Galjart, N., Grosveld, F., Vorobjev, I., Tsukita, S., and Akhmanova, A. (2005) CLASP1 and CLASP2 bind to EB1 and regulate microtubule plus-end dynamics at the cell cortex. *J Cell Biol 168*, 141–153.
20. Lee, T., Langford, K. J., Askham, J. M., Bruning-Richardson, A., and Morrison, E. E. (2008) MCAK associates with EB1. *Oncogene 27*, 2494–2500.
21. Jiang, K., Wang, J., Liu, J., Ward, T., Wordeman, L., Davidson, A., Wang, F., and Yao, X. (2009) TIP150 interacts with and targets MCAK at the microtubule plus ends. *EMBO Rep 10*, 857–865.
22. Rogers, S. L., Wiedemann, U., Hacker, U., Turck, C., and Vale, R. D. (2004) Drosophila RhoGEF2 Associates with Microtubule Plus Ends in an EB1-Dependent Manner. *Curr Biol 14*, 1827–1833.
23. Jaworski, J., Kapitein, L. C., Gouveia, S. M., Dortland, B. R., Wulf, P. S., Grigoriev, I., Camera, P., Spangler, S. A., Di Stefano, P., Demmers, J., Krugers, H., Defilippi, P., Akhmanova, A., and Hoogenraad, C. C. (2009) Dynamic microtubules regulate dendritic spine morphology and synaptic plasticity. *Neuron 61*, 85–100.
24. Grigoriev, I., Gouveia, S. M., van der Vaart, B., Demmers, J., Smyth, J. T., Honnappa, S., Splinter, D., Steinmetz, M. O., Putney, J. W., Jr., Hoogenraad, C. C., and Akhmanova, A. (2008) STIM1 is a MT-plus-end-tracking protein involved in remodeling of the ER. *Curr Biol 18*, 177–182.
25. Smith, P. A., Tripp, B. C., DiBlasio-Smith, E. A., Lu, Z., LaVallie, E. R., and McCoy, J. M. (1998) A plasmid expression system for quantitative in vivo biotinylation of thioredoxin fusion proteins in Escherichia coli. *Nucleic Acids Res 26*, 1414–1420.
26. Beckett, D., Kovaleva, E., and Schatz, P. J. (1999) A minimal peptide substrate in biotin holoenzyme synthetase-catalyzed biotinylation. *Protein Sci 8*, 921–929.
27. de Boer, E., Rodriguez, P., Bonte, E., Krijgsveld, J., Katsantoni, E., Heck, A., Grosveld, F., and Strouboulis, J. (2003) Efficient biotinylation and single-step purification of tagged transcription factors in mammalian cells and transgenic mice. *Proc Natl Acad Sci USA 100*, 7480–7485.
28. Grosveld, F., Rodriguez, P., Meier, N., Krpic, S., Pourfarzad, F., Papadopoulos, P., Kolodziej, K., Patrinos, G. P., Hostert, A., and Strouboulis, J. (2005) Isolation and characterization of hematopoietic transcription factor complexes by in vivo biotinylation tagging and mass spectrometry. *Ann N Y Acad Sci 1054*, 55–67.
29. Rodriguez, P., Braun, H., Kolodziej, K. E., de Boer, E., Campbell, J., Bonte, E., Grosveld, F., Philipsen, S., and Strouboulis, J. (2006) Isolation of transcription factor complexes by in vivo biotinylation tagging and direct binding to streptavidin beads. *Methods Mol Biol 338*, 305–323.
30. Kolodziej, K. E., Pourfarzad, F., de Boer, E., Krpic, S., Grosveld, F., and Strouboulis, J. (2009) Optimal use of tandem biotin and V5 tags in ChIP assays. *BMC Mol Biol 10*, 6.
31. Chapman-Smith, A., and Cronan, J. E., Jr. (1999) The enzymatic biotinylation of proteins: a post-translational modification of exceptional specificity. *Trends Biochem Sci 24*, 359–363.
32. Stepanova, T., Slemmer, J., Hoogenraad, C. C., Lansbergen, G., Dortland, B., De Zeeuw, C. I., Grosveld, F., van Cappellen, G., Akhmanova, A., and Galjart, N. (2003) Visualization of microtubule growth in cultured neurons via the use of EB3-GFP (end-binding protein 3-green fluorescent protein). *J Neurosci 23*, 2655–2664.
33. Driegen, S., Ferreira, R., van Zon, A., Strouboulis, J., Jaegle, M., Grosveld, F., Philipsen, S., and Meijer, D. (2005) A generic tool for biotinylation of tagged proteins in transgenic mice. *Transgenic Res 14*, 477–482.
34. Lansbergen, G., Grigoriev, I., Mimori-Kiyosue, Y., Ohtsuka, T., Higa, S., Kitajima, I., Demmers, J., Galjart, N., Houtsmuller, A. B., Grosveld, F., and Akhmanova, A. (2006) CLASPs attach microtubule plus ends to the cell cortex through a complex with LL5beta. *Dev Cell 11*, 21–32.
35. Maney, T., Hunter, A. W., Wagenbach, M., and Wordeman, L. (1998) Mitotic centromere-associated kinesin is important for anaphase chromosome segregation. *J Cell Biol 142*, 787–801.
36. Rogers, S. L., Rogers, G. C., Sharp, D. J., and Vale, R. D. (2002) Drosophila EB1 is important for proper assembly, dynamics, and positioning of the mitotic spindle. *J Cell Biol 158*, 873–884.
37. Coquelle, F. M., Caspi, M., Cordelieres, F. P., Dompierre, J. P., Dujardin, D. L., Koifman, C., Martin, P., Hoogenraad, C. C., Akhmanova, A., Galjart, N., De Mey, J. R., and Reiner, O. (2002) LIS1, CLIP-170's key to the dynein/dynactin pathway. *Mol Cell Biol 22*, 3089–3102.

Index

Anne Straube (ed.), *Microtubule Dynamics: Methods and Protocols*, Methods in Molecular Biology, vol. 777,
DOI 10.1007/978-1-61779-252-6, © Springer Science+Business Media, LLC 2011

MIX
Papier aus verantwortungsvollen Quellen
Paper from responsible sources
FSC® C105338

If you have any concerns about our products,
you can contact us on
ProductSafety@springernature.com

In case Publisher is established outside the EU,
the EU authorized representative is:
Springer Nature Customer Service Center GmbH
Europaplatz 3, 69115 Heidelberg, Germany

Printed by Libri Plureos GmbH
in Hamburg, Germany